Write of Passage

This book, originally published by the Panama Canal Museum, is reprinted in honor of the Museum's history and that of the people who lived and worked in the Panama Canal Zone.

LIBRARY PRESS@UF

AN IMPRINT OF UF PRESS AND
GEORGE A. SMATHERS LIBRARIES

WRITE OF PASSAGE

Stories of the American Era of the Panama Canal

Panama Canal Museum

Library Press @ UF
Gainesville, Florida

Copyright 2017 by Library Press @ UF on behalf of the University of Florida
Original copyright 2008 by the Panama Canal Museum, transferred to the Smathers Libraries in 2012
This work is licensed under a modified Creative Commons Attribution-Noncommercial-No Derivative Works 3.0 Unported License. To view a copy of this license, visit http://creativecommons.org/licenses/by-nc-nd/3.0/. You are free to electronically copy, distribute, and transmit this work if you attribute authorship. *However, all printing rights are reserved by the University of Florida Press (http://upress.ufl.edu). Please contact UFP for information about how to obtain copies of the work for print distribution.* You must attribute the work in the manner specified by the author or licensor (but not in any way that suggests that they endorse you or your use of the work). For any reuse or distribution, you must make clear to others the license terms of this work. Any of the above conditions can be waived if you get permission from the University Press of Florida. Nothing in this license impairs or restricts the author's moral rights.

Names: Panama Canal Museum, publisher. | George A. Smathers Libraries, publisher.
Title: Write of passage : stories of the American era of the Panama Canal / presented by the Panama Canal Museum.
Description: Gainesville, FL : Library Press @ UF, [2016] | Originally published in 2008 by the Panama Canal Museum. | Summary: An account of life, work and play in the former Canal Zone and in Panama during the American Era of the Panama Canal. There are accounts of the family histories by 92 storytellers that include reflections, historical perspectives, and family photos.
Identifiers: ISBN 9781944455033 | ISBN 1683400070
Subjects: LCSH: United States. Panama Canal Commission. | Panama Canal (Panama)—Officials and employees. | Panama Canal (Panama)—History. | Canal Zone—History.
Classification: LCC F1569.C2 W75 2016

Library Press @ UF is an imprint of the University of Florida Press.

| LIBRARY PRESS@UF |

AN IMPRINT OF UF PRESS AND
GEORGE A. SMATHERS LIBRARIES

University of Florida Press
15 Northwest 15th Street
Gainesville, FL 32611-2079
http://upress.ufl.edu

Series Foreword

The Panama Canal Museum, formerly located in Seminole, FL, closed in 2012 and transferred its collection to the George A. Smathers Libraries, greatly enhancing the University of Florida's holdings on Panama and the Canal. The Museum's mission was to document, interpret, preserve, and articulate the leadership role played by the United States in the history of the Panama Canal, with emphasis on the construction, operation, maintenance, and defense of the Canal and the contributions to its success by people of all nationalities. This mission continues to guide the preservation of the Panama Canal Museum Collection. As such, the Smathers Libraries have reprinted this book, originally published by the Panama Canal Museum, in honor of the Museum's history and that of the people who lived and worked in the Panama Canal Zone.

The Panama Canal Museum Collection (PCMC) in the Department of Special & Area Studies Collections at the George A. Smathers Libraries is the leading research collection in the United States for the study of the American era of the Panama Canal. The Collection documents the U.S. experience in the Panama Canal Zone and Panama, and to a lesser degree, it also preserves historical information about the Canal prior and subsequent to U.S. construction and operation.

The Smathers Libraries preserve and provide access to the historically significant and distinctive materials about Panama and the Canal in order to facilitate knowledge creation and dissemination and to support and advance the commitment of the Smathers Libraries to excellence in education and research and contribute to the University of Florida's standing as a preeminent public research university.

Visit the Panama and the Canal Digital Collection at: *http://ufdc.ufl.edu/pcm*.

Judith C. Russell, Dean of University Libraries, University of Florida

... running across stickers in your bare feet to see who wins ... waving at the guys on ships ... Golden Altar ... dancing at the Panama Hilton ... chasing the rain ... eating michas ... ¡Carn at Summit Gardens ... Saturday afternoon at Amador Beach ... riding the Panama Railro ... submarine races out on the Causeway ... a "real" tropical rainstorm ... dances at the Ti French fries and gravy at the Gatun Clubhouse ... ring-a-levio at dusk ... heading home wh Margarita hill on the sides of the tennis courts ... sunset over the Caribbean ... the Gatu Conjunto ... sleeping to the sound of rain on a tin roof ... riding the train to football game races ... Christmas tree burns ... battleball in the Margarita Gym ... jeeps full of G.I.'s whis Bravo's band floating across the bay from Colon ... onion rings at the American Legion ... t ... banyan trees on Roosevelt Avenue ... steak sandwiches at Gamboa Golf Club ... swimmin ... Boy Scout Camp Chagres ... meat on a stick ... the lottery ... Balboa Police Station (enc ditches after a big rain ... Chorrera Falls ... BHS letters on Sosa Hill ... cases of soda deli Canal Zone at the Civil Affairs Building ... Draft Board Local #1 ... collecting snakes ... Frida in the big sleigh at Mr. Townsend's garage on Santa Claus Lane ... the American Legion floor Lake ... swimming races for movie tickets during summer vacation at Gatun Pool ... walki the pond ... fishing on the Chagres River ... skydiving over CocoSolo/France Field ... ceviche docked at Pier 1 ... visiting the Leper Colony ... spaghetti night at Amador Officers' Club ... carnivalito ... midnight mass at St. Mary's ... drag races at France Field ... meatball sand ...taking your life into your hands on the Transisthmian Highway ... Jamboree parties at th trails of Mindi Acres and Ft. Davis ... fried corvina — anywhere ... swimming parties at the through Balboa whenever Sosa Hill burned ... the old guy with the swagger stick who walked Panama at the beach ... hot feet on the black sand at San Carlos beach ... 5-cent ice crea low-flying planes over BHS ... plantain chips ... blocks of ice delivered to your home ... sea the Seventh Fleet at the Atlantic breakwater ... Chinese plums ... the morning flag-raising — anywhere ... driving up the Admin. Building hill during your driver's license test ... drinkir dinners at Balboa Yacht Club ... fishing trips to Isla Perlas ... gathering around the plaque mud ... beer and pizza at Ft. Davis snack bar after Friday night football games ... putting wall parties, Coco Solo ... hunting lobster at night on the reefs outside the breakwater wal Whiskey-a-Go-Go in Panama City ... the natural slide at Goofy Falls ... weekends at Santa Gatun Lake ... watching the flying fish from the Taboga launch ... tree frogs in El Valle ... r top of the hill in Curundu ... watching the U2s take off from Albrook Field ... picking berries Curundu ...the manual pinsetters at Balboa Bowling Alley ... eating raspadilla snow cones the best pizza in the world from the Napoli restaurant ... snorkeling around Taboga Island . the flats at Albrook AFB ... sound of the bullfrogs ... cruising the Causeway at night ... the

...they pass by on the Canal... splashing at Gooty Lake... attending mass at the Church of the ... riding the chivas with your friends... a kiss under the palm trees at Rio Del Mar... picnics ... wading in the Goethals Memorial... sunrise from Gold Hill the morning after graduation ... skinny-dipping at Cocoli Lake... hiking the Las Cruces Trail... weekends at Santa Clara... the streetlights turned on... sliding down the Admin. Building hill on palm fronds—and the spillway with all gates open... ginnup season... eating fresh limes with sugar... Lucho y Su ... "the other side" ("the other side" being the opposite side from where you lived)... cayuco ... at the girls... warm rain... mango season, from green to ripe... the sound of Colegio Able ...uto Cine... panning for gold in Rio Pecorar... Fourth of July celebrations at Balboa Stadium ...K-9... the DDT truck... leaf-cutter ants... weekends at Taboga Island... 25-cycle electricity ... said)... green police cars... Nancy trees... Farfan Beach... teen clubs... swimming in the ...d to your home for a dollar... Chinese Gardens... Cocoli Gun Club... the rubber map of the ...ght at the skating rink — then walking fast to make the Owl Show at Diablo Theatre... Santa ...ping with the music... Saturday speedboat races at the Cristobal Yacht Club and Miraflores ...he railroad tracks from the Aids Building to the Gatun Yacht Club... catching guppies at ... the Police Lodge... 25-cent rum and cokes at the drive-in theater... sneaking aboard ships ...day nights at El Rancho... Johnny Mazetti from the clubhouse... skipping school to go to ...es from the Napoli... toad hunting at night... Margarita's "Snob Hill"... France Field crabs ...i Docks... chop suey and wontons from the Balboa YMCA... riding horses through the back ...tel Washington saltwater pool... catching the ferry to Ft. Sherman... wild animals running ...nd wearing a WWII Nazi uniform... the Blue Angels flying over Panama Bay... drinking Cerveza ...nes at the YMCA... veterans of the Spanish American War at Memorial Day ceremonies... ...es taking off and landing behind Cristobal High School... watching the sun come up behind ...Guardia Nacional stations... gooey cinnamon buns at Margarita Clubhouse... empanadas ...rple Passion at the drive-in... throwing up Purple Passion on the way home... $1.50 lobster ...HS, trying not to step on it... Little Theatre plays... football games in torrential rains and ...ies on the train tracks... hidden beaches past Piña Beach... Mine Dock keg parties... sea ...seeing people off on the S.S. Cristobal... The Snake Pit in Curundu... the Sombrero and the ...ra... the spider monkey at Rio Mar... camping at the Taboga cross... fishing for bluegill in ...g horses at Santa Clara... camping with the scouts at Rio Hato... the rope swing at the ...erro Punta... watching Carnival parades on Via España in Panama City... chasing iguanas in ... the Canal Zone bus terminal... the sound of an old palm branch falling from a palm tree... ...ting monkey meat on a stick... the sound of the approaching rain coming from the jungle... ...wooden Balboa Clubhouse... the GAP (later the BRIDGE) Clubhouse for Balboa teenagers...

Write of Passage

STORIES OF THE AMERICAN ERA OF THE PANAMA CANAL

VOLUME I

Presented by the Panama Canal Museum

© 2008 Panama Canal Museum. No part of this book may be copied by any means without written permission from the publisher, the Panama Canal Museum.

Printed and bound by Rose Printing Company Inc.
2503 Jackson Bluff Road
Tallahassee, Florida 32304
www.RosePrinting.com

Designed and illustrated by McGuirePro Marketing of Safety Harbor, Florida
www.McGuirePro.com

About This Book

THE COVER: In designing the cover, Gary McGuire has captured graphically the symbolism contained in the phrase, "Write of Passage." "Write" is depicted by the Canal Zone-postmarked envelopes on the front and by the letter and photographs on the back, representing the American era stories contained in the book. "Passage" is conveyed in several ways:

- By the two ships in various states of passing through the Canal, with the one on the front cover beginning its transit on the Pacific side of the isthmus, and the other on the back cover completing its journey through the Canal as it leaves Gatun Locks headed toward the Atlantic Ocean.

- By the remnants of the old Thatcher Ferry used in the early days of the Canal, shown in the foreground, contrasting with the Bridge of the Americas which replaced the ferry in 1962. The background photo on the facing page shows the ferry with the bridge under construction. Both the ferry and the bridge served as the primary means for passing from one side of the Canal to the other since the Canal's opening, and their juxtaposition reflects the passage of both time and history.

- By the spectacular sunsets on both covers, which are symbolic of the setting of the sun for the last time on the American Era of the Panama Canal.

ROOSEVELT MEDAL HOLDERS: Those stories that reference a Roosevelt Medal recipient or holder are identified with an image of the Roosevelt Medal at the top of the family story. The medal refers to the person(s) about whom the story is written and/or to individuals mentioned in the story who are also recipients. Names, medal numbers, and number of bars received are mentioned at the end of those relevant stories.

FAMILY HISTORIES: These pages list alphabetically all stories by story title. If there are two family names in the title of the family story, the first family name is used to alphabetize the story. There is also a **Story Contributors** page that lists the story writers' names alphabetically, followed by their family story title.

"SEE ALSO . . ." REFERENCES: Many family stories have at the end of the story a "See also the John Doe family history." This represents a tie between stories where story writers asked us to reference other family histories. During the review and editing stage, the editors noticed family ties between stories that don't mention other family histories in the body of their story. With those story writers' permission and where evident, references between those stories are made.

LANDMARK PHOTOGRAPHS: There are also related "landmark" photos selected and inserted by the editorial team at the end of some of the family histories as space permitted.

FAMILY PHOTOS: Submissions included a wide range of sizes, formats, resolution, clarity, and condition. We weighed these factors, as well as available space and related written content, to determine the size and placement of photos and captions.

PHOTO COLLECTION AT END OF BOOK: This section contains photos that take the reader on a pictorial journey through the American Era of the Panama Canal. It includes a poem fondly remembering the Canal Zone as well as photos that continue the theme of honoring the Roosevelt Medal and descendants mentioned throughout the book.

Dedication

This book is dedicated to the thousands of men and women
of all races and nationalities who participated in the construction,
operation, and defense of the Panama Canal.

HISTORY OF THE PANAMA CANAL MUSEUM

The Panama Canal Museum is the only museum in the world to preserve a unique experience in American history — the construction of the Panama Canal and role played by the United States in the history of Panama. Anticipating that after the Panama Canal was turned over to Panama, and that the history of the Americans working and living in the Panama Canal Zone would most likely be lost to future generations unless a concerted effort was made to preserve and document it, several former Panama Canal residents founded the museum on March 10, 1998.

The museum records and honors the participation of the many organizations, businesses, and individuals of all nationalities that provided support to the Canal effort over the years. The museum's Board of Trustees, staff, and volunteers have focused on building a firm financial foundation and its collections during these early years of operation. In 1999 the museum opened its doors to the public in a small office center suite and two years later moved to a larger, ground floor suite in the same building. It is sustained by memberships, generous donors, and museum store revenue.

In 2006, the Charles W. Hummer, Jr., Research Library, housing books, pamphlets and other historically significant materials relating to Panama and the Panama Canal, was dedicated in honor of President Emeritus and founder, Chuck Hummer.

Three galleries interpret the history of the Panama Canal through permanent and rotating exhibits. The permanent exhibits explore the history of the Panama Canal from the 16th Century Spanish Period, through the French failed attempt in the 1880's, and on through the American era of construction and operation. The temporary and rotating exhibits have included the history of:

- The Panama Railroad Company
- The American Military in Panama and Operation Just Cause
- 75 Years: The Panama Canal Society
- Theodore Roosevelt's Contribution to the Building of the Panama Canal
- The Role of the West Indians: 1904-1914.

The Museum has many other programs that feature guest speakers and programs about the history, culture, and current affairs of the Panama Canal.

The Panama Canal Museum
7985 113th Street, Suite 100
Seminole, FL 33772-4785
(727) 394-9338
www.panamacanalmuseum.org

MESSAGE FROM THE PRESIDENT

A Journey Through History

The building of the Panama Canal by the United States from 1904 to 1914 at the time was the most significant and monumental engineering achievement the world had ever seen. Its completion, despite incredible obstacles, changed the lives of thousands of people in nations around the world and brought to realization a centuries-old dream of connecting the Atlantic and Pacific oceans.

The brave and adventurous workers who went to Panama during the construction period faced unimaginable hardships on a daily basis; death was a constant companion as yellow fever, malaria and other tropical diseases took their toll. In 1915, Ira Bennett, in his *History of the Panama Canal*, wrote:

> *Thousands came and hundreds stayed. It was a great place for trying out a man; the weakling soon lagged behind and dropped out. Only the upstanding, right thinking, energetic and industrious man could make good in a country where the climate bored itself into the very soul of every individual and put the acid test upon his nerves, where diversions were few and occasions for homesickness many.*

American workers who went to Panama and stayed were rewarded for their loyal service: a medal, ordered by President Theodore Roosevelt, was issued to every American worker who remained on the job for two consecutive years. A bar, attached to the medal, represented an additional two years. There were 7,404 medal recipients; but only 41 of those received all 4 bars, representing the full ten years of construction work. While the Panama Canal remains as a great and lasting memorial to the skill and perseverance of those who shared in the glorious task of its construction, the Roosevelt Medal will always be considered as their badge of courage.

The family histories contained in this book provide a fascinating glimpse of life in Panama and the former Canal Zone during the American Era of the Panama Canal, from 1904 to 1999. Many of the stories document the incredible hardships faced by those early construction workers and proudly acknowledge the receipt of a Roosevelt Medal by one or more family members. Those stories are specifically indicated by an image of the medal imprinted on the title page. Other stories cover various periods from the opening of the Canal to the end of the 20th Century, when the U.S. transferred the Canal to Panama.

On behalf of the members of the Board of Trustees of the Panama Canal Museum, it is a privilege to bring you these stories; we hope they convey the pride and dedication of the many men and women who contributed to the success of this great American enterprise. Efforts have been made to preserve the rich flavor of each individual story. Minimal editing for grammatical accuracy, flow of content, and consistency has been made, but the stories are actual accounts as recalled by family members, written in their own words.

Joe Wood

Message from the President Emeritus

Reflections of a Zonian

The Panama Canal Museum and this project, "The Write of Passage," have caused me to reflect on the history of my own family and how it was woven into the fabric of the tapestry of the Panama Canal. I reflected on those first days in 1904 when my grandfather began our story by living in a tent drenched with the wetness of heat or torrential rains. How this brave man endured through those early years of diseases like yellow fever, malaria and tuberculosis. How he repeatedly renewed his contract to see the Canal's construction completed and to remain a part of its operation for another quarter century. I hope that you will join me celebrating his story and those of the others in this volume.

Early on, it seemed that if a few of us did not take the initiative to preserve the history of America's role in Panama's history, it might be lost or at least saved in a very superficial way. It is the confluence of people, places and events that gives color and depth to this history of the Panama Canal. It is up to those of us who lived during this period fulfilling one role or another to furnish the scraps of our stories and photographs of our lives during this special era of social experimentation and technological innovation and application. The Panama Canal Museum is a fitting way to preserve as much as possible through our contributions of funds, artifacts, photographs and personal support. For those of us who lived there, this history will fade into the darkness of time, but the museum can keep this history alive for future generations.

It is up to each of us to be the face reflected in the mirror that captures the images of our lives and those who preceded us. It always seemed somewhat strange to me that more people who participated in this special time and place have not supported the Panama Canal Museum to preserve their history and the memories and friendships that so many express with such warmth. We are delighted that this volume shows the commitment of nearly a hundred authors to record their own histories and the willingness of several outstanding photographers to enhance our histories with their photographs. Perhaps the success of this initial volume will stimulate more interest, more stories and even, more volumes. For it is these personal reflections that give an authentic description of so many aspects of what it meant to live this history of the Panama Canal.

A page or two seems a mighty small snapshot of what for many were the stories of generations. But these stories will provide those in the future with a glimpse of our lives and stories with a flavor that no novel or history book can capture.

I, for one, thank the Panama Canal Museum and those who labored to make this "Write of Passage" a reality. For those who read these stories in the future, I hope you find them as exciting to read and reflect upon as it was for us to live them.

Charles W. Hummer, Jr.

THE ROOSEVELT MEDAL

On November 16, 1906, President Theodore Roosevelt gave a speech to a large gathering of Canal workers in Cristobal during his official visit to the Canal Zone to inspect the progress of the Canal's construction. He was so impressed with their work that he told them, "I feel that to each of you has come an opportunity such as is vouchsafed to but few in each generation. I shall see if it is not possible to provide for some little memorial, some work, or some badge, which will always distinguish the man who did his work well on the Isthmus, just as the button of the Grand Army distinguishes the man who did his work well in the Civil War."

With those words, Theodore Roosevelt initiated the plan which was to honor the participants of the great Panama Canal undertaking. The badge of distinction took the form of the Roosevelt Medal, or, as it was called in the early days, the Canal Medal. It was presented to employees of the Isthmian Canal Commission and the Panama Railroad working in Panama during the construction of the Canal, who were American citizens and who served their Government for two consecutive years of satisfactory service between May 4, 1904, and December 31, 1914.

The medal was designed by artist Francis D. Millet, who later was to lose his life in the sinking of the *Titanic*. The medals were made from nearly 1,000 pounds of staybolts taken from old locomotives, excavators and railroad cars left behind in Panama by the French when they abandoned the Canal project.

On each medal was engraved the name of the recipient, the dates of service for which the medal was awarded, and a number which placed the recipients in order of their employment date. In addition to the medal, a bar was issued for every two years of additional service. At the end of the construction period, there had been issued 7,404 Roosevelt Medals; 3,883 First Bars; 1,865 Second Bars; 636 Third Bars; and 41 Fourth Bars, indicating only 41 Americans worked the entire ten-year construction period.

Theodore Roosevelt summed up the contribution of construction-day workers and the great achievement that was the Panama Canal when he said, "It is not the critic who counts, not the man who points out how the strong man stumbled, or where the doer of deeds could have done them better. The credit belongs to the man who is actually in the arena; whose face is marred by dust and sweat and blood; who strives valiantly; who errs and comes up short again and again; who knows the great enthusiasms, the great devotions and spends himself in a worthy cause; who, at the best, knows in the end, the triumph of high achievement; and who, at the worst, if he fails, at least fails while daring greatly, so that his place shall never be with those cold and timid souls who know neither victory nor defeat."

The Roosevelt Medal was highly cherished by the thousands of men and women who toiled and sacrificed under the most adverse conditions to make the Canal a reality. Those brave, courageous and dedicated individuals entered the arena where few dared to go; survived the hardships and the death and disease that everyone faced in the unknown jungles of Panama; and built a legacy for generations of their descendants who can proudly proclaim their connection to that incredible achievement.

Acknowledgments

To Jim Deslondes for proposing that the Board of Trustees of the Panama Canal Museum undertake a family history book project.

To the members of the Board of Trustees of the Panama Canal Museum for their vision, determination and initiative in bringing together this collection of rich family histories during the American era of the Panama Canal.

To the Museum leadership of Joe Wood, Chuck Hummer, Kathy Egolf, Dick Cunningham, Elizabeth Neily, Paul Glassburn, and Gerry DeTore for their oversight, encouragement and direction throughout all phases of the project.

To Pat Kearns, Anne Magee Severy, Betty Ann Allen Hansen, Kathleen Steiner Bennett, and Joanne Steiner Robinson for contributing their family story early so the Museum could use it as the example for participants to follow.

To Barbara (Bonanno) Marshall for leading and managing tirelessly and skillfully this book project through its every detail at every stage.

To the editing team of Peggy Huff, Sue Robbins, Kathy Egolf, and Barbara (Bonanno) Marshall who read the family histories with enthusiasm and love of reading and learning about personal Panama Canal histories.

To Paul Glassburn for overseeing the financial responsibilities of this book project.

To Panama Canal Museum staff members Marcy Corrigan, Helen Morris, and Marilyn White for their suggestions and administrative support.

To the photograph research team of Dick Cunningham, Chuck Hummer, Joe Wood, and Barbara (Bonanno) Marshall for their countless hours of reviewing many unique landmark photos, depicting life in the Canal Zone, that were selected and are illustrated throughout the book.

To several photographers and organizations for contributing the "landmark" photos used throughout this book: The Panama Canal Authority, the Panama Canal Museum, and the U.S. Department of the Army; Don Goode and Kevin Jenkins, former Panama Canal Commission official photographers; Captain Wilbur Vantine, retired Panama Canal pilot; Dennis White, son of P.A. White, retired Chief, Dredging Division; Allan Hawkins V., retired from HSBC Bank (Panama), S.A.; Bill McLaughlin, from his website, www.czimages.com; David Wright of David Wright Photography & Productions (www.digitalimageproduction.com); and Sue Robbins and Andra Nash English, story contributors.

To the Panama Canal Authority for providing the photo on the front cover; Allan Hawkins for providing the background photo on the copyright page; Kevin Jenkins for providing the photo on the back cover; and Don Goode for providing the photo that starts the color photo section.

To Lesley Hendricks, former member of the Panama Canal Museum Board of Trustees, whose website, www.czbrats.com, provided the script on the inside pages of the front and back covers.

To Nellree Berger for her scenic postcards from the well-known Panama and Canal Zone postcard albums she assembled, circa 1970's.

To Celina Barkema Vargas for her poignant poem about life in the Canal Zone.

To the design team of Gary and Jean (Medinger) McGuire of McGuirePro Marketing for the design, illustration, and layout of the book's cover and content.

To the story writers — without them, there wouldn't be a *Write of Passage: Stories of the American Era of the Panama Canal*.

FAMILY HISTORIES

Allen, William J., Family	1
Armbruster, Edwin and Jean, Family	4
Ashton, William Francis, Family	7
Baker, Donald Thompson, Family	10
Barnard, William Andrew, and Alvin Monroe Rankin Families	13
Berg, Curtis L., Family	15
Bliss, Gerald DeLeo, Sr., Family	18
Bloemer, Robert J., Family	21
Boggs, Reinhard A. and Max Reinhard	24
Bonanno and Gately Family	27
Bowman, Audrey Benoit, Family	30
Brayton and Wertz Families	33
Butler, James, and Charles William Dawson Families	35
Carter, Harold, Family	37
Catanzaro, Anthony J., Family	39
Cockle, George Dean, Family	42
Coffin, Joseph W., Family	44
Compton, Harry and Rebecca, Family	46
Conley, William H., Family	48
Corin, Theodore S., Family	50
Corrigan, John Paul, Sr., Family	52
Corrigan, Peter Tiernan "Pete," Sr., Family	55
Crooks, Robert, Family	58
Davis, John M., and Michael Kenny Families	60
Deakins, Fred B., Family	63
Dennis, Sylvester P.; Arthur E. Baker; and William C. Bain Families	65
Dolan, Edward V., Family	67
Edwards, Thomas, Family	69
Falk, Henry Edwin, Sr., Family	71
Foley, Thomas Paul and Mary Denn, Family	73
Furlong, Ralph Edward, Jr.	76
Geddes Brothers	78
Godfrey, William B., Family	80
Grimison, Thomas I., Family	83
Hall, Peter A., Family	86

Hill, William G. and Bernice A. (Sanders), Family	88
Hollander, Charles Sanford	90
Hollander, Eleanor Freund	92
Hollowell, David Cooper, Family	94
Hood, Jack B. and Patricia C. Perry, Family	96
Huber, Charles Clarence, Family	97
Huff, Mercer Blanchard, Family	100
Huff, Maenner Blanchard	103
Huff, Maenner Blanchard, Family	106
Hummer, Charles D., Family	109
Hunter, Oscar R., Family	112
Hurst Family	114
Jackson, John James and Grace Norris, Family	117
Johnston, William C. Caley, Family	120
Kane Family	122
Kapinos, Andy, Family	124
Keenan, William Henry (Harry), Family	126
Kelly, Captain Dean and Joyce, Family	128
King, Anna Jacobs, Family	130
Kleefkens, Hermanus, Family	133
Krziza Family — Esther, Ethel, Leo	135
Lindsay, Walter R., Family	138
Maenner, Ludwig Theodore, Family	140
Mann, Tony and Anna	143
Mayo, Francis T., Family	146
McArthur, William David, Family	148
McCartney, James, and James Lewis Phillips Family	150
McCracken, Irene	137
McCullough, Maurice Lee, Family	152
McMillan, Robert R., Chairman of the Panama Canal	154
McNatt, John A.	156
Mellander Family	158
Morland, Gilbert	161
Morland, Virginia	163
Nash, Milton Lee, Family	165
Orr, Elmer Franklin, and Simon Butler Jones Families	168
Peterson, Charles H. and Margaret R., Family	171
Peterson, Thomas C. and Barbara D., Family	173

Potter, Russell B., and Ross Cunningham Families	174
Quinn, Patrick Joseph and Jane "Jennie," Family	177
Raines, E. L., Family	180
Ramsey, Erwin and Dorothy, Family	183
Rankin, Alvin M., and Paul D. Thompson Families	186
Reece, Roy D., Family	188
Richardson, Gladys Houx and Joe, Family	190
Robbins/VanderWeg Family	193
Rowley, Samuel Harvey, Sr., Family	195
Ryter, Clifton W., Family	197
Salter, Earl, Family	199
Sanders, Bruce Gordon and Grace Aloise (Meister), Family	202
Schmidt, John E., Family	205
Schmidt, Louis "Pop," Family	207
Schroeter, H.C. (Bert), Family	210
Sill, Fred and Ruth, Family	212
Smith, Harry Grant, Family	214
Smith, J. Bartley and Mercedes Alegre, Family	217
Smith, William Charles	220
Stevens, John Frank, Chief Engineer, 1905 to 1907	222
Stillwell, Ellis Dayre	225
Stilson, Charles H., Family	228
Stockham, Roy C., Family	231
Sullivan, Thomas C., Jr., and Joe Ridge Families	233
Taylor, William D.	236
Thomas, Hugh M., Family	238
Walker, Murry, Family	240
Wertz, Joseph Anderson, Family	242
Westberg, John Emanuel, Family	245
Wood, Joseph J., Family	247
Worsley, Robert C.	250
Zeeck, William Carl and Christine Peterson, Family	253
Zent, Llewellyn "Lew," Family	256

Story Contributors

Allen, Karl D., and David Hollowell
   David Cooper Hollowell Family................94

Armbruster, Ed
   Edwin and Jean Armbruster Family 4

Barr-Ausnehmer, Betty
   John Emmanuel Westberg Family...............245

Berg, Inez, Carl, Franz and Janice
   Curtis L. Berg Family........................15

Berger, Nelliee Baker Smith
   William Charles Smith.......................220

Bloemer, Angela
   Robert J. Bloemer Family......................21

Bowman, Bob
   Audrey Benoit Bowman Family..................30

Brandt, Dorothy (Godfrey) and Ira
   William B. Godfrey Family....................80

Bruce, Shirley Brayton Wertz
   Brayton and Wertz Families...................33

Carter, Lece and Family
   Harold Carter Family 37

Clary, Karen Husum, and Lorrie Husum Allen
   Thomas Paul and Mary Denn Foley Family 73

Clinton, Frances Thompson
   Alvin M. Rankin and Paul D. Thompson Families..186

Cockle, George R.
   George Dean Cockle Family....................42

Collins, Anita
   Reinhard A. Boggs and Max Reinhard Boggs Family.24

Conley, Olga Johnston
   William C. Caley Johnston Family.............120

Conley, Roger
   William H. Conley Family.....................48

Corin, Ted and Georgia
   Theodore S. Corin Family 50

Corrigan, David
   Gilbert Morland.............................161

Corrigan, Tim, and Rosie Corrigan
   John Paul Corrigan, Sr., Family...............52
   Peter Tiernan "Pete" Corrigan, Sr., Family 55
   Charles Clarence Huber Family.................97

Corrigan, Taffy Koepke
   John James and Grace Norris Jackson Family ...117

Crooks, Bob
   Robert Crooks Family 58

Cunningham, Lynne (Carolyn) Coffin
   Joseph W. Coffin Family......................44

Cunningham, Dick
   Russell B. Potter and Ross Cunningham Families..174

Deakins, JoElla
   Fred B. Deakins Family.......................63

Dolan, Bonnie Davis
   Edward V. Dolan Family 67
   J. M. Davis and Michael Kenny Families.........60

Dolim, Virginia Ridge
   Thomas C. Sullivan, Jr., and Joe Ridge Families....233

Dunbar, Bonnie Bain
   Sylvester P. Dennis, Arthur E. Baker, and William C. Bain Families................................65

English, Andra Lee Nash
   Milton Lee Nash Family......................165

Falk, Murray
   Henry Edwin Falk, Sr., Family.................71

Fitzgerald, Mickey, and Emily Bliss and grandchildren
   Gerald DeLeo Bliss, Sr., Family...............18

Furlong, Brenda
   Ralph Edward Furlong, Jr. 76

Garner, Marge Zent
   Llewellyn "Lew" Zent Family 256

Girand, Juanita Jones
   Elmer Franklin Orr and Simon Butler Jones Families 168

Goldstein, Suzanne Schmidt
   Louis H. "Pop" Schmidt Family 207

Grimison, Richard and Tom, and Janice Scott
   Thomas I. Grimison Family 83

Gross, Beverly Phillips
   James McCartney and James Lewis Phillips Family. 150

Haugen, Joann Hummer, and Chuck Hummer
   Charles D. Hummer Family 109

Hawks, Frank Stevens
   John Frank Stevens. 222

Hill, Robert
   Bruce G. and Grace Aloise (Meister) Sanders Family 202
   William G. and Bernice A. (Sanders) Hill Family ... 88

Hood, Jack and Patricia
   Jack B. Hood and Patricia C. Perry Family 96

Huff, Peggy A.
   William Francis Ashton Family 7
   Thomas Edwards Family 69

Hunter, William R. and Dorothy E.
   Oscar R. Hunter Family 112

Kane, Joseph and Lucille
   Kane Family. 122

Kearns, Pat and Family
   William J. Allen Family 1

Keenan, Charles
   William Henry (Harry) Keenan Family. 126

King, Tara
   Anna Jacobs King Family. 130

Krziza, Leo
   Krziza Family – Esther, Ethel, Leo 135

Lanfranco, Nancy Kaufer, and Jane Kaufer Cochrane
   Peter A. Hall Family 86

Lindsay, Judith R.
   Walter R. Lindsay Family 138

Loera, Helen Hurst
   Hurst Family. 114

Mann, Tony
   Tony and Anna Mann. 143

Marsh, Annette Kelly
   Capt. Dean and Joyce Kelly Family 128

Marshall, Barbara (Bonanno), Sherry (Bonanno) Dunlap, and John Bonanno
   Bonanno and Gately Family 27

Martin, Norma Stillwell
   Ellis Dayre Stillwell. 225

Mayo, Francis L., Lt. Col. U.S. Army (Ret.)
   Francis T. Mayo Family. 146

McArthur, David
   William David McArthur Family. 148

McCracken, Irene
   Personal Observations 137

McCullough, Judi
   Maurice Lee McCullough Family. 152

McMillan, Robert R.
   Robert R. McMillan, Chairman of Panama Canal .. 154

McNally, Gail Dawson
   James Butler and Charles William Dawson Families 35

Mellander, Gus
   The Mellander Family 158

Mizrachi, Robert
   Virginia Morland 163

Neu, Carmen Smith
   J. Bartley and Mercedes Alegre Smith Family ... 217

Newman, Betty Lockwood Skow
    Harry Grant Smith Family 214

Oberholtzer, Anita
    Anthony J. Catanzaro Family. 39

Peterson, Thomas C. and Barbara
    Charles H. and Margaret R. Peterson Family......171
    Thomas C. and Barbara D. Peterson Family 173

Quinn, Bruce and Marc, and Elaine Lombard Newland
    Patrick J. and Jane "Jennie" Quinn Family........177

Raines, Fred
    E. L. Raines Family 180

Ramsey, Steve and siblings
    Erwin and Dorothy Ramsey Family............183

Rankin, Ginny Kleefkens and other grandchildren
    Hermanus Kleefkens Family 133

Reece, Jim and Janet
    Roy D. Reece Family 188
    Roy C. Stockham Family 231

Richards, Roberta, and Jacque Williams
    Robert C. Worsley 250

Richardson, Gladys
    Gladys Houx and Joe Richardson Family........190

Risberg, Pat Geddes, and Barbara Geddes Tung
    Geddes Brothers............................78

Robbins, Sue
    Robbins/VanderWeg Family 193

Rowley, Skip, Dorothy Rowley Gerhart and June Rowley Stevenson
    Samuel Harvey Rowley, Sr., Family 195

Ryter, Virginia (Wennik) and Don
    Clifton W. Ryter Family 197

Salter, Bobby
    Earl Salter Family 199

Schafer, Catsy Taylor
    William D. Taylor 236

Schield, Blanca McNatt
    John A. McNatt. 156

Schmidt, Jr., John E.
    John E. Schmidt Family 205

Schroeter, Val and Tina
    H. C. (Bert) Schroeter Family 210

Sill, Fred
    Fred and Ruth Sill Family 212

Smith, Carol Kapinos
    Andy Kapinos Family. 124

Smith, Pauline Sue Pincus
    Charles H. Stilson Family 228

Spelman, Margaret Annette (Peggy) Rankin
    William A. Barnard and Alvin M. Rankin Families ..13

Thomas, III, Hugh M.
    Hugh M. Thomas Family.....................238

Wagenbrenner, Jane Compton
    Harry and Rebecca Compton Family 46

Walker, Lessie Platt
    Murry Walker Family 240

Williamson-Musco, Roberta, and Rosemary McCorkle
    Charles Sanford Hollander....................90
    Eleanor Freund Hollander 92

Willoughby, Edie Huff, and Christine Baker Huff Fewell
    Donald Thompson Baker Family 10
    Ludwig Theodore Maenner Family 140
    Maenner Blanchard Huff 103
    Maenner Blanchard Huff Family 106
    Mercer Blanchard Huff Family................100

Wood, Jr., Joseph J.
    Joseph J. Wood Family 247

Zeeck, Charles E.
    William Carl and Christine Peterson Zeeck Family..253

Zornes, Ginny Lee
    Joseph Anderson Wertz Family 242

The William J. Allen Family

FOND MEMORIES FROM PROUD GRANDCHILDREN — FOR OUR GRANDCHILDREN

William James Allen was born to Irish immigrant parents in Gananoque, Ontario, Canada, on November 2, 1870. His parents came with their families to Canada in the 1840's; his father is said to have been born at sea. Upon the death of his mother, William Allen moved to New York City to look for work.

Will Allen was a gifted carriage painter by trade and was known for his artistic talent with gold leaf detailing which he applied to expensive carriages. While living in a boarding house in New York around the mid-1890's he met Margaret Breheney, whose mother Mary worked as a cook in the same boarding house. Mary and her daughter were recent immigrants from Roscommon, Ireland.

In April 1896 William Allen married Maggie in New York City. They were blessed with several children within the next few years, including Suzanne (July 19, 1900), Margaret (August 23, 1902), Walter (July 19, 1904), William B. (July 5, 1906), and Marian (Dolly) (February 5, 1909). Dolly was born four months after Will left New York for a position in Panama, where he and thousands of others went to help construct the Canal and earn money to send home to feed their families.

William James Allen

According to Panama Canal employment records, William Allen was hired as a painter, arriving in Panama on October 15, 1908. He worked for the Canal continuously until his retirement as painter, leadingman, in the Mechanical Division, on November 30, 1932, when he reached the mandatory retirement age of 62. He and Maggie never left Panama after retirement, living in 12-family quarters and then with several of their children. They are both buried in Corozal Cemetery.

After his first "tour of duty" in Panama, Will went back to New York for vacation in 1910 and saw his youngest daughter Dolly for the first time. He returned for another two-year tour, again without his wife and children. When Will returned to New York for another vacation in 1912 and told Maggie there were "many lovely young 'señoritas' in Panama who would love to have an American husband," she decided it was time to pack up the children and join Will in Panama. The entire family arrived in Panama in June 1912.

By 1912 living conditions for families with children were significantly improved, but still an adventure. Oldest daughter Susie remembered that upon disembarking at Cristobal pier they had to board a very dark train with only a kerosene lantern at each end of the train coach. They were met by a horse-drawn wagon and taken in the rain to their old French quarters in one of the construction-day towns. These quarters were built three to four feet off the ground and had a wide staircase going up the front. All night long they were terrified at the sounds of the animals scampering across the corrugated tin roof of their new home. The next morning after the rain had stopped, Maggie walked out to the top of the stairs and called all the children to come see the beautiful country, so fresh and green. Then they all noticed a huge snake stretched across the front steps, which caused an immediate panic and a stampede into the house.

Each house had an ice box and a coal stove. Ice and coal were delivered daily in a covered wagon drawn by horses and driven by a big man. Provisions were sold at the daily-stocked commissary, and to assist the ladies with their "commy" shopping, a horse-drawn wagon made trips throughout town delivering the ladies to and from the commissary. Later, when things were a little easier, Maggie would occasionally send her oldest daughter to the store to do the shopping. One day as Susie was walking home from the commissary up the hill past the Administration Building, a fancy carriage with a driver stopped, and the passenger, a nicely dressed gentleman, asked her if he could give her a ride home. Susie was delivered with her groceries right to her door by the Governor of the Canal Zone!

The Allen children passed on to their children stories about how on Sundays after church all the older children would climb down rope ladders to explore and run along the still-dry Canal bed collecting pretty stones and agates. After the Canal opened, another Sunday afternoon pastime enjoyed by the older folks was walking over the hill to "the limits" and having chili with a cold draft beer.

Will and Maggie encouraged all their children to participate in music and sports. Susie refused to learn to swim and would not put her face under the water. One afternoon Will took her down to Pier 18, tied a rope around her, threw her in and told her to sink or swim. That story was one we all heard and remember quite well. Youngest son Willie was an excellent swimmer and champion diver. He was on the red, white, and blue team coached by Henry Grieser, and he had the opportunity to compete as a diver in the U.S. Olympics; but Maggie wouldn't let him go because she did not want him missing school. Willie's children still have his gold medals won primarily for his diving ability. Susie, Margie and Walter were the performers who loved to sing and dance, and Dolly played the piano.

Once the Canal was completed, the Allen family lived exclusively on the Pacific side. However, for a short time before then, the family lived on the Atlantic side, where Edna was born in Colon hospital in June 1913.

The family was aboard the *SS Ancon* when it made the first transit through the Canal on August 15, 1914. There were chairs at the front of the ship for the ladies and children to sit in while the men stood on the front and both sides of the ship so all would have a good view of the Canal they had helped build. This episode made an impression on the family; they could see and truly appreciate the work of all those brave men who helped create this wonderful engineering marvel.

> *Our family church, St. Mary's Catholic Church, sat on top of a hill in Balboa with a long stairway up the hill to the front of the church. At the top of the church was a large cross to which Grandpa would apply gold leaf whenever it began to look shabby.*

During World War 1 when the son with whom she was living in New York joined the Army and went off to Europe, Grandma's mother, Mary, came to Panama to live with her daughter and son-in-law. Mary (Nana) remained in the Canal Zone with the Allen family until her son Thomas (Tom) Breheney returned to New York City from the war and was discharged in May 1919. After Nana died, Tom then decided to join his sister and family in Panama, where he eventually went to work for the Dredging Division.

Our family church, St. Mary's Catholic Church, sat on top of a hill in Balboa with a long stairway up the hill to the front of the church. At the top of the church was a large cross to which Grandpa would apply gold leaf whenever it began to look shabby. He volunteered his time to do the work long after he retired and indeed until he was almost blind from cataracts. Years later one of his grandchildren (Willie's daughter) found some sheets of Grandpa's precious gold leaf in their home.

The vacations from the Panama Canal Company were wonderful two-month visits to the United States. Travel was on the Panama Line ships with an always exciting stopover in Port-au-Prince, Haiti. Talk about luxury — our Panama Line ships could stand right up there with today's cruise lines! These vacations were terrific morale boosters for Canal employees and a wonderful way to allow families to see and experience what life was like in the U.S.

Of the six Allen children, only two of them left the Canal Zone and raised their families in the United States. The four who remained in Panama brought up their children there and educated a total of 16 children in the Canal Zone school system. Many of those children also married and raised their children in Panama. Suzanne (Susie) Allen married Charles F. Magee, had three sons and three daughters, and lived in Panama until retirement. Margaret (Margie) Allen married Francis C. Pepe, had one son, and lived in New York City. Walter Allen married Mary Salvato, had one son, lived in Panama until he died while employed by the Panama Canal Company, and is buried in Corozal Cemetery. William (Willie) Allen married Louise Kerr, had four daughters, and lived in Panama until retirement. Marian (Dolly) Allen married Jerome E. Steiner, had three sons and three daughters, and lived in Panama until retirement. Edna Allen married James A. Slusher, had five daughters, and lived in Panama for a number of years before eventually moving to the States.

From 1908 to the present, almost 100 years, there have been Allen descendants living in Panama.

William James Allen
Roosevelt Medal No. 6491

PHOTO COURTESY OF PANAMA CANAL MUSEUM

Opening of the Panama Canal, SS Ancon in east chamber of Pedro Miguel locks, August 15, 1914.

The Edwin and Jean Armbruster Family

AS TOLD BY ED ARMBRUSTER

Ed was born in March 1937 in Washington, D.C., and grew up in Arlington, Virginia. In his senior year of high school, his family moved to New Port Richey, Florida, where Ed graduated in 1955 and where he met Jean Morris, also born in Washington, D.C. After Ed and Jean were married in December 1956, Ed joined the Air Force, had a two-year tour in the Philippines, and finished out enlistment at Keesler AFB at Biloxi, Mississippi, before moving the family to Cocoa, Florida. Ed was employed by the Martin-Marrietta company at Cape Canaveral.

In 1963 Ed and 35 others from the company were sent to Panama to conduct tests on the Pershing missile at Fort Sherman. During their four months, the Armbrusters rented vacation quarters in Gatun and then drove their Volkswagen back to Cape Canaveral via Central America. With Ed and Jean were three daughters, Margie (6), Judy (4), and Debi (2). While in the Canal Zone, Ed applied for and one year later got a job as an electronic technician with the Panama Canal Company. He later transferred to the Meteorology/Hydrographic Branch as a hydrological technician at the Madden Dam field office. Ed retired after 25 years of service. Here are a few of the Armbrusters' fondest remembrances.

Ed arrived on the isthmus two weeks before Jean and the girls. He was met at Tocumen airport and was driven to duplex 168B Williamson in Gamboa, which was the family residence for seven years. Ed was pleasantly surprised to see that this three-bedroom, one-bath, wood-framed quarters was temporarily furnished. There was a welcome package that included sheets, pillows, toilette articles and even groceries, which made an indelible impression on this new employee.

Jean and the girls traveled to Panama on the *SS Cristobal*. They quickly settled in the tropical paradise, excited about their new life. Margie was registered for second grade and Judy was enrolled in kindergarten, held in a side room off the Gamboa gymnasium. The kindergarten curriculum included swimming lessons three days a week, and in a few weeks Judy (5) was swimming the length of the pool and awarded her B badge. Margie also earned her B badge at that time. However, of all the Armbruster girls, it was Connie who earned her B badge at the youngest age of four. Connie was the only Armbruster daughter born in the Canal Zone, at Gorgas Hospital.

Jean and Ed

One of the family's favorite activities was to attend movies at the Gamboa Theater. Most Friday or Saturday nights, the Armbruster girls would attend a movie. After buying their tickets, they would climb the cement stairs to be greeted by a door keeper, who not only accepted their tickets, but made sure the movie patrons obeyed the rules. After the movie started, bats would fly out from the projector toward the screen, then turn and head back at head level toward the patrons. Kids would yell and throw popcorn at the bats, only to be reprimanded by the "rule enforcer." Another popular movie activity was sitting so hard on the seats that the steel balls in the seat ball-joints would break loose and roll down the hardwood floors while the kids would anxiously try to pocket them before being caught.

These steel balls, or steelies as they were called, would be killers in the game of marbles played before school.

The Armbrusters loved living in idyllic Gamboa. They had grown used to the rattling of the window panes whenever a ship passed through the Canal just yards away from their home. The family would spend lazy evenings watching the ships and enjoying the warm breeze before the DDT trucks came by spraying for mosquitoes. It was not unusual to see wild animals around our house including anteaters, coatimundis, sloths, nikkis, parrots and monkeys. The arrival of Christmas trees, the parties and Christmas tree burns were fun times. The Gamboa Country Club was a favorite hangout on Saturdays. Margie had a horse that was kept at the Gamboa stables. The girls often treated their friends to hamburgers and sodas, charging all to their dad. One payday Ed came home with a two-week paycheck of $7.21; the girls' charging privileges were suspended. Gamboa was like Mayberry, USA. The school, clubhouse, commissary, post office, clinic, train station, theater, swimming pool, police and gas station and gym were all within a five-minute walk from home. Also terrific was home leave, putting the car on the *SS Cristobal*, the four-day cruise to New Orleans, fun and games on board, great food, six weeks' traveling around the States, returning to New Orleans the night before sailing, and sailing back to Cristobal every other year.

Most days, Ed had a pleasant five-minute walk to work at the electronics shop next to the Gamboa train station. An exciting event took place only months after Ed's arrival on the isthmus. Ed boarded the tug *Taboga* to be a radio operator on a three-day mission in the Caribbean to rescue a gun-running converted PT boat (the *Jane Dee*) that was dead in the water and sinking. The U.S. Air Force from Howard AFB directed the rescue mission. The sea was so rough that the helmsman was tied to a post by the wheel, and the heavy seas prevented moving the 11-member crew, including one woman. The *Jane Dee* was towed to the Cristobal Yacht Club where she was impounded by a U.S. marshal and later sold at auction. Another memorable assignment was the annual trip to Punta Mala in the interior of Panama by the town of Pedasi to service the lighthouse and radio beacon at an old U.S. Coast Guard station.

The Armbruster daughters, top L-R: Judy, Margie; inset L-R: Connie, Debi.

In 1965 Ed organized the Chagres Aero Club, acquiring an 11-acre land lease from the PCC on a strip of land one-half mile east of Gamboa along the Canal, and with a rented bulldozer and grader from the U.S. Navy cleared and leveled a 1,700-foot-long runway. Ed was the first pilot to land there in 1966 and the last to fly out in 1984. He obtained an FAA flight instructor certificate and gave instructions to over 400 students covering 12 years. In the heyday of Bohio International, there were 31 planes based there. Occasionally, Ed would be called on to fly special missions for the Canal Zone Police, Smithsonian Institute, Military Intel units, and Central Intelligence Agency.

In 1971 the family moved to Bougainvillea Street in Balboa after Ed was transferred to the Balboa electronics shop. The girls were transferred to Balboa schools and a new chapter began. They were active in school, BHS Select Chorus, community activities, Girl Scouts, parades, Rams cheerleading, and dance filled their lives. The family often would take trips to the beaches or to the mountains of Cerro Punta.

In 1972 Ed transferred to the Meteorology/Hydrographic Branch field office by Madden Dam. It was near the field office where Ed overturned a truck and was rushed to Gorgas Hospital, where his ruptured spleen was removed, saving his life. Not only did Ed love his job and the adventures it brought, but he loved teaching flying to many Canal Zone residents. Scuba diving and spending hours probing for old bottles on the ocean floor off Taboga Island contributed to his love of adventure. He was active in his church, The Church of Christ of Latter-day Saints (Mormon). He served for two years as President of the Los Rios Civic Council. Also because of his unique job with the PCC and his flying business, quite often Ed would meet with famous people, including General Omar Torrijos; General Manual Noriega; movie greats John Wayne and Sophia Loren; *General Hospital's* Dr. Rick Weber and nurse Barbara Spencer; and motivational guru Tony Robbins, who led Ed and many others to walk barefoot over hot coals at a seminar at Curundu Junior High School.

Jean enjoyed being a stay-at-home mom until the four girls were in school. Her career was first with the Air Force and then with the Army in the intelligence field as a political analyst for the USA 193rd Infantry Brigade at Fort Clayton, and as a collection manager for the USA 470th Military Brigade in Corozal. Jean and the girls were involved in horseback riding, scuba diving with the Balboa Dive Club, and various community activities. Her favorite memory is being part of the annual Ocean-to-Ocean Cayuco Race in the boat *Slave Gallery*. In 1989 Jean and Ed moved to Fort Bragg, North Carolina, where Jean transferred to the U.S. Army Special Operations Command Headquarters. All four girls graduated from Balboa High School.

There has never been and quite possibly never will be an equal to the utopian existence in the Canal Zone, a near perfect society in a near perfect environment. We were truly blessed to be a part of the American era, playing a significant role in the maintenance and operation of the Panama Canal and residing in the Canal Zone, truly a "Paradise Lost."

The Armbruster home, 1971-1979.

PHOTO COURTESY OF PANAMA CANAL MUSEUM

Aerial photo of the Canal at Gamboa Reach showing the Dredging Division and the towns of Santa Cruz and Gamboa on the right.

The William Francis Ashton Family

As Told by Peggy A. Huff

William Francis Ashton was born in Plymouth, England, February 14, 1872, and immigrated to Calumet, Michigan, with his family at the age of six. He was an amateur boxer in his teens, known as the "Fighting Cornishman." In 1889 he went to work for the Tamarack Mining Company and from 1892 to 1898 for the Ashland Mining Company and the famous Calumet Hecla in Ironwood, Michigan. He then moved to Sault Ste. Marie and was superintendent of the Copper Queen mine.

William met Anna Holmberg while working for the Salvation Army. She was born in Trollhattan, Sweden, April 23, 1872, and immigrated to the United States at the age of 18; after arriving, she taught herself to speak and write English. William and Anna were married in 1892 and had ten children — Emma, Herbert, Stanley, Gladys, William, Ellen, Harry, Robert, James and Edward, who died at the age of four.

William went to Peru to put in a water system and then to Brazil to work on a water purification system; he contracted and survived yellow fever while in South America and lost three fingers in a dynamite explosion while working on a mining project. When returning to the U.S., he passed through the isthmus and on April 23, 1905, accepted a position as a pump man at the Mount Hope pumping station. His reputation as a workman who could handle anything landed him the position of general foreman of all pumping stations and pipe lines in the Atlantic division, and he supervised the construction of the water purification plant built in Mount Hope. Anna and their five children joined him the following year. They lived in a little frame building on a hilltop behind Mount Hope. Anna operated a boarding house for bachelor workers, sometimes feeding as many as 20. Her young daughters helped by kneading bread and setting tables. The Ashtons could not resist a man who was homesick or lonesome, and their home became known as "Bums' Haven." They gave money to those down and out on their luck. One such handout was $20, a huge sum in those days; the recipient paid it back years later. William opened a saloon in Panama, but was forced to close it because it was against U.S. government policy. When President Roosevelt visited the isthmus in 1906, he dropped in on the Ashton family home. Anna was upset that she had to serve lunch on tin plates. The President is reported to have said, "Now don't apologize, Mrs. Ashton. What's good enough for the working man is good enough for me." Earlier in the day William had been at the controls of the train that carried the presidential party. When he placed his handkerchief on the train seat for Mrs. Roosevelt, she stopped him; "I won't need that," she said. "I'm no better than the rest of them."

William F. Ashton

The family moved in 1921 to California, where William prospected for oil. He formed the Blue Streak Gas Production Company, which was later sold to Sunoco. From California they moved to Syracuse, New York, where

William Ashton controls the Presidential train; Emma Ashton at far right, standing.

he worked for Solvay Process, which later became Allied Chemical. He transferred to Hopewell, Virginia, in 1927 as superintendent of pipe fitters and retired from Allied Chemical in 1948. He and Anna returned to the isthmus to live with their daughter Ellen and Kenneth Edwards in Pedro Miguel. William died in November 1950 and Anna on August 26, 1953. He was a member of the American Order of Foresters and The Society of the Chagres.

William and Anna's sons, Herbert and Robert, had brief careers in Panama. Herb worked as a tool boy and pump operator; Robert was a fireman and played semipro baseball in the Panama league. He was a cutup and loved to joke; his friends called him Joe Palooka, after the comic strip character. Their three daughters raised families in the Canal Zone.

Emma Elizabeth was born July 21, 1893, in Ironwood, Michigan. During President Roosevelt's visit she appeared in a photo of the presidential party on a railroad car. She married Thomas Joseph Ebdon on September 20, 1913. On July 16, 1914, Thomas, known as "Pop," hired on with the Locks Division and served 43 years of continuous service. He was a locomotive and crane operator. He and son, Joe, Jr., achieved national notoriety in the 1989 "this is not your father's Oldsmobile" television commercial, and Pop was also featured on the back cover of an issue of *Philip Morris* magazine at 100 years of age. Pop and Emma had four children — Thomas Joseph, Jr.; Frederick; William; and Doris.

Gladys was born in 1902 in Michigan. One of her fondest memories was of the time President Roosevelt lifted her up when he visited the family home in Mount Hope. She recalled his bushy mustache and shiny eyeglasses. Gladys married Edward "Jake" Sullivan, a paint contractor foreman, and lived in Panama City most of her life. They had three sons — William, David and Eddie, who died of appendicitis at the age of 18. After David's death she raised her grandson, David, Jr. Gladys, like brother Bob, was known for her great sense of humor.

Ellen Doris was born in Trollhattan, Sweden, July 17, 1907. She married Kenneth Morris Edwards July 17, 1929, in Virginia. In 1937 she joined him in the Canal Zone with their two daughters, Anne Elizabeth, born May 25, 1930, in Petersburg, Virginia, and Helen Virginia.

Anne lived all her early isthmian years in Pedro Miguel. She graduated from Balboa High School in 1948 after spending most of her senior year in Gorgas Hospital with malaria, contracted on the lower Chagres River while fishing for tarpon. After graduating, she worked for the Army. She married Hugh Douglas Hale on December 3, 1949, in Pedro Miguel. Hugh was born at Gorgas Hospital on November 30, 1927, to Philip and Edna Beall Hale. Anne was a housewife and served her family in the traditional fashion — she was a room mother and Girl Scout leader. She and the girls in Troop #77 were the first females invited to use the Boy Scout facilities at Camp Chagres on Madden Lake. Hugh was a farm and little league coach and was one of a few dads active in the Girl Scouts. He retired from the Fire Division in 1976, after serving assignments at almost every Canal Zone fire station, including those on military sites. They moved with 10-year-old daughter Glenora to Kerrville, Texas, and later to Tarpon Springs and Yalaha, Florida. Hugh, a Navy veteran, died January 16, 1991, while visiting his children in Panama and is buried in the American Cemetery in Corozal.

William and Anna Ashton with eight of their ten children.

Anne and Hugh's daughter Peggy Anne was born October 19, 1950, at Gorgas Hospital. She attended Gamboa, Diablo and Balboa elementary schools, and Diablo Junior High School. In 1968 she became a third generation graduate of Cristobal High School. She married Dennis Rex Huff on September 4, 1970. As newlyweds they lived in Honeymoon Hotel — old Building 41, Gatun — and later in Coco Solo, New Town and Old Town in Gatun, Diablo and Los Rios. She received an associate of science degree from Panama Canal College in 1984 and was a Department of Defense budget analyst intern. When the Canal reverted to Panama in 1999, Peggy terminated her 16-year federal career and Dennis, a Panama Canal pilot, retired with 32 years of service. After living 49 years on the isthmus, they moved to Florida. William Ashton referred to his mother in a written document as Mrs. E.E. Huff. Five generations later, her great-great-great-granddaughter, Diana Rae, was born on April 3, 1971, to Peggy and Dennis at Coco Solo Hospital. Diana graduated from Balboa High School in 1989. She married Thomas Richard Grimison on May 25, 1996. They have four children — Carly Anne, Thomas Dennis, Samantha Rae, and Megan Elizabeth.

Anne and Hugh's son Douglas Edward was born on November 8, 1952, at Gorgas Hospital. He attended schools in Diablo, Balboa, Margarita and Coco Solo; he graduated from Balboa High School in 1970. Doug began his Canal career as an electrical apprentice on Gatun Locks on July 7, 1975. In 1976 he became a towboat mate apprentice and in April 1981 a towboat master. He entered the pilot-in-training program in March 1987 and retired in 1999 as a Panama Canal pilot. Daughter Lauren Nicole was born March 18, 1984, at Gorgas Hospital. She graduated from Coronado High School, Coronado, California, in 2002 and received a bachelor's degree in political economics from the University of California at Berkeley in 2006.

Anne and Hugh's daughter Glenora Rae was born at San Fernando Clinic in Panama City, Panama, on April 6, 1966. She graduated from Tarpon Springs High School, Tarpon Springs, Florida, in 1984 and has two children — Todd Douglas McInerney, born on May 22, 1992, in Charleston, South Carolina, and Dillon Nicole McInerney, born on March 9, 1995, in Tullahoma, Tennessee.

William Francis Ashton
Roosevelt Medal No. 471 with Three Bars

See also the Thomas Edwards family history.

The Donald Thompson Baker Family

AS TOLD BY CHRISTINE BAKER HUFF FEWELL

An American Adventurer

Donald Thompson Baker was born in Lewisburg, Pennsylvania, on September 14, 1873, and attended Bucknell University there. He later went west and graduated from the University of California at Berkeley as a mining engineer. He engaged in mining activities in Nevada, California and Mexico before going to Central America in 1910 to spend the next few years mining in Nicaragua, Costa Rica, and Panama.

Donald had met a young woman named Sarah Evangeline Weeks of Yaphank, Long Island, and become engaged. However, his mining activities in Central and South America kept him away for long periods of time. After five years of engagement, Sarah Weeks's parents summoned Donald to a meeting in Costa Rica where they confronted him with the need to settle down and fulfill his promise of marriage to their daughter. On April 17, 1913, they were married in the Church of the Good Shepherd in Costa Rica in the presence of her parents and two witnesses. Donald decided to settle down in Panama City and established a lumber business exporting exotic tropical woods which were used at that time for inlay work in marquetry.

In 1913, Donald Baker and his new wife, Sarah Evangeline Weeks, born October 30, 1877, in Yaphank, took up residence at La Marina, 12 Avenida Norte in Panama City on the waterfront near the Presidential Palace and within walking distance of his lumberyard.

The Bakers had two daughters: Sarah Antoinette Baker, called Antoinette or Toi, born August 16, 1914 (later married to Maenner B. Huff), and Edith Weeks Baker, born January 31, 1917.

The Baker family home from 1912-1947, left of Marina Building, waterfront, near Presidential Palace, Panama City.

Donald managed his lumber business, exporting exotic woods and other native products such as ivory nuts, or tagua, for 25 years. He was known for his environmentally sound methods of selecting trees for removal. His business acumen was reported in a book that described his negotiations with the owners of 90 square miles in the upper basin of the Rio Chagres. He located the heirs to these holdings dating back to Spanish colonial times and obtained the timber rights after convincing the owners that their valuable timber was being poached with no return to them. He also devised a way to stop the poaching. By posting a guard at the mouth of the Rio Chilibre, all logs driven down the river, the only means of egress, could be monitored. Every man passing by with logs was offered as much for the timber as it would cost the company to log it. If the offer was refused, the logs were held under embargo pending a court decision. He also had timber rights to remove selected trees from Canal Zone territory near Madden Lake.

Donald Baker was a member of the National Rifle

Association and won numerous medals for expert marksmanship with his pistol. He believed that it was important for his two daughters to know how to handle a pistol; he took them out in the jungle and gave them lessons in marksmanship.

Donald died on September 20, 1939, at Panama Hospital of cancer of the sigmoid, following an illness of one week. He was buried in the Baker family plot in Lewisburg, Pennsylvania.

Sarah Evangeline Weeks Baker was very involved in the social life of the Canal Zone community. She was a dedicated member of St. Luke's Episcopal Church and an active member of the Daughters of the American Revolution. She was a leader and sponsor of the Children of the American Revolution, which her grandchildren and other children attended regularly. She was an accomplished writer and had many people with whom she carried on active correspondence, including letters written to and responded to by Edith Roosevelt, widow of Theodore. She also took great pleasure in engaging in conversations with newcomers and introducing them to others she knew.

Both Antoinette and Edith Baker attended Canal Zone schools. They were driven to school from Panama City by their father's chauffeur and then took the bus home. The two sisters were very athletic and participated in outdoor activities in the tropics. They were both accomplished equestrians and went riding every weekend. They often paddled their cayuco, a carved native wooden canoe, across Panama Bay to Bella Vista and even the 12 miles from their waterfront home to Taboga Island. They made numerous trips with their father to the Darien jungle when he went there for matters related to his lumber business. On one occasion, after traveling up the Chagres River on a boat called the *Argo*, they arrived at a small native village on the shore near Brujas. Antoinette

Donald Baker, award-winning marksman, 1914.

and Edith spent the night in a house belonging to some Swedes who owned a banana plantation. While their father went further upriver to conduct business, the girls entertained themselves by playing Swedish records on an old Victrola. They became alarmed when the Swedish sailors staying in the village got boisterous as their drinking continued into the night. The girls stayed awake guarding the locked door with the pistols their father had taught them to shoot.

Antoinette graduated from Balboa High School in 1933 and attended Canal Zone Junior College for a year before leaving for New York to study dress design at Pratt Institute. While in New York she ran into a high school classmate from the Canal Zone, Maenner B. Huff, who was there attending Pace College; the couple began a courtship. They were married twice — on November 1, 1935, secretly, and again on January 25, 1936, when Antoinette's mother arrived in New York and insisted that she be present to witness the marriage. Antoinette and Maenner returned to Panama in 1939 with their two year old son, Donald, with the intention of Maenner joining his father-in-law's lumber business. When Donald was diagnosed with cancer and died a week later, Maenner obtained employment with the Panama Canal Company.

Antoinette and Maenner had three more children, all born at Gorgas Hospital: Christine Baker Huff was born on October 12, 1942; Edith Louise Huff was born on June 6, 1944; and Linda Antoinette Huff, later called Toni, was born on March 3, 1947. All four children attended school in Balboa and graduated from Balboa High School.

From April 14, 1943, to April 1944, Antoinette was a member of the Women's Army Service Platoon (WASP) and was trained for work on the plotting boards and other jobs in the bombproof tunnels in the fortified islands of Fort Amador during World War II. The meetings, requiring her to don her uniform and leave for training, left a vivid impression of the importance of her task on the two young children she had at the time. She worked for a short time as a clerk/stenographer at Fort Clayton and Quarry Heights. However, as her family grew, she devoted her time to being a full-time homemaker and mother to her four children. She achieved great satisfaction from returning to Canal Zone Junior College to finish her degree, graduating in 1955, the same year that her oldest child Donald graduated from Balboa High School. She was an accomplished gourmet cook and seamstress. She sewed clothes for the family and taught her children to sew at a time when the variety of ready-to-wear clothing was still scarce in the Canal Zone. She spoke fluent Spanish and it was always an exciting adventure to accompany her on trips to the fabric stores in Panama City to select patterns and fabrics for a new outfit. Antoinette was involved in Canal Zone life as an active member of the Daughters of the American Revolution, of which she served as regent. She was a member and president of the Balboa Woman's Club and an active member of St. Luke's Episcopal Church.

> *She achieved great satisfaction from returning to Canal Zone Junior College to finish her degree, graduating in 1955, the same year that her oldest child Donald graduated from Balboa High School.*

Antoinette Baker Huff left the Canal Zone in January 1975 when her husband Maenner retired. They made their home in Brevard, North Carolina, until her death on January 5, 1990.

Edith Baker graduated from Balboa High School in 1935 and Canal Zone Junior College. She left the Canal Zone in 1936 to attend her father's alma mater, Bucknell University. She earned a master's degree in library science from Columbia University. Edith was married to Harrison Hampel, an artist, from 1944 to 1963, and made her home for most of her life in Philadelphia where she worked as a research librarian for the University of Pennsylvania for 25 years, *TimeLife* in New York for five years, and Temple University in Philadelphia for 20 years. For 40 years, Edith was a docent at the Philadelphia Museum of Art. She died on June 17, 2002, after a brief illness with cancer.

See also the Ludwig Theodore Maenner; the Mercer Blanchard Huff; and the Maenner Blanchard Huff (two) family histories.

St. Luke's Episcopal Church, Ancon.

THE WILLIAM ANDREW BARNARD AND ALVIN MONROE RANKIN FAMILIES
AS TOLD BY MARGARET ANNETTE (PEGGY) RANKIN SPELMAN

My maternal grandfather, William Andrew Barnard, a two-bar Roosevelt Medal holder, was born February 14, 1884, in Lanarkshire, Scotland. His family and that of my maternal grandmother emigrated through Ellis Island and landed in Brooklyn, New York, as young children. The families became lifelong friends. While working as secretary for the Health Department in New York, William requested and received a transfer to work on Canal health problems as secretary to Colonel William Gorgas.

In 1916, my grandfather traveled to Brooklyn and wed his lifelong friend Jennie; he stayed long enough to leave my grandmother with child and returned to Panama. Months later he escorted my grandmother and my three-month old mother, Margaret E. Barnard, to Panama.

When Colonel Gorgas left Panama, he asked William to leave and work with him; but my grandfather stayed with the Canal as an admeasurer until my grandparents returned to New York after his retirement on May 31, 1944. It was a sad day when they left; but thanks to the Panama Railroad ships, my wonderful parents, Margaret and Thomas Lee Rankin, along with my sisters Janice (1935) and Colleen (1945) and me (Peggy, 1938), sailed every other year on the Panama Railroad ships from Panama to Truesdale Lake in South Salem, New York, for a three-month winter visit, attending schools in the United States. We found our schools in Panama to be ahead of and harder than our schools in New York.

Canal workers hiking in Panama. Far R front – William Arthur Barnard, maternal grandfather, secretary to Col. Gorgas. Then became a ship admeasurer — the first to retire from the Canal in that position.

My paternal grandfather, Alvin Monroe Rankin, a three-bar Roosevelt Medal holder, was born December 23, 1879, in Knoxville, Tennessee, of Scottish descent. He arrived to work on the Canal on September 6, 1906, as a mechanic. He could fix just about anything, but specialized in the train systems that pulled the ships through the locks. He met and married my grandmother, Agnes Walker, who lived in Panama.

Alvin Monroe Rankin and Agnes Walker Rankin had four children who all worked with the Canal as did some of their offspring: Alvin Rankin (wife Margaret Batchelder) and their children: Billy, Robert, Bonny and Richard; Anita Rankin (husband Paul Thompson, a Canal policeman) and their children Alvin and Frances; Carlos Rankin, a Holy Cross priest who was ordained in Balboa, Canal Zone, and besides being my first love as a young girl, was Father Superior at Moreau Seminary at Notre Dame during the time of the design and building of the new Seminary; and the youngest, my father, Thomas

L-R: Frances Thompson Clinton, Janice Rankin Craighead, Margaret (Peggy) Rankin Spelman, Alvin Thompson, at the farm of Aunt Blanche Walker McIntyre, Gamboa.

Lee Rankin (wife Margaret Barnard), whose first job was loading bananas onto trucks for shipment, but advanced to head up the Mechanical Division of the Panama Canal Company. My mother worked in the censorship department in Balboa in a wooden building behind the post office along with my Aunt Anita during World War II.

We left the Canal Zone in the mid-1950's with many tears shed by Janice, Colleen and me when my father joined the American Bureau of Shipping as a surveyor in Michigan, Ohio and Spain, later retiring to Florida and Colorado.

William Andrew Barnard
Roosevelt Medal No. 6867

Alvin Monroe Rankin
Roosevelt Medal No. 2043 with Three Bars

See also the Alvin M. Rankin and Paul D. Thompson; the Gerald DeLeo Bliss, Sr.; the Reinhard A. Boggs and Max Reinhard Boggs; and the Hermanus Kleefkens family histories.

Front L-R: Peggy Rankin Spelman, Janice Rankin Craighead, Back L-R: Margaret Barnard Rankin (mother); Thomas Lee Rankin (father), FarFan Beach.

PHOTO BY DON GOODE

Panama Canal towing locomotive used for all transiting ships.

The Curtis L. Berg Family

As told by Inez, Carl, Franz, and Janice

Panama's Best Friend from North Dakota

Curtis Leslie Berg was born October 20, 1905, on a small farm near Perth, North Dakota. His father, Carl Johan Berg, emigrated from Norway; and his mother, Helen Nelson, also of Norwegian stock, was born in Lake Park, Minnesota. They homesteaded in North Dakota in the 1890's.

Curtis graduated from North Dakota State Agricultural College in Fargo in 1928 with a degree in civil engineering. Adventure and work in the Central American tropics beckoned, so he accepted a position as a junior engineer with the United Fruit Company of Boston. He traveled by train from Fargo to San Francisco. From there he sailed on a banana freighter of the Great White Fleet to Puerto Armuelles, in the Provincia de Chiriqui, Panama, that port being the nation's thriving banana capital on the Pacific coast.

Curtis L. Berg

He worked as a "banana cowboy" for United Fruit in Panama, mostly surveying (banana plantations require many miles of railroad) from 1928 through 1939 except 1932. That year he roughed it as a cat skinner (Caterpillar tractor operator) for an American contractor building Madden Dam, an important auxiliary structure for Canal operations. During that decade he became familiar with bananas, coffee, many other species of flora, the fauna, the people, the language, the national spirit, and the culture of Panama. He became quite proficient in conversational Spanish. For all his curiosity and excitement about the host country, he should have sailed and soldiered with Hernando Cortes or Francisco Pizarro.

In 1939 he met eighteen-year-old Margarita Nunez Gomez, a house maid, from the small town of San Martin in the El Volcan highland region of Chiriqui Province. Her father was Juan Gomez from the Galicia region of Old Spain; her mother was Lorenza Munoz Nunez of Chiriqui, claiming Spanish-French-Basque-Guaymi heritage. After a brief and clandestine courtship, Curtis and Margarita were married in the provincial capital of David in October 1939. Though she never finished grade school, over the years she acquired English as a second language, educated herself to a high degree, excelled in cooking and sewing and provided the children's initial religious instruction, all while being the exemplary housewife.

As war in Europe raged and spilled over national boundaries, America prepared for the possibility of direct involvement. In 1940 Curtis worked as a heavy equipment operator with an American contractor, making improvements to vital military installations in the Canal Zone. Defense of the Panama Canal was paramount. In this capacity he worked at France Field, Coco Solo, Albrook Field, Rodman, Naval Air District, and other sites. He excelled with the shovel, dragline, and D-8 bulldozer. On the night of December 7, 1941, he "defended" Rousseau on the West Bank single-handedly against sabotage with just a flashlight. After the U.S. entered the war, he worked as a machinist and U.S. Navy fuels specialist at Rodman Naval Station for another contractor.

Near the end of the war, Curtis began continuous employment as a U.S. Navy civilian, a "sand crab" in Seabees slang, until retirement in July 1970. First, he worked as a gauger at the Arraijan Tank Farm; his next two positions placed him in the Public Works Department at Rodman as a field foreman for the Streets and Roads Division, then to the air-conditioned office of the Planning and Estimating Section in the Rodman administrative building. Upon retirement as a Navy civilian, he was credited with about 43 years of federal service. Rarely did he miss work, and he proudly sold hundreds of hours of sick leave back to the U.S. Government.

Residences: After a brief stay in David, he and his new wife moved in early 1941 into a two-story apartment in the Casco Viejo section of Panama City near the Presidential Palace. Sometime in 1942, they moved into Navy family quarters at Locona; then into a rental in Arraijan in 1945; then, again, to their small, owned finca in Pedregalito in 1946; and, lastly, to stay in Pan-Canal quarters in Cocoli in the summer of 1947. This townsite transferred to Navy control sometime in the early 1950's. The family lived in two separate fourplex quarters, 2396 and 2398 Avocado Place, in "The Hollow." The final move was to the duplex at 2726 Nicobar Avenue sometime about mid-1956.

His family: By the time of the final relocation to Cocoli in the Canal Zone, there were six children: Inez Helen, born 1940 in David; Carl Nelson, born 1941 in Casco Viejo; Stanley Clayton, born 1943 in Panama City; Franz Gilbert, born 1945 in Arraijan; and Janice Madelyn, born in 1947 in Pedregalito. The last act of the family play was Elizabeth Ann, born at Gorgas Hospital in 1954, the only child not born in the Republic and, thus, not needing registry at the U.S. Embassy on Avenida Balboa in Panama City. All attended Cocoli School and Balboa High School.

In the summer of 1951, the entire Berg family made its first trip ever to the United States, visiting relatives in North Dakota and Oregon. One year later Curtis took a 90-day leave of absence from the Navy to work at Cerro Bolivar in Venezuela. There he worked as a heavy equipment foreman for Bechtel Corporation, subcontractor under U.S. Steel Company.

Hobbies and recreation: In Panama he was enthusiastic about lake fishing, ocean fishing, netting lobster, and searching for rare seashells and gemstones and learning about plants, animals, geology, and geography. His reading covered many subjects: engineering, science, mathematics, astronomy, anthropology, biology, botany, history, poetry and literature classics; medicine held a superior interest. In his private jungle garden patch in Cocoli, he raised bananas, plantains, pineapples, sugarcane, and yuca. He hoisted cold beer regularly with his best friends. On his supper plate he preferred the food of the tropics. Chess, Chinese checkers, and cribbage were his favorite games. In music he liked Strauss waltzes, polkas, Jimmy Rodgers blue yodels, Bing Crosby, and Panama's own Lucho Azcarraga. His favorite piece of music was "The Third Man Theme" by Anton Karas. Occasionally he argued the good and bad of organized religion. Long, stirring poetry recitations were his special domain, particularly the poetry of Rudyard Kipling. Politics held little interest for him, especially that of Panama.

In his private jungle garden patch in Cocoli, he raised bananas, plantains, pineapples, sugarcane, and yuca.

His character: He fixed his own flats, cleaned his own sparkplugs, changed his own oil, repaired his own transmission, replaced his own brakes, and aimed his own headlights. In short, in everything he always did as much as he could for himself. He always finished the meal on his plate, decrying waste. When finding himself in unexplored surroundings, he bragged, "I'm an adaptable animal." Young children warmed to him for his affectionate attention; dogs, cats, and parakeets did not panic at his approach; friends sought him out for his companionship, experience, knowledge, instruction, and trust. Mosquitoes, he said, never bothered him. Blue-eyed and five-feet-eleven tall, the Panama sun bronzed him over the years.

After normal work hours at Rodman, he would change into his old khakis and cut away weeds or harvest plantains in his jungle garden, or read for an hour, or work on his 1950 Chevy. On weekends it was the same, but there was also fishing at the Third Locks Cut, weaving shrimp and lobster nets, collecting seashells at Vera Cruz, rock collecting at the mud flats, netting lobster at

Chame, checking shrimp traps at the Rousseau bridge, hoisting with friends at the NCO Club or VFW, and doing odd jobs at the quarters. He enjoyed swinging a machete, planting, gathering clams, and telling stories.

He and Margarita retired in Tampa, Florida. In early 1970, an unfavorable civil court case in Arraijan concerning his modest and undeveloped retirement property there soured him forever on the Panama judiciary. The little republic then lost one of its best American friends. The retirees, however, made various trips back to the Canal Zone to see family and old friends. They reported on changes occurring under the Canal-neutrality treaties.

On September 13, 1996, in Tampa, he passed away instantly after a heart attack at the age of almost 91. His marriage to Margarita had lasted almost 57 years. When that sad and sudden end came, she expressed great satisfaction in her life's partner. He was a constant provider, a loving husband, and a good father.

Their children and grandchildren: In 1957 Inez (BHS 1959) married U.S. Army 1SGT Fred Clark of Andalusia, Alabama. Their union produced Tramel, 1961; Pamela, 1968; and Trina, 1969.

In 1969 Carl (BHS 1960) married Patricia Bartels of Elgin, North Dakota. Their union produced Jennifer, 1971; Carolyn, 1973; Jason, 1975; and Christian, 1979.

In 1973 Stanley (BHS 1961) married Diane Christiansen of Glendale, Arizona. Their union produced Jacqueline, 1974.

In 1981 Franz (BHS 1963) married Christine Karavas of Staten Island, New York, having two young daughters from her first marriage.

In 1966 Janice (BHS 1965) married Patrick Swanstrom of Duluth, Minnesota. Their union produced Patrick, Jr., 1967; Elizabeth, 1974; and Erika, 1981.

In 1983 Elizabeth (Plant HS, Tampa, Florida, 1972) married Abbas Tajiani of Kurdish Iran. Their union produced Farah, 1984.

Rodman Naval Station.

PHOTO COURTESY OF PANAMA CANAL MUSEUM

The Gerald DeLeo Bliss, Sr., Family

PROUDLY TOLD BY EMILY BLISS, MICKEY FITZGERALD, AND BLISS GRANDCHILDREN

I'm so glad my grandfather went to the Canal Zone - he sure laid the foundation for the great childhood & life I have - every one should have had this opportunity. Jeanne Walker Wagner, Granddaughter, Carole Walker Miller, Mickey Fitzgerald 7/4/09

Gerald DeLeo Bliss (1882-1957), Postmaster in the Canal Zone, 1905-1934, was born on April 30, 1882, at Sherman, New York. He entered the postal service in Sherman after high school and was later appointed Assistant Postmaster at Chautauqua Assembly, New York. So it came about that he was appointed Postmaster at Pedro Miguel when he arrived in the Canal Zone in December 1905. He received a merited promotion and was appointed Postmaster at Culebra in 1911 and then at the Cristobal Post Office, where he remained until his retirement in 1934.

Gerald Bliss had great influence in establishing mail delivery and trading with the residents of Pitcairn Island (famed descendants of the mutinous crew of the Bounty) and the rest of the world. In 2002 Pitcairn Island issued a postage stamp honoring his contribution to their early days. In February 1929 he received the first airmail delivery from Miami by Col. Charles Lindbergh. He was a popular Canal Zone public figure and was a member of several organizations, including the Elks, Freemasons, Shriners, and Rotary, and was among the invited dignitaries aboard the *SS Cristobal* on August 3, 1914, as the first ship to transit the Canal, prior to the opening official transit on August 15, 1914. Gerald and Mabelle were well known and loved there.

2002 Pitcairn Island stamp honoring Gerald D. Bliss.

Gerald D. Bliss, Postmaster, Cristobal, receives first airmail delivery from Charles Lindbergh, February 1929.

Mabelle Hart Bliss (1886-1960), also born in Sherman, New York, arrived in the Canal Zone in early 1906 with their first born, Gerald D. (Budd), Jr., in her arms. She was one of only two women on the ship from New York to Panama and was the first American woman resident of Pedro Miguel. They lived in railroad box cars until housing could be built. She was active in community and church, playing piano at Cristobal Union Church. Friends and visitors to the Canal Zone frequently gathered at their home.

Gerald D. Bliss, Jr. (1905-1992), was the first of seven children born to Gerald D. Bliss, Sr., and Mabelle Hart Bliss. Affectionately referred to as "Budd," he grew up in the Canal Zone. Budd was quite adventurous — he rode motorcycles across the isthmus and owned and flew several airplanes. He entered into a career with the Panama Canal Company at the Administration Building as an accountant. Budd, like his father, was active in the Masonic Order and achieved 32nd degree and Past Master. He married Eleanor Lawson of Atlanta, Georgia. In 1950 the family moved to California, where Budd started another career as a private business entrepreneur and later a vending machine supply company. Nola, their eldest daughter, was a baby when they moved to the Canal Zone. She remembers Sunday lunch at the Tivoli Hotel, driving through the tree-lined Roosevelt Tunnel near the train station, watching ships being pulled through the locks, swimming at Far Fan Beach, picnicking at Summit Gardens, hiking in the mountains of the interior to see the golden frogs, exploring ruins of Old Panama, and riding on the train to the "other side" to visit relatives.

Gerald D. Bliss III (Jerry) was born in the Zone, attended Balboa Elementary School, and was 12 years old when the family moved to California in 1951. He remembers riding his bike all around the Ancon-Balboa side of the Isthmus. The police must have picked him up a dozen times; they all knew Mom and Dad's phone number, and they would promptly return him home. He knew every road and alley in the area, and he is glad the Zone was such a safe place to grow up with everyone knowing the Bliss family.

Another adventure was entering the empty elevator at the Hotel Tivoli and cranking the handle. Jerry went up a few feet before the operator jumped in and saved his life. He loved the rocking chairs on the Tivoli veranda. Marjorie was also born in Ancon, and Carolyn was born after the family moved to California.

Six more children were born to Gerald and Mabelle in the Canal Zone.

Manola Faye (1907-1925) graduated from high school in 1925, went to New Jersey to attend college, and died soon afterward from meningitis.

Zonella Louise (Zonie) (1910-1993) was very musical and while a teenager played the violin at the silent movies in Cristobal. If the horses were running fast, she played fast; if the scene was sedate or romantic, she played the same tune, only slowly. Zonie graduated from Oberlin College and after moving to Los Angeles became a registered nurse and was head nurse to the U.S. Veteran's Hospital in Sylmar, California. She and husband, Jack Field, lived in San Fernando, California.

Ramona Eugenia (Tinsie) (1912-2001) was a dancer and played in the all-girl saxophone band while in high school. She married J.O. Barnes (Barney), who was a dedicated butterfly collector; Treasurer for AAA Panama; an avid chess player, including playing by mail with others throughout the world; and head of the Panama Canal Company Payroll Department. They moved to Florida upon his retirement, and Tinsie retired from the Sarasota Post Office.

Gladys Evelyn (1914-2006) was born in the French house near the Cristobal Fire Station and the Cristobal Commissary; thus she was destined to marry a fireman, B. Donald Humphrey, on January 21, 1938, and she was an avid shopper 'til the day she died 92 years later. During her school days she was a lettered athlete in a couple of sports and participated in others. She was a member of the local chapter of the Atlantic Order of Eastern Stars and was very active in supporting her daughter, Donna, and son, Donald, in any and all of their activities. She was Mother Advisor to the Rainbow Girls and both den and troop mother in the scouts. Gladys was endeared by *all* of her children's friends, many referring to her as their "second mother;" these people made special efforts always to visit with her at reunions of the Panama Canal Society, of which she was the Sarasota, Florida, reporter for over 25 years. She retired from Gatun Locks as the Time and Leave Supervisor in February 1964 to her and Don's home in Sarasota, Florida.

Mabelle Jean (1916-2001) (Mayno) and the love of her life, George Walker, met in elementary school in Cristobal and after graduating from high school married in California. Mayno retired from the Records Division at the Administration Building, and George retired from communication equipment repair with the Navy. The family lived in Cocoli for many years and loved being involved with the horses at Kobbe Stables and out-of-town shows, aquaplaning in the 3rd locks, skiing at Madden Dam and "Walker" Lake by Miraflores Bridge and the Island in Gamboa, school activities, Rainbows, ping pong, the Finca and Santa Clara.

Mayno and George's children were: Mickey (BHS '59), who married while attending college in West Vir-

> *Another adventure was entering the empty elevator at the Hotel Tivoli and cranking the handle. Jerry went up a few feet before the operator jumped in and saved his life.*

ginia. They loved living in Heidelberg, and their three children have great memories of living in the Canal Zone. Fred (BHS '61) (1944-1998) graduated from the apprenticeship program and worked in the telephone division with the Panama Canal Company until his retirement. He and his wife Kay (Mills) (BHS '63) have two sons, Scott (BHS) and Shane (BHS). The family loved outdoor activities, jeep mud runs, participating in motorcycle racing, camping, fishing, exploring the beautiful country of Panama, snorkeling, netting saltwater fish for their aquarium, and family fun. Fred was a loving husband, father, and friend — always positive in his thoughts and spreading cheerfulness to all with whom he came in contact. Jeanne (BHS '63) graduated from X-Ray school in California, worked at Gorgas Hospital and then married Jack Wagner of Curundu. They lived in the Zone with their two daughters, Gayle and Jeannine, until 1977, when they moved to Washington State for four years before moving on to Alaska. Nowhere else has the ocean, beach, lakes or life style been like the Zone. Life in the Zone was certainly special! Carole (BHS '66) moved to Virginia after attending Canal Zone Junior College and later settled in Florida. She has retired from SunTrust Bank and now operates a roofing company with her two sons.

Canal Zone Post Office Cristobal, Gerald D. Bliss, Postmaster (center in white suit with bow tie), with employees.

Curtis Hart (1921-2005) was thirteen years old when his father retired from Canal Zone service and moved to Miami, where he attended junior and senior high school. Curt served in World War II in Europe, 1945-1947, as Lieutenant in the U.S. Army. Returning home, he married Emily Jackson of Miami; they moved to the Canal Zone, where he worked in the Engineering Division and also the Oil Handling Plant. In 1952 they returned to Miami, where they graduated from the University of Miami; then they moved to Brevard County, where Curt was an aerospace employee with the U.S. Air Force Transportation Department at Port Canaveral until his retirement in 1982. He played French horn in high school, and he and Emily both played in the Brevard Symphony. His favorite hobbies were growing red papayas and flying his own plane.

Gerald DeLeo Bliss, Sr.
Roosevelt Medal No. 1074 with Three Bars

See also the William Andrew Barnard and Alvin Monroe Rankin; the Alvin M. Rankin and Paul D. Thompson; the Reinhard A. Boggs and Max Reinhard Boggs; and the Hermanus Kleefkens family histories.

The Robert J. Bloemer Family

AS TOLD BY ANGELA BLOEMER

Panama Canal Pilot, 1965-1983

November 1965 is when husband Bob received a telegram for the position as Panama Canal Pilot. We were anxious to move to Panama to have Bob home with our family rather than travel the seas. We flew from New Jersey to New Orleans to board the *SS Cristobal*. At the time our sons Robert and Donald were 10 and 4 years old, respectively, and Barbara was 9. We arrived in Cristobal on November 9 and settled temporarily in a Coco Solo apartment. Mr. Camolich, a Canal pilot, met us and escorted us to our apartment. December 1965 was our first Christmas away from our home. Since our furniture was not due to arrive until January, to celebrate Christmas we bought cheap tree ornaments from the commissary, and our children made popcorn streamers from color construction paper. We had to stand in line at a warehouse to purchase our Christmas tree as trees were shipped from the U.S. for all the Panama Canal employees. It was a social way to meet others in the community. On Christmas Eve day a truck with Santa Claus and adults singing Christmas carols passed by our apartment. Santa gave each child a gift. And on Christmas day Captain Meeker and his wife invited us for dinner. They had a gift for all of us, and a festive dinner was enjoyed; it was the best Christmas I had experienced — sad to be away from our family back home, but it was great to have had what all we were given.

Our furniture arrived in January; it was exciting to receive all our belongings. On arrival to Coco Solo, Barbara started school in fifth grade, Robert in sixth, and Donald was placed in nursery school. Bob had started his 18-month pilot training period in the Canal. He would take a ship to Balboa and stay overnight at the Tivoli Hotel, the only hotel in the Canal Zone. On occasion I would ride the train to Balboa and have dinner with him and stay overnight. The Tivoli Hotel's front porch had many rocking chairs, and I spent time with older folks there listening to their stories of the past.

Bob with Angela, receiving his license, June 18, 1967.

Bob received his pilot's license on June 18, 1967. He always said that being a Panama Canal pilot was the very best position he ever had. He met with captains from all over the world and piloted ships of all sizes through the Canal. He was always amazed about commanding large ships through Canal locks with minimum clearance on either side. This was a great challenge. On one of his assignments he had to take a launch to the ship anchored in Canal waters, and on arrival the stern of the launch caught the accommodation ladder causing the perches to break; both Bob and the accommodation ladder landed in the water — he still boarded the ship soaking wet, and I was called to have a complete set of dry clothes for the jitney driver when he arrived at our home to be delivered to Bob.

Many years we enjoyed the annual pilots' party held alternate years at the Elks Club on the Atlantic side and at the Amador Officers' Club on the Pacific side. Each annual party was sponsored by four pilots and their wives. Our transportation to the Pacific side for these parties was via the Panama Railroad train. These train rides were festive, and these pilot parties were the best.

After several months when the Canal organization discovered I was a registered nurse, I was asked to serve as a relief nurse in the Public Health Division, which I accepted. I worked three years as a relief nurse during which time I was able to meet with many people who worked in the Canal Zone on the Atlantic side. In order to travel outside Panama, vaccinations were required; and we nurses in the clinics were able to give the immunizations to all.

Every year the Governor gave one person in every division an opportunity to sail on his boat for three days on a fishing trip. It so happened one year I was chosen from our division, and Bob and I went. The trip was along the Pacific coast; food and service on board was provided, along with equipment for fishing. We were able to catch dolphin which were simply beautiful with green and blue coloring, and I was able to take one home to show our children. This was a lifetime experience for me even though we both were a bit seasick.

When working at Rainbow City's Health Center, a center for Panama Canal local Panamanians and their families, one had to know Spanish. I decided to take an advance course in Spanish to accommodate the patients. There was a health center in every Atlantic side community: Coco Solo, Margarita, Gatun, and Rainbow City to name a few. I relieved the nurses whenever I was asked to do so at all the Atlantic side centers. As a relief nurse I did not receive any compensation or civil service time, so after three years I decided to stop and stayed home with our children. I also spent time swimming and twice earned a 50-mile badge, which was an accomplishment. I joined the Sweet Adeline chorus group; we performed one year in a competition in Hollywood, Florida. I was a member of the College Club and also served as president one year. One time I was in charge of the country fair for the College Club and earned $3,000 for a scholarship to give to one of our graduating students — a rewarding project.

All three of our children were on swim teams. Robert

Bloemer Family, Casarina Street, Margarita, 1971. L-R: Robert, Donald, Angela, Barbara, Bob.

was very active on the football team in high school and Barbara in swimming and cheerleading and sports. One year Robert and his cayuco team won the annual cayuco race sponsored by the Boy Scouts and the next year placed second. Robert played drums in a private band; one year they competed in Colon and won first award earning big bucks. After Robert and Barbara graduated from Cristobal High School in Coco Solo, they both attended Baylor University in Waco, Texas. Robert studied pre-medicine and Barbara studied nursing.

In 1975 Bob and I moved to the Pacific side to Balboa where we lived for 10 years, after having lived 10 years on the Atlantic side. Donald attended Balboa High School. He was in a private band, too, and was very active with the music department at school as well. He performed in the musical stage production *Pirates of Penzance* as the Sergeant, and at times he was asked to conduct the school band. He graduated as the "best talented" voted by his classmates. One summer Mr. Carwithen, Cristobal High School's music director, conducted the musical *Camelot*, and Barbara, Donald and I were asked to be included in the performance—a great challenge which we thoroughly enjoyed. After graduation, Donald attended North Texas State University.

While living in the Canal Zone, I took golf lessons at Amador golf course. On one occasion I met the Egyptian Ambassador who then invited Bob and me to his home where we met other ambassadors and wives from different countries; I was surprised to have had this opportunity to be with such elegant people. Times following, Bob and I were invited to Ambassadors' homes on their National Day and to our U.S. Ambassador's home for a farewell dinner for the Egyptian Ambassador.

When we lived on the Pacific side, Bob and I decided to build a home in Coronado. We spent many wonderful times there and also joined and played golf at the Coronado Golf Club.

The sad day for us was when there was a reduction in force of the ship pilot force. Since Bob, a senior pilot, had taken an early retirement and had been rehired, he was one of the pilots who left the force as a result of the reduction. As luck would have it, Bob was asked by a pilot in Chiriquí Grande to work with Petro Terminales in Panama, which he accepted; we moved to and lived in Boquete during the two years that Bob was employed with Petro. He then was called for a position with Loop, Inc., in New Orleans which he accepted and worked

Bob and Angela's 50th Anniversary (2005). L-R: Robert, Bob, Angela, Barbara, Donald.

for another 10 years until his retirement in 1995. In the meantime we built a home in the community of Eldorado in McKinney, Texas. We joined the Eldorado Country Club and now live in Stonebridge, another community in McKinney.

Transiting ship with linemen at bow in heavy rain.

The Reinhard A. Boggs and Max Reinhard Boggs Family

AS TOLD BY ANITA COLLINS

Father and Son and Their Descendants

Reinhard and his son, Max Reinhard Boggs, were of German descent, but came to Panama from California. Reinhard A. Boggs, a carpenter by trade, worked on the Canal during the construction days from 1909 to 1911. He was a Roosevelt Medal recipient, but did not receive a first bar because his second year of service was interrupted when he left the Canal Zone to work in South America. His son, Max Reinhard Boggs, worked on the Canal from 1910 to 1914, was a recipient of Roosevelt Medal No. 5577, and also received a first bar No. 3457.

Max was very athletic. He was an excellent swimmer and an expert at pistol and rifle marksmanship. He enjoyed hiking in the jungle and taking cayuco trips up the Chagres River. Max met and married Victoria A. Walker, whose father was British and worked for the All American Cable Company in Central and South America. Max and Victoria had three daughters, Stella, Anita, and Zona, all born in Panama and graduated from Cristobal High School.

Anita shares her fond memories of life in the Canal Zone living on the Atlantic side:

As a true "Atlantic sider" it is the best place to live in Panama although those from the "other side" wouldn't think so. Cristobal was divided into Old Cristobal where all the businesses and activity took place and New Cristobal including the residential and school area, the hospital, and the Hotel Washington's salt water pool. The high school was practically in a Coconut Grove on the beach. Old Cristobal was where the clubhouse was located. It was like a city of its own. It included a movie theater, restaurant, soda fountain, bowling alley, pool tables, reading area, barber and beauty shops. Sometimes as youngsters we would get together and walk through Colon to go to an early movie, then walk back home down "Front Street" acting like tourists looking at all the beautiful items in the store windows. If you went to a late movie, it was fun to ride a coach (carametta) back home.

New Cristobal was a perfect place to roller skate. It had very little traffic. A big thrill was "hitching behind" a carametta and being pulled along. If there were too many of us hanging on, the coachman would swish his whip as a signal for us to let go, or else. Often as small children we would pool our nickels and dimes together to have the coachman take us for rides around the area. There were large concrete horse watering troughs all around the area which were tempting for us for many reasons — dunking each other in the water, a play area for sail boating, or just taking a dip to get refreshed during a hot day. It was a wonderful time to explore and enjoy your freedom. One never worried about the children walking or riding their bicycles down along the Colon Beach Road to the Washington Hotel swimming pool. Much of the community center activity took place at the play shed. It was a wonderful gathering place for us to play. At night the adults would take advantage of this feature as well and take part in activities such as volleyball, basketball, handball and even fencing.

My father, Max, obtained a Canal Zone Government land lease for a small island on Gatun Lake. A bohio was built until he could complete a proper cottage. The cottage had a front and back porch with two big rooms inside. The back porch was used as the kitchen and dining room while the front porch was the living area. As typical for island accommodations, the facilities were outside. There was a privacy fence of bamboo on three sides with the lake on the fourth side — a beautiful view but care was taken to time your retreat around the passage of the occasional cayuco.

To get to the island you either rented a launch leaving from the Gatun docks or rode the train to Frijoles and paddled cayucos. The trip would take you past Barro Colorado Island. At night on the small island, I could hear the howler monkeys on the far off island of Barro Colorado. Their eerie howl would give me goose bumps. It was interesting to see dead trees stretching their limbs up out of the water as if they were reaching to drag you under. Passage through the channel had to be made with care. There might be a short tree or stump just below the water line. You could get stuck on it or even be overturned. A channel had to be marked and cut before a launch could get to the island. Outweighing the task of making the channel through the trees was the benefit of easily gathering orchids and other air plants because now they were all at your level. At night as teens we would go out in cayucos with flashlights and count how many alligators we could see. Then the next day we would swim near the same areas. As teenagers, groups of us would plan trips to the island on weekends or holidays with our parents or school teachers as chaperones. It was great fun.

My sisters, Stella and Zona, and I met the men we would marry in Panama. Stella married Colonel James D. DeMarr and had four sons, James, Jr.; Glen; Barry; and Victor. I married Albert B. Collins and had three children, Brenda, Mark, and Alita. Zona married Dr. Paul Harry Dowell, and they had five children, Paul Harry, Jr. (Pablito); Albert Clay; Harry Winship; Richard; and Vicki.

When the government turned New Cristobal over to Panama, we all moved to Margarita or Coco Solo, a former naval base. The Dowell and Collins families moved to Coco Solo into a three-bedroom four-family house. The Dowells moved downstairs with the Collins living upstairs and all of us sharing the same basement (one big family).

My children were able to grow up in the same type of freedom and security that I experienced as a child. They also experienced the fond memories of hours at the play shed or as youngsters who climbed up to the high open window of the high school auditorium peeking down on the "big kids" practicing for an upcoming thespian production or visit to the Chinese 5 & 10 store for a treat. There was always excitement over the homemade candy that looked like an umbrella on a toothpick. What an awesome childhood. Can you imagine going to school at Coco Solo elementary and peering out the window watching the ships' passage in and out of the Canal? That must have been a real challenge for the teachers who needed to keep the kids on task.

The seasons in Panama may have been limited to rainy season, dry season, football season, top season, yoyo season, skin diving season, marble season; but even without a winter season, no one missed out on the spirit of Christmas. There might not have been sleigh rides and chestnuts roasting on an open fire, but nothing beat the fun of a Christmas tree burn. What a thrill it was to throw your tree into the fire and jump back while the embers flamed up into the sky and the air was engulfed with the sharp crackles of the limbs as they ignited into a ball of flames. Your face would glow from the heat of the fire. It was a community gathering and time for all to have great fun.

As a mother, how do you calm the wows of a child who has heard the tales of tuli vieja or the stories of how the migrating crabs from France Field would raise up their claws to puncture the tires of the cars who dared to encroach on their path? It was a beautiful place and a safe place. From the excitement of carnival to the private gathering on the Dowell family owned Santa Rita Farm, there was no better place to nurture and raise your chil-

> *At night on the small island, I could hear the howler monkeys on the far off island of Barro Colorado. Their eerie howl would give me goose bumps.*

dren. I can look back and say there is no place like our home in the Canal Zone. Thank you to Max and Reinhard for bringing us to this land to experience such wonderful memories.

See also the William Andrew Barnard and Alvin Monroe Rankin; the Alvin M. Rankin and Paul D. Thompson; the Gerald DeLeo Bliss, Sr.; and the Hermanus Kleefkens family histories.

Reinhard A. Boggs
Roosevelt Medal No. 5370

Max Reinhard Boggs
Roosevelt Medal No. 5577 with One Bar

PHOTO COURTESY OF PANAMA CANAL MUSEUM

Cristobal Clubhouse, Canal Zone.

PHOTO COURTESY OF PANAMA CANAL MUSEUM

The Hotel Washington swimming pool was a salt water pool with sea water pumped in from Limon Bay. It was a favorite place for guests of the hotel and residents of the Canal Zone to go for a day of recreation.

The Bonanno and Gately Family

AS TOLD BY THE BONANNO SIBLINGS

Captain Justin and Jeanne Bonanno, twins John and Barbara, and younger daughter, Sherry, were living in New York before the family moved to the Canal Zone in 1958. Our father was working as a sea captain for the United States Lines, and had been spending very long periods of time away from home at sea. He wished to come ashore to have more family time, so he applied both to the U.S. Coast Guard and to the Panama Canal Company as a Canal pilot. The Coast Guard offer came through first; however, about a month later, he received an offer from the Panama Canal and asked for and received a release from his Coast Guard commission to accept the offer. The Bonanno family arrived on the Isthmus of Panama in March 1958.

We three children were 4 and 2, respectively. We lived in multiple family quarters in Coco Solo for six months until we moved to Margarita's Sixth Street. The next six years on Sixth Street were a kid's delight as there were 100 plus of us who lived there during that time. There was always great fun and activities going on in Margarita and other townsites close by: watching movies under people's houses, playing street games such as ring-a-levio and four square, butterfly catching, sliding down hills on cardboard, Cub Scouts, Brownies, Christmas tree bonfires with a full day of sack races, picnics, games at the Red Barn, Saturday morning movie matinees at Margarita theater, baseball games at the ballpark behind Margarita Clubhouse and theater (Who doesn't remember Ivan, the popcorn man, at those games?), Brazos Heights's annual fair, and so many more activities organized by parents and other people who were active community members.

Jeanne and Justin Bonanno, 1964, Los Rios.

While living in Margarita, our father built us a beautiful life size playhouse, a roller coaster, and a two-seater soap box derby which we loved to enjoy with our friends. We remember the man on the bicycle who would blow his flute to let neighbors know he was in the neighborhood to sharpen knives, and we remember running with scores of friends behind the DDT truck until our parents reprimanded us for running through a chemical spray which we thought of as clouds.

The Bonanno family moved to the Pacific side in 1964 to Boquerón Street in Los Rios where Dad lived until he retired in 1988. Our father built an in-ground swimming pool, and he paved the way for other families to seek permission from the Canal administration to build similar pools. Our weekends were filled with many friends at home enjoying our backyard pool.

We attended Los Rios Elementary School through sixth grade; then attended the brand new Curundu Junior High School for seventh through ninth grades (John and Barbara were in Curundu's first graduating class.); then Balboa High School, where Barbara and John graduated in 1971 and Sherry graduated in 1973. We all loved our high school experiences — ROTC (all three of us), drill team, a quality education, dedicated and caring teachers, many friendships with kids coming in from all over the world with a lot of different life experiences. Brother John remembers many gatherings of friends at the home of Mr. Belden in Ancon, who welcomed many Balboa High School and Canal Zone Junior College students in his home. After school we loved to explore Panama City and its interior locations and learn another country's culture.

John and Sherry attended Canal Zone Junior College before heading to the United States for more advanced education and remained in the U.S. thereafter. During Sherry's first year at Canal Zone Junior College, she served as the CZJC Director for the "Straight Scene" drug education program which was overseen by Clarence Payne, the Governor's Youth Advisor. In that post, she was responsible for training high school students to speak to elementary school students about drugs. After Barbara attended college in New York and worked a year in Washington, D.C., she returned to Panama to work for the Panama Canal Commission from 1975 to 1997 — the last 12 years in the Washington, D.C., Office of the Secretary, where she retired from federal government employment.

Our father spent many years as a Panama Canal ship pilot, and he later transitioned into marine management positions to include Chairman, Board of Local Inspectors, and his last post, Deputy Marine Director, before he retired. During our father's years in Panama, he was decorated by the President of Panama with the Vasco Núñez de Balboa medal, the highest Panamanian civilian award, for his efforts to initiate a program for young Panamanian students to obtain U.S. maritime academy appointments. He also received the Panama Canal organization's Gold Medal award for public service and was awarded the Distinguished Service Award in recognition of his professional work and contributions to the isthmian community. He served 10 years as Kings Point's field representative and also on the U.S. Governor's selection committee for both U.S. and Panamanian students applying for appointments to U.S. maritime academies.

Our mother Jeanne volunteered in many capacities in the Canal Zone instead of seeking employment as a nurse which she was by profession. She served on the Girl Scout Board; was a Brownie and Cub Scout leader; and used her nursing skills in a variety of community health activities, to include assisting with polio immunizations, and helping disabled children with physical movement and therapy when we lived on the Pacific side. She volunteered with the Pilots' Wives Association, especially its hospital visiting committee which visited foreign sailors convalescing in Gorgas Hospital, providing conversation,

Frank Gately and Margaret (Gately) Bonanno and Justin Bonanno, 1994, Bradenton, Florida.

reading materials, and writing letters to their families. She was active with the community's Tuesday Club, a monthly social club of women who met on both sides of the isthmus. Our mother died on the Transisthmian highway after a day with Tuesday Club ladies. Her heart attack then took her life at the young age of 50 in 1975.

In 1977, our father married Margaret Marie Gately, which pleased us greatly. Margaret arrived in the Canal Zone in 1957 from Boston to teach. She began her teaching career on the Atlantic side at Cristobal High School, teaching biology and general science. She taught four years at CHS before transferring to the Canal Zone Junior College on the Pacific side. Margaret continued to teach biology at the college, but later transitioned into Assistant Dean at the college and also served as a counselor. Before she met our father and while living in the Canal Zone, Margaret was the recipient of three National Science Foundation grants to Brown University; University of Southwestern Louisiana; and University of Southern California, which took place at the University of Costa Rica. She was very active in the Catholic Church in the Canal Zone and also taught catechism for many years.

Margaret's brother, Frank Gately, arrived in 1961 to teach (U.S. history and American Institutions and occasionally Economics and American Government) at Balboa High School, which he did until he departed in 1973. Frank's first recollection of Panama is reflected in this memory: "It was 7:45 a.m. in August 1945 when Japan surrendered to the United States, thus ending World War II. I was at the wheel of the *SS Galrid Franchere* just finishing the 4 to 8 a.m. watch. After waiting our turn to enter the Panama Canal, I was again at the wheel when our ship crossed Gatun Lake. It was a night transit. When the *Franchere* sailed into the Caribbean Sea, I thought I would never see Panama or the Canal Zone again. By August 1961, I had returned."

Frank was also junior class sponsor at BHS for several years, and he was assigned to oversee the student association, too. He recollects: "Until my sister Peggy married Captain Justin Bonanno, we had connecting apartments in Williamson Place. One attraction to Williamson Place was its attractive rent. Gasoline was 15 cents to 17 cents a gallon (no state or federal taxes). On Saturday mornings, I would walk to the Balboa clubhouse for a five cent cup of coffee and two cent donut. "

Twins Barbara and John, and Sherry Bonanno, 2001.

Living for 30 years in the Canal Zone was a wonderful experience for our whole family. Our father was able to pursue a career that he loved, our mother became a vital part of the community and was able to stay at home to raise her children, our stepmother Margaret loved her students, and for the three of us, everyday was an adventure where we made close life long friends, and appreciated our country's endeavor with the building of the Panama Canal and overseeing its stewardship for almost 100 years. It's a source of pride for our whole family that we lived during, and were a part of, this unique slice of American history.

PHOTO BY KEVIN JENKINS
Aerial view of Corozal and Los Rios.

The Audrey Benoit Bowman Family

As Told by Bob Bowman

In 1917 Bert J. Benoit went to the Canal Zone as a machinist on a two-year contract with the Mechanical Division. He was enticed to go to Panama by his brother, Otave, who was working on the Canal, and his sister, Bertile Casanova.

Bertile's husband, Charles M. Casanova, a recipient of the Roosevelt Medal for his construction-day service, was employed on the Atlantic side of the Canal. Known affectionately as "Queen," Bertile was well known in the Canal Zone for the wide variety of flamboyant hats she wore. People would comment: "If you didn't see Queen coming with her hat, the scintillating fragrance of Channel No. 5, which she did not use sparingly, would announce her presence." Charles and Queen had two sons, Alton and Roland. Roland graduated from Cristobal High School, worked his entire career with the Panama Canal's Maintenance Division, and retired to Slidell, Louisiana. He died in 2006. Alton died in Balboa in the mid 1950's.

In 1919 after completing his two-year contract with the Canal, Bert Benoit returned to the United States. One of his six sisters, Lucy, and his brother-in-law, Alfred, owned a bakery shop on Second Street in New Orleans, where Bert began working. While there, Bert met Edna Luft, the youngest of eight children born to Frederick G. Luft and Henrietta Schmitt of Germany. Edna, born on January 1, 1901, and Bert Benoit were married in New Orleans on March 17, 1920. They had two children, Audrey, born April 1923 in a private residence in Algiers, Louisiana, and Burton J., born in July 1930 in Gorgas Hospital in the Canal Zone.

After several years, Bert decided to return to Panama, so on March 25, 1925, he embarked with his family on a United Fruit ship from New Orleans to Cristobal, Canal Zone. His job was as a machinist in the Mechanical Division in Ancon. When that division later became the Industrial Division and moved to Mt. Hope on the Atlantic side, Bert and Edna took up residence in Margarita where they became nightly Bingo players at clubhouses and Knights of Columbus, VFW and Elks clubs. Bert's retirement in 1959 resulted in his and Edna's return to New Orleans, where he died in January 1961. Audrey and her son, Ron, visited Edna every summer when school was recessed. Their transportation was supplied by the Canal Zone via the steamships *Ancon*, *Panama* and *Cristobal*, departing from Cristobal and arriving in New Orleans after a four-day journey. Edna passed away in January 1987.

Edna and Bert Benoit in front of their house on the Prado.

Audrey grew up in Balboa, living first in the Flats, then moving to concrete quarters on the Prado. She attended first grade at the Balboa School located at the end of the Prado which, at that time, served first grade through high school. Because of overcrowding due to the unexpected influx of children, Audrey was sent to the Knights of Columbus building on Balboa Road for second, third, and fourth grades. The K of C served drinks and meals to adults in the evening and housed children during the day.

For fifth and sixth grades, Audrey was sent to what was affectionately known as the "Monkey Plum Tree School" (named for the Monkey Plum tree in front of the building) on Tavernilla Street near the Balboa clubhouse and swimming pool, overlooking Balboa Road. One of Audrey's fondest memories of her early years was taking bicycle rides with several of her classmates to Pedro

Miguel swimming pool, a distance of seven miles. (This was no small feat as the bicycles of those days had no gears and were not exactly built to minimize wind resistance.) After elementary school, Audrey attended junior high in the old wooden building located next to the Junior College. After two years there, she returned to the building where she originally started first grade, graduating from Balboa High School in 1941. The next year, in 1942, the high school moved to its new building next to the Junior College.

Audrey began working in the summer of 1941 as a student assistant at the Information Desk in the Administration Building when she met L.D. Bowman, Jr. He showed up to take a Civil Service exam — carrying his own typewriter — since the testing office did not supply the machines. After a short conversation with Audrey, L.D. took his test and mentioned to the test supervisor on the way out that "he was going to marry that girl at the Information Desk." (It appears more than one soldier in this era used this line and ended up marrying young ladies in Panama and the Canal Zone.)

L.D. successfully passed his test and was hired in the Transportation Section. He eventually went on to become a Supervisory Marine Traffic Controller in the Port Captain's building in Balboa. L.D. (also known as Larry), had gone to the Zone in 1935 with the 14th Infantry Brigade at Fort Davis on the Atlantic side. When his second enlistment was up in 1940, he went to work for the U.S. Army at Fort Amador. Asked about his job at the time, Audrey's answer was that L.D. was a spy working for the Office of Strategic Services, the forerunner of today's CIA. Audrey and L.D. courted for a year and a half and were married in a civil ceremony in Colon in 1942 and later had a church wedding at St. Mary's Mission in Balboa.

L.D. and Audrey went on to raise three kids, Bev, Bob, and Ron. While having her first child in Gorgas Hospital, Audrey's maternity ward mates were Mary Egolf (having daughter Kathy) and Peggy Mitten (having twins Sue and Sheila). They shared many stories and had the added excitement of blackouts and darkened windows due to the war. There were no official visiting hours, so the ward was constantly filled with all their friends dropping by to chat.

Audrey's mother and father lived in Margarita on the Atlantic side, so it was a common occurrence for Bev or Bob, while they were growing up, to hop the Panama Railroad train at 4:40 on Friday afternoon to Mt. Hope, play bingo that evening with their grandparents, and drive back with them on Saturday via that wonderful pot-marked roadway from Cristobal to Balboa. On Sunday morning, before returning home, they would go to the Chinese Garden near the Corozal Cemetery to pick up vegetables.

L.D. "Larry" Bowman, 1935, Ft. Davis.

In addition to his job as a Marine Traffic Controller, L.D. was one of the founders of the Pacific Softball League. Initially, the games were played at the Ancon Laundry ball field until moving to the parks by the Balboa Railroad Station. L.D. served as president of the league for many years. His kids would later remark that they thought his real job was to pick up the keg of beer before games, deposit it after the game in the brewery, and then drop the write-up and score of the game at the *Star & Herald* for the morning paper.

L.D. died in 1962 due to heart valve replacement complications, requiring Audrey to enter the workforce to raise Ron and provide for Bev's and Bob's further education. She proceeded to take every civil service exam being offered, even traveling to Paraiso, La Boca, and Red Tank. One of her favorite interviews was with the Post Office, where she had to demonstrate the ability to throw a 50 lb. sack of mail. Her endeavors were rewarded when she was hired in the Accounting Division, working until her retirement in 1985. Most of her work location was in the old Balboa Shoe Section building.

Retirement has been a joy for Audrey, settling in New Orleans in the very same house her mother and father, Edna and Bert, had bought when they retired from the Canal Zone. She successfully survived Hurricane Katrina and her evacuation although her family didn't hear from her for several days. Her house, fortunately, was not damaged by the hurricane.

Audrey's three children have presented her with six grandchildren and four great-grandchildren and have gone on to successful careers. Bev returned from college to the Canal Zone and worked for the Marine Bureau. She later married Joseph Wood, who rose to the position of Deputy Administrator after the invasion of Panama in 1989. Bev and Joe have three sons, Craig, Brian, and Scott, and, upon retirement, moved to Tallahassee, Florida, in 1995, where they reside today. Bob married Mary J. Wells and had a career as an Air Force fighter pilot and later an airline captain at US Airways. They have two sons, Wade and Michael, and retired to Cocoa, Florida, in 2004. Ron, the best athlete in the family, being named to several All-Zone sports teams, is still actively employed as the city manager of Boerne, Texas, and has a daughter, Kristan.

Bev, Bob, and Ron will be forever thankful to their mother and father, Audrey and L.D., for their dedication and perseverance to the Panama Canal Zone way of life.

See also the Joseph J. Wood family history.

Audrey B. Bowman

PHOTO COURTESY OF PANAMA CANAL MUSEUM

El Prado, 1915, from Sosa Hill. Shown are Balboa Elementary School on the far right at the end of the Prado and Tavernilla Street houses (front right foreground) where "Monkey Plum Tree School" was located.

The Brayton and Wertz Families

As told by Shirley Brayton Wertz Bruce

The Brayton/Wertz families were in Panama during the construction era. Between the two families, there are six Roosevelt Medal holders listed at the end of this story.

During the first transit of the SS Ancon in 1914, my future husband Fred's parents, Fred and Lillie Wertz, Uncles Carl and Harry, and his Uncle Robert and Aunt Lou Lumby were invited guests on the ship.

My grandfather, Charles Ross, arrived in Panama in 1910 with his wife and three daughters. He was hired as a conductor on the Panama Railroad. They lived in Bas Obispo and later moved to Empire. My dad, Rodman Brayton, also went to work with the Canal in 1910. He married my mother, Marian Ross, and they had six children. After the birth of Donald and Dorothy, it was back to the United States. My brother, Jack, and I were born in Providence, Rhode Island. We returned to Panama in 1927. Our first home was a four-family wooden house on Las Cruces Street in Balboa. The second home was a four-family cement house on Barnebey Street. My sister Alice and my brother Ross were born in Panama. Ross died in infancy.

My life in Panama started when I was five years old. I stayed close to home and played games of ring-a-levio, mumble peg, kick the can, red rover, baseball, jacks and jump rope with our neighbors. In the late 1920's, we

Margarita Recreation Association dance, New Year's Eve, 1949. Standing, L-R: John Frensley, Mr. Barfield, Dave Sink, Fred Wertz, George Wertz; Seated, L-R: Mrs. Barfield, Gladys Wertz Brayton, Mrs. Frensley, Ann Sink, Shirley Brayton Wertz, Edna Wertz.

spent many Saturday hours at the "play shed" playing and doing arts and crafts. After-school hours were spent between our house and the neighbor's house. We were a total of 19 kids. Curfew was when the streetlights went on around 6 p.m. Bedtime was at 8:30 p.m. with homework done, baths taken, and clothes laid out for the next day. Some Saturdays we would take the trolley car at the YMCA and visit Central Avenue generally coming home empty handed. There was catechism on Saturday mornings, and afternoons we would go to a 15-cent matinee.

In 1937, we moved to the Atlantic side on Colon Beach and could see ships coming and going through the breakwater. My dad would herd all of us to the front steps for a lecture on the different types of ships, the flags they flew and what type cargo they were carrying. Dad's lectures became very useful to me when I worked at the Cristobal Port Captain's Office in the 1970's.

In 1941, I married Fred Wertz, Jr., and we had two daughters, Donna, born in 1947, and Carol in 1954.

Summers were spent in Santa Clara at our adobe cottage. Daily we visited the beach and went horseback riding. I would stay the week with my nephews, Jack and Rod. Fred, my brother Don, and his wife Gladys would return to work and make the trip back on the following weekend. Because gasoline was rationed during the war, car trips were made using half gasoline and half kerosene.

Fred loved anything having to do with water. One stateside vacation during the 1950's, we stopped at Cypress Gardens and watched a waterskiing show. We bought skis to take home. We bought a speedboat for skiing from surplus and spent hours refurbishing it. Fred was the first one to ski in front of the Cristobal Yacht Club and enjoyed teaching others how to ski. Our friend, Bill Brooks, owner of a seaplane, dared Fred to ski behind his plane in Colon Bay. Fred took the dare; we watched from shore with open mouths, but no mishaps. Later we bought a cabin cruiser for family outings going to White Beach, Isla Grande, or the San Blas Islands where swimming, fishing, and lobstering at night was the norm. For breakfast, we dined on scrambled eggs and fresh lobster.

In the late 1950's, I began working at the Cristobal Yacht Club and hired student assistants for various office jobs. My husband Fred spent all his off time on his boat.

When scallops were running, my brother-in-law, Robert Wertz, sent 16 gunnysacks of live scallops to us by railroad. My nephew Rod figured out how to remove and clean the scallops from their shells. The debris was taken to the Mindi Diversion, a small inlet alongside the road to Gatun. The Dredging Division had recently dredged the area to widen it.

We made many family vacation trips to the United States on the Panama Line ships. One trip we picked up a new car and drove to California and back. We bought an ice chest, which was new on the market, and we were able to picnic our way across the United States.

Prior to my retirement, the Canal notified us that the Mount Hope Cemetery was being turned over to Pan-

Gladys Wertz Brayton and Fred Wertz in front of the Panama Canal Yacht Club, circa 1950.

ama. I requested all my families' remains be moved to Corozal Cemetery, which was to become a national cemetery. Grounds Maintenance employees went to each gravesite, dug down six feet, and picked up six shovels full of dirt that was later transferred to the new gravesites with the headstones. When my father-in-law died, embalming was not available, so his sons built a cement tomb. When work began at my father-in-law's grave, they encountered a solid wall of cement and needed a jackhammer to break it up.

In 1984, I not only left my adopted home, but I also left part of my family. After being a widow for more than 20 years, I married my best friend, Don Bruce, and moved to Florida. Don died in 1996. I am again a widow and continue to reside in Florida with my daughter and son-in-law, Donna and Wendell Sasso.

Rodman Brayton
Roosevelt Medal No. 5982

Robert Lumby
Roosevelt Medal No. 630 with Three Bars

Charles Ross
Roosevelt Medal No. 6256 with One Bar

Carl Wertz
Roosevelt Medal No. 2523 with Two Bars

Fred Wertz, Sr.
Roosevelt Medal No. 2430 with Two Bars

Harry Wertz
Roosevelt Medal No. 1553 with Two Bars

The James Butler and Charles William Dawson Families

AS TOLD BY GAIL DAWSON MCNALLY

The early 1900's brought notice across the country of the beckoning adventure of the building of the Panama Canal. We can assume that one fine day this enticing news reached Kalamazoo, Michigan, where one James Butler lived with his wife and eight children. The recruiting propaganda was obviously irresistible. Workers were needed on the Panama Railway; and James Butler, a railway engineer, heeded the call; incredibly, packed up his wife and large brood of children; presumably, boarded a train; and then, evidently, set sail for Panama. From where? We don't know, but we guess it was either New York or New Orleans. What amazing courage! For awhile the family lived in Empire, a temporary townsite near the banks of the Canal construction where the children traveled on horseback and flourished in the bustling, but hot and insect-infested, construction atmosphere of the "big dig." Along with many others, James's effort earned him the Roosevelt Medal.

In 1924 the youngest Butler daughter, the popular Frances Gabriel, became the bride (see photo) of Charles William Dawson, an Irish immigrant who, along with his brother Frank (at left), also heard the call from across the "pond" in County Cork and arrived on the Isthmus to seek work with the All American Cable Company. Two of Frances Gabriel's sisters, Marie and Mary Margaret McCormick and son Bobby, are in the wedding party. The smallest child is "Lal" Dawson, daughter of Frank Dawson. Even in the 1920's elegance was evident.

The marriage of Gabriel Butler and Charles Dawson produced three sons, Michael James, Herbert Charles, and William Henry Dawson. The family lived in a variety of locations, including Panama City, Pedro Miguel, Diablo Heights, Curundu Heights, and even, for awhile, Cali, Colombia. The boys all graduated from Balboa High School, were active in sports, and enjoyed strong popularity and friendships.

Michael graduated from Georgetown University, and, after time in the Air Force, returned to Panama to become eventually General Manager of Gulf Oil. Two of Michael and Maritsa (Uribe) Dawson's five children, Gabriella Dawson de Caprilles and Mark Dawson, remain in Panama. The others, Dr. Michael Dawson, Charles Dawson, and James Dawson, reside in the United States.

Herbert (Herb) attended the University of Buffalo and was then awarded an appointment to the U.S. Merchant Marine Academy at Kings Point. As a Lieutenant (j.g.), Herb returned to Panama as Aide to Admiral George Wales, Commandant of the 15th Naval Air District, where he met and later married Gail Blank, a young intelligence officer. In the decades from the 1960's to the

Wedding of Frances Gabriel and Charles W. Dawson, 1924.

1980's, Herb was involved in several projects and formed Webco, Panama, a subsidiary of a worldwide military brokerage firm serving the U.S. military retail operations. Herb was a founding member of the United States Navy League and served as its president twice. He remained in the Navy Reserve and retired as a Captain.

Herb and Gail Dawson's five children, Brian, Sharon, Erin, Kelly and Herb, Jr., reside in Miami, Virginia, Boston, and New York. Herb's youngest daughter, Kelly Frances, was the last U.S. member of the Washington office of the Panama Canal Commission with ties back to the Canal's construction days.

William (Bill) Dawson attended the Colorado School of Mines and then accepted an appointment at the U.S. Naval Academy. He completed a 27-year naval career to include ship command and retired as a Captain in Falls Church, Virginia, with wife, Connie. Connie Glassburn was also reared in the Canal Zone, and her grandfather, Daniel Eggleston, was a long-time Panama Canal pilot. Bill and Connie Dawson's three sons and five grandchildren live in California and Virginia.

In addition to the three Dawson sons, Frances Gabriel's siblings, their spouses and children and all the generations to follow were sewn forever into the fabric of Panama and the Canal Zone. The oldest Butler daughter, Aileen, married a Sousa musician, Joe Flynn, and returned to Kalamazoo. Her sister, Marie, met a bachelor doctor, James Gallagher, who traveled to Panama by ship; after marriage Marie settled in Buffalo, New York. James Butler, Jr., resided in Colon; a successful businessman and developer, he married Aurora, a beautiful Colombian, who resided at the Hotel Washington in Colon after her husband's death. Macel Marie Butler married Arthur Goulet of French extraction who became General Manager of the Atlantic side Panama Canal commissaries during World War II. Mary Margaret Butler married William McCormick, who assumed management of the Canal Zone's premiere hotel, the Tivoli Guest House, so beloved and memorable to so many in the Canal Zone. Michael Butler remained a bachelor overseeing operations of the Coco Solo tank farm which stored fuel for various Canal operations.

The legacies of courageous human beings who fulfilled the dream of long ago visionaries, who successfully put the "path between the seas," were numerous and proud. As their numbers dwindle in the years to come, their unique heritage should never be forgotten. The Butler and Dawson families are proud to be part of this Panama Canal Museum historical project. The children and grandchildren of Mike, Herb and Bill are world citizens with a privileged history.

James Butler
Roosevelt Medal No. 1322 with Three Bars

Fifteenth Naval District Building, Ft. Amador.

PHOTO BY CAPT. WILBUR VANTINE

The Harold Carter Family

FOND MEMORIES OF THE CANAL ZONE FROM THE CARTER FAMILY

The Carters came to Panama for the first time on their way to Ecuador. The first sign of what was to come arrived as soon as the curved doors of the MATS turbo-prop Constellation swung open at Albrook Air Force Base on a warm evening in the spring of 1961 — a rush of air — the rich, dense, humid and exotic smell of the jungle and ocean breeze. We were seduced.

We returned in 1963, Harold, Virginia, Chris, Marc, Barbara and Lece, who was adopted while we were in Ecuador. This was Harold's second tour of duty with the Inter-American Geodetic Survey. A two-year tour turned into 15 years with retirement in 1978. Barbara married Richard W. Bacot, Jr., and was the only one to remain in Panama until July 1999.

Remembrances include wooden-framed houses on stilts on North A Street in Curundu, watching the thunderstorms rage through the night from the second floor of the house, seeing the "ballbat" birds' nests swing in the wind, and hearing the chatter of those beautiful birds.

Traveling to the interior to go to the beaches was a favorite. On the way, you could always smell the wood-burning stoves. In Gorgona, near the Ocean-Blue Cottages, we viewed the eclipse of the sun through a pinhole in a pie pan and later slept in our VW bus and hammocks. Who could forget the hot black sand, burning our feet, requiring Father to carry Barbara and Lece as well as our Great Dane puppy from the beach to the cottage walkway. Listening to the great music from the distant cantina in the evenings was part of the experience. Returning from the "interior," we bought fresh cheeses, michas and guandú along the way.

There was the often nearly deserted Río Mar beach of that time — the black volcanic sand that burned your feet, the jellyfish, and the radical tides that moved the water's edge hundreds of feet from shore up to the hills, then back down again. As a previously landlocked 12-year-old, Chris was overwhelmed with awe and wonder at this limitless ocean and the power and magic of those six-eight foot waves — this primal, eternal, mystical energy.

The Carter family, 2007. Back row L-R: Hal, Marc, Chris. Front row L-R: Lece, Virginia, Barbara.

While spending the day learning to body surf, time would stop, and you could feel at one with the sun, sky, water and earth — it was heaven. At night one would still feel the glow of the sun, the ocean spray, and the rising and falling, pulling and pushing of those glorious, endless waves.

One day after surfing and after walking along the shore with his surfboard up to his chest in the ocean for a quarter of a mile, Marc saw that he was bleeding profusely from his foot, and he came to the realization that sharks really don't like people!

We remember Nuevo Vigiá, a favorite interior town on Madden Lake, where our friends were Toto and family and Eusabia, a Darién Indian woman, and family, who made baskets for us. We remember the first time we used our trolling motor to power the cayuco. A motor boat with two men was heading in the opposite direction from us. They turned off their motor and stared, mouths open in awe as our boat silently made waves along the other side of the inlet!

The VW (Volkswagen) Club trip to the Darién was our first time into "the jungle." Fording rivers was exciting. The single beautiful red puff flower, seen at a distance from the primitive road, was impressive; and the cool-off in a stream was welcome.

Curundu, 2003-A West 1St Street, was another home sweet home with a mango tree in the back yard and orchids and where trash men feared our Great Dane, Bunny. For several years we were shaken by the roar of the U-2 take-offs at precisely 7:00 a.m.

Curundu Clubhouse, bowling alley, and theater were neighborhood places where you could get popcorn with real, hot-melted butter. We enjoyed horseback riding at the Curundu stables with our teacher, Esther Moxan, and galloping the horses up Curundu 4th Street and down to the clubhouse to enjoy the best empanadas ever.

We remember no need for much clothing or need for shoes at all, resulting in calloused feet that were tough enough to put out a burning cigarette.

Curundu Pool, the Home of the Curundu Cougars, was a place to learn to swim and hang out daily. The Balboa Skating Rink, where we all learned to skate, and enjoyed a night out as young teenagers, was another great hang out.

We remember Panama, the only place where you could go fishing in the big ditches just using a net and capture all sorts of tropical fish, but mostly guppies.

We have fond memories of going to Penonomé with Barbara's best friend, Arlene Price; staying at her grandmother and aunt's house; and going to the local theater to see movies.

These are some of the Carter family memories of the Panama Canal Zone, a wonderful place to live and grow up.

PHOTO COURTESY OF WWW.CZIMAGES.COM

The town of Curundu was built in the early 1940's.

The Anthony J. Catanzaro Family

AS TOLD BY ANITA OBERHOLTZER

'Right' Passage: Dad's Decision

Although my family does not date back to Canal construction days, after drinking water from the Chagres River, Panama will always be home in our hearts.

Dad, Anthony J. Catanzaro, married Mother, Jane V. Filichia, in 1920 in Chicago, Illinois. Their union produced four offspring — Rose in 1921, John in 1925, Anita in 1928, and Tom in 1936.

Having accepted a construction position with M.P. Severins Company, Dad traveled to Panama in 1942 and began working at Howard Air Force Base. After completing that contract, he started as a machinist with the Panama Canal Company's Mechanical Division.

In 1944 Mother, Tom and I made our first trip to the isthmus. Our prop plane landed at Albrook Air Force Base, which was open to commercial air traffic. Post-war, Rose and her husband Richard Brogie moved to the Canal Zone. They had four children — Janis (Eckel), Rick, Ken and Steve. After John's Navy discharge, he traveled to Panama and began working at the Army Transportation Corps. During a trip to New York to upgrade his engineer's license, he met Olga Romero, who was born in Panama but was working in Manhattan. They married; and Jane Ann (Pattwell), Grace (Makowski), and John Anthony are their children. Anita married Rusty Oberholtzer, and Judy (Hebble) and Joey followed. Caught up in the draft, Tom landed at Fort Bragg, North Carolina, then blew further north for his doctorate in clinical psychology at the University of Chicago; he continues to work in the "Windy City."

Mother and Dad's wedding day, September 25, 1920.

Our first living quarters, built in a hilly area on stilts of different heights, was a multi-family, old French-style building on Pyle Street in Balboa. You entered from street level, where short stilts supported the building; as you walked through the apartment, the all-wood edifice extended over the lower hill level and was held up by much longer stilts. In those days it seemed everyone had the same door keys. Lost yours? Just borrow the neighbor's to unlock your door.

Dad told Mother not to bring many clothes since everything we would need was available at the "no-money-only-coupon-book" commissary. It was wartime and there were meager pickings in the clothing department. Thank God Mother sewed clothes — by hand — until we found inexpensive dressmakers aplenty. Balboa Commissary was a two-story concrete building; downstairs was mostly food and upstairs, apparel. The "annex," a one-story wooden building further down Balboa Road, sold shoes and hardware.

In late 1944 Dad transferred to the Army Transportation Corps as chief engineer on the commanding general's 100-plus foot ship. Six years later, he returned to the PCC to work for the Mechanical Division in Mount Hope. He transferred to Gamboa's Dredging Division in 1953 and remained there until his early retirement to Gorgona in 1958.

Varsity teams from Balboa and Cristobal high

schools, and eventually Canal Zone Junior College, traveled via the Panama Railroad to "de 'udder side' fer de games dem." We were fun-loving and full of school spirit; our boisterous cheers and songs allowed little rest for adult passengers on those ocean-to-ocean trips aboard the P.R.R. School dances took place at the elegant company-owned Tivoli Hotel in Ancon and the Washington Hotel in Colon; these soirees continued until those buildings were surrendered.

St. Mary's study club met on Monday evenings. We enjoyed trips to Gorgona beach staying at the Ridge family's cottage. Since Dad was the "Chief," we occasionally went deep-sea fishing or transited the Canal on the general's ship. Much free time was spent at the pool, Saturday matinee, bowling alley and gym; the Thatcher Ferry ride to Far Fan Beach was always a treat. After playing hard we indulged ourselves by consuming Klim shakes and French fries with gravy at the "Clubbie" soda fountain or scarfed down the best hot dogs in town at the Balboa bowling alley. What a life!

Dad, our "Chief."

Living and growing up in the Zone was a wonderful and unique experience. It was a great place to raise children. Everybody knew almost everybody; people watched out for one another, and kids were safe. Zonians enjoyed the best of two worlds — life in the tropics and frequent vacations in the States, first via company-owned ships and later on chartered flights. It was like life in Archie comics — year-round outdoor swimming, fishing, and water skiing, free activities at the playshed, dodgeball, tennis, little league baseball, scouting, the cayuco race, Christmas tree burns, Holy Family's and St. Luke's annual fairs, horseback riding, street games, chasing DDT trucks and dreading the "7:20 p.m. bedtime train." When the drive-in opened at the Civil Affairs Building, previously the Albrook AFB terminal, to sell soft-serve ice cream, jumbo fried shrimp, hamburgers, hot dogs and other sandwiches, we thought we were big time and almost on a level with the States. Long before designer jeans were being marketed, Canal Zone kids fashioned their own with bargain-priced mola patches. Though never merchandised, grade-school kids dabbled in making perfume from fragrant ilang-ilang flowers. The icing on the cake was the thrill of seeing celebrities. Queen Elizabeth and Prince Philip visited, and we were close enough to see her peaches-and-cream complexion. President Eisenhower, very much the Commander-in-Chief, rode down Balboa Road. President and Mrs. Nixon made a whistle stop at the Gatun railroad station and waved to us. Hollywood "celebs" stopped by, too — from glamorous Lana Turner to the real *Miracle on 34th Street* Santa, Edmund Gwenn.

We added tropical produce and native cuisine to our own Italian and American favorites. While Dad kept us supplied with an abundance of fresh seafood, the Knights of Columbus sold the best 10-cent hamburgers. We all enjoyed Mary Jane cookies, Princess Pat bars and poppy seed braided bread from the company bakery. Remember the patacones, bananas, roseapples, ginneps, star fruit, Chinese plums, carne en palito, Jamaican bun, empanadas, piva, hot buttered micha, raspaduras and paletas on the "local economy?" The Chinese gardens supplied us with fresh fruits and vegetables. Dining at the Sunday evening Tivoli buffet was a steal at $2.00 for adults and $1.00 for children. We climbed and picked fruit from our favorite mango trees. Sure, there were a few broken arms and some mango poisoning from eating too many green ones; but, oh, they were good. Having our

> *We all enjoyed Mary Jane cookies, Princess Pat bars and poppy seed braided bread from the company bakery. Remember the patacones, bananas, roseapples, ginneps, star fruit, Chinese plums, carne en palito, Jamaican bun, empanadas, piva, hot buttered micha, raspaduras and paletas on the 'local economy?'*

own avocado tree in the backyard was a dream come true; what we wouldn't give for that now.

In the mid 1950's, the PCC brought on the colossal electric current conversion from 25 to 60 cycle which put us on par with the military bases. Replacing wringer-type washers with automatic ones and purchasing air conditioners for the first time was exciting. Shortly thereafter another milestone was reached when Caribbean Forces Network was born. It launched the Zone into the world of television. How anxiously we awaited those few hours a day beginning with the 3:00 p.m. test pattern and sign-on — "Channel 8, Fort Clayton; Channel 10, Fort Davis!"

Where else could you paddle a cayuco in so few hours from the Atlantic Ocean to the Pacific Ocean? Dad and his friend, Richard Biava, often volunteered with the Explorer Scouts. When the Explorers held the first cayuco race in 1954, Dad organized the very first spaghetti dinner for race participants. He and Mother prepared their delicious sauce with meatballs. The dinners continue in conjunction with what is now the "Annual Ocean-to-Ocean Cayuco Race." My parents would be very proud to know their two great-grandsons, Tom and Alan Eckel, together with *Snafu* crew members Wendell Sasso and David Cohen, retired the rotating first place trophy for the open class male category in 2007 — that's coming in first place for three consecutive years. Wow!

Family gatherings were many and held at various locations, including Brazos Brooks police range and Gorgona, where John owned property with two furnished cottages and a four-star bohio, complete with a designer tile dining table and matching built-in barbecue. Swimming, riding horses on the beach and hammock time were additional attractions. One year we celebrated Judy's and Joey's common birth date (their 18th and 16th birthdays) with a water skiing party on a houseboat in Gatun Lake. We were surrounded by anchored ships from many nations awaiting Canal transit.

Here's another coincidence Judy shares: During her 1967 tenure as Governor of Girls State, the last one held in the Canal Zone, Harold R. Parfitt was the lieutenant governor of the Canal Zone. Between 1975 and 1979 he was the last governor, so they share the distinction of being the last two governors in the Canal Zone!

Dad always said he wanted to live the rest of his days in Panama. So he did, until his death March 13, 1959, at Gorgas Hospital. Mother died April 29, 1996, in Chicago. They are buried together at Corozal American Cemetery. Rose died October 21, 1989, and is buried at Corozal with Richard, who died April 8, 1978.

All these wonderful times are etched in our hearts forever.

PHOTO BY KEVIN JENKINS

Annual Ocean-to-Ocean Cayuco Race sponsored by the Explorer Scouts, with the Gatun Yacht Club in the background.

The George Dean Cockle Family

AS TOLD BY GEORGE R. COCKLE

George Dean Cockle, the youngest of eight children, was born in Galveston, Texas, on August 4, 1901. His mother died while delivering her ninth child, who also died. George's early years were spent with his oldest brother on a farm near York, Nebraska. Later, returning to Texas, he attended Houston Business School, which prepared him for administrative employment. Upon graduation in 1923, he accepted employment in Antafagasto, Chile, with commissary services at the nearby copper mines.

While transiting the Panama Canal he ran into several old friends who convinced him to apply for employment there. He did and was promptly hired as Chief Clerk, Commissary Division, at the Mount Hope facilities. His "metal check number" (an old mining term used in the early Panama Canal days) was 301, a number he bought every week for Sunday lottery drawings and which never won big.

While on a 1927 vacation visit to his stepmother living in the San Francisco bay area, he met Caroline Elizabeth McKellar whom he later married. Her father, an industrial artist in San Francisco, had created the Sun Maid Raisin Girl, Beach-Nut and the Morton Salt Girl trademark images among others and co-drew "Maggie and Jiggs," a popular cartoon strip of that era.

They were married in Fruitvale, California, and made their first home on 3rd Street in New Cristobal. They had two children: George Robert (July 29, 1929) and Dale Sherwood (September 22, 1936) — both born in Colon Hospital. They later resided on 8th Street and after that on Margarita Street until 1941.

With the increased service demands placed on the Panama Canal by the hostilities preceding World War II, George was transferred in late 1941 to the Pacific side, becoming the manager of the Balboa grocery and cold storage facilities. He oversaw the rapid expansion of these services both to provide the local needs and to handle transiting maritime demands. The family resided on Tavernilla Street and later on Pendleton Street next to the All American Cable Company. He continued in this management role until returning to the Atlantic side.

Again he resumed commissary duties at the Mount Hope facilities. The family returned to their previous home on Margarita Street. This was the only up-and-down two-family house in New Cristobal with stairs into the attic — where our extensive Lionel electric train layout operated.

George Dean Cockle

George retired in 1956 with 33 years of service, moving to Pharr, Texas. He and his wife enjoyed doing volunteer work and gardening. She died in 1965, and he followed in 1971. They are both buried in nearby McAllen, Texas.

Son George Robert was active in scouting on both sides of the isthmus. A WeBeLoS Cub, an Eagle Scout and Order of the Arrow, he represented the Canal Zone youth during General of the Army Eisenhower's Panama visit. George attended elementary school in New Cristobal and junior and senior high in Balboa (BHS 1947). He attended his freshman and sophomore years at Carson Long Military Institute in Pennsylvania.

Joining the Army Air Corps at Albrook Field, George flew with the 91st Reconnaissance Squadron which was photo-mapping all the South American countries. Later he became aide to Brigadier General Blackshear M. Bryan, Caribbean Command, Quarry Heights. Transferring from the new Air Force to the Army, his military career spanned 22 years. As a combat infantry officer and a rated Master Army Aviator, his service included most South American countries, Trinidad, Alaska, France, Germany and Vietnam. His various military citations include the Distinguished Flying Cross with Oak Leaf Cluster, 21 Air Medals, 2 Bronze Stars and other merit and service awards. He retired in 1969 as a lieutenant colonel.

In 1952 George married Mary June Smalley of Macon, Georgia. She died in 1977. They had one son, John Robert, now deceased. He later married Roberta Sue Watts Dempsey of Clinton, South Carolina.

In 1972 George began a public relations and writing career in the transportation field, authoring 21 books on railroad and military aircraft subjects. He currently resides in Tallahassee, Florida, where he is working on another book.

Son Dale Sherwood attended Canal Zone schools, class of 1954, Cristobal High. Active in Junior ROTC, he was the Battalion Commander of his class. Gaining the Governor's appointment to the U.S. Military Academy in 1954, he graduated and was commissioned in 1958 in the U.S. Army Signal Corps. He received a Master of Science Degree in Electric Engineering in 1964 from Northeastern University, Boston, Massachusetts. His military career included tours in Central America, Taiwan, Spain and Vietnam as a Signal Officer. His various citations include the Legion of Merit, Bronze Star and other merit and service awards. He retired in 1981 as a colonel.

In 1959 Dale married Joanna Cragun of Alexandria, Virginia. They had two children: Linda Jo (June 6, 1959) and John David (June 21, 1960). He later married Betty Jane Ross in 1967 at Fort Leavenworth, Kansas.

In 1981 Dale entered the high technology communication field with Eagle Signal and later 3M, both in Austin, Texas. After retiring from 3M, and moving to Burnet, Texas, he began substitute teaching in both their middle and high schools. This later culminated in his teaching eighth grade mathematics on a full-time basis and then instructing computer science at Central Texas College in Burnet County.

Active as an Amateur Radio Operator (K5JIC), Dale served in a leadership capacity in several computer clubs. His other community activities include playing in the community band and emu ranching. Trips back to Panama included the "Last Graduation" at Balboa High on May 15, 1999, and celebrating his 40th wedding anniversary there in 2007.

PHOTO COURTESY OF PANAMA CANAL MUSEUM

Industrial Plant, Mount Hope. A busy day at the Commissary Division delivery platform, November 25, 1936.

PHOTO COURTESY OF PANAMA CANAL MUSEUM

Industrial Plant, Mount Hope. South-end Commissary Division warehouse, November 25, 1936.

The Joseph W. Coffin Family

AS TOLD BY LYNNE (CAROLYN) COFFIN CUNNINGHAM

Early in 1910 Joseph Wilbur Coffin II arrived to play baseball in Panama from California, where he had played semi-professional ball. Panama at the time was well known for its baseball. Prior to traveling to Panama, he seemed to live a footloose and fancy-free life traveling around the United States. Employment, other than playing baseball, didn't seem to be a concern. "Big Joe" as he was known on the Isthmus was 6' 6" tall, and his fame as a southpaw pitcher had preceded him from the west coast. He played on several different teams and lived up to his reputation. Life in the tropics appealed to Joe, so at the age of thirty he took a job and stayed for over 30 years.

Joe was born in 1880 in Providence, Rhode Island, the youngest child of a well-to-do New York City family. His first job was for the Panama Railroad Commissary as a clerk with pay of $100 per month. After ten years with the Commissary Division, Joe transferred to the Marine Division at Cristobal. His new duty was as a dispatcher. Later he was promoted to Chief Clerk and served under many port captains. He was awarded the Roosevelt Medal (No. 6373) with one bar (No. 3831) for four years' Canal construction employment from 1910 to 1914.

During his time in Panama, Joe was involved in many service clubs — Elks, Shriners, and Masons, to name a few. He held many offices in the various clubs and was well known as the close second in his Masonic lodge for honors as the champion greeter of Cristobal and the isthmus. He was also a member of the University, Century, Strangers and Washington Cotillion Clubs.

Joe married Kate Hurst in November 1917. They lived in Colon and then Old Cristobal. Katie May Hurst was born in 1893 in York, Pennsylvania, the youngest child of a Pennsylvania Dutch family. She completed what was then called "practical nursing school" and arrived in Panama in 1917 to work at Panama Hospital.

Kate and Joe on the "Panamanian Riviera."

They had two children — Joseph W. Coffin III and James Coffin. Both "little Joe" and Jim completed high school at Cristobal. Among our family keepsakes are letters dated 1917 from Joe's parents in New York City to their new daughter-in-law. They caution Kate about Joe's frivolous ways, write about the politics going on in New York City at the time, and remind Joe to keep in touch with his family. Panama seemed so far away to the older Mr. Coffin and his wife. "Big Joe" retired from the Panama Canal in April 1943. Joe and Kate returned to her birthplace of York, Pennsylvania. Their youngest son, James, left the Zone with them and never returned.

Joseph W. Coffin III was employed first by the Commissary Division as a butcher apprentice and then hired as a fireman for the Cristobal district. He married Carolyn Geddes (daughter of another Roosevelt Medal holder, Thomas Geddes) in 1942. Early in 1944 Joe traveled to New York City to bring Carolyn and their first child to live with relatives. He joined the U.S. Navy and served in the South Pacific on a floating dry dock. Joe and Carolyn returned to Panama at the end of the war, and Joe resumed his job as a fireman. They lived first in Old Cristobal and then moved to Gatun, where they lived most of their married life. Joe was a fireman on the Atlantic side during his career and retired as a Lieutenant.

Joe and Carolyn had nine children — Carolyn (Lynne) Coffin Cunningham, Joseph (Bud) W. Coffin IV, Thomas Geddes Coffin, Jon (Sid) David Coffin, Marjorie (Honey) Coffin, Gerald Coffin, Marcia (Candy) Coffin Wheatley, Kathi Coffin Reca, and Nannette Coffin Halliwell.

Joe and Carolyn retired in 1973 also to York, Pennsylvania, to be near his younger brother, Jim. Their first six children had graduated from Canal Zone schools before they left and two of the boys (Joe and Tom) had attended Canal Zone Junior College. After finishing school in the Canal Zone, Lynne, Joe, Tom, and Jerry all moved to New York City, where their mother had family. After Lynne's marriage she moved to New Hampshire and then retired to North Carolina. Joe finally settled in New Jersey and Tom in Massachusetts. Jerry is now a Lieutenant with the New York Fire Department. Marjorie and Nannette live in York, Pennsylvania, and Marcia and Kathi live in Virginia. Jon (Sid) Coffin continues to work on the Panama Canal as a tug boat captain and lives in Margarita.

"Big Joe" and Kate's decision to make the trip and life in Panama have produced two children, nine grandchildren and ten great-grandchildren. Their descendants are all proud of their heritage and ties to the Panama Canal.

Joseph W. Coffin II
Roosevelt Medal No. 6373 with One Bar

See also the Geddes Brothers family history.

PHOTO COURTESY OF PANAMA CANAL MUSEUM

10th Street, Colon, Masonic Parade, 1914.

The Harry and Rebecca Compton Family

AS TOLD BY JANE COMPTON WAGENBRENNER

My Heart Remains in the Land of My Birth

Harry and Rebecca Compton, young missionaries from Ohio, first landed in Panama in 1884. They took the train across the isthmus and stayed ten days in Panama City waiting for a boat to take them to Valparaiso, Chile.

Years later Rebecca wrote, "I well remember how my very soul wept the first time I witnessed the terrible heat, sickly faces, bad smells, filth, gambling, and other vices and degradation of Panama. I did not eat a thing but oranges, and I was so disgusted with them that I could not eat them any place again for about ten years."

In 1892 Rebecca returned to Panama with her son William Taylor Compton who was born in 1889. Matters in Panama had not changed a great deal. Rebecca wrote, "Oh the mosquitoes! We were in Central Hotel Panama and Taylor could not sleep. The mosquitoes must have reminded him of the mewing of kittens. 'Mama, estos gatitos no me dejan dormir!'"

After teaching assignments in Coquimbo, Chile; Montevideo, Uruguay; Mendosa, Argentina; and Quito, Ecuador, the Compton missionaries were assigned to the Church of the Seawall in Panama City, Panama. They lived there from 1911 to 1916, when they retired to Delaware, Ohio.

Living in Panama those five years, Rebecca wrote:
All the wonderful cleaning up was going on, ice was being manufactured and comfortable homes, too, with baths. Schools were in process of erection as well as the Y.M.C.A., plants, libraries, and best of all community centers where religious services were conducted. Who did it? The Missionary Societies? No. Uncle Sam—bless his old heart. I have never felt ashamed of that cry in my soul for help for those poor people. Neither have I doubted that God felt much as I did and enlisted the interest of the only person who was rich enough and at the same time willing to perform the task in God's own way.

Taylor Compton, center, boarding ship, circa 1930.

In 1900 at age 11, Taylor Compton was sent to relatives in Ohio to learn English and to be educated. When he graduated from Ohio Weslyan in 1912, he returned to Panama and worked in payroll in the Clubhouse Division. He later joined Canal Zone Customs, a job he enjoyed because he loved the sea, working outdoors and boarding ships from all around the world.

In 1931 Taylor married Ethel Jane DeGroot from Richmond Hill, Long Island. The missionaries, Harry and Rebecca Compton, had stayed with the DeGroots when they went to the States to preach and solicit money for their missionary work. Taylor won the lottery and bought a new Buick roadster and a diamond engagement ring. After the wedding on Long Island, the newlyweds drove across the United States and took a Chinese ship from Los Angeles to Panama. They settled into House 27 on Fifth Street in Cristobal, where they lived until World War II.

Three children were born to Ethel and Taylor Compton in Colon: Jane in 1934, Harry in 1935, and Rebecca in 1942. When World War II started, Taylor felt very patriotic, left his employment in Panama and joined the Army. He was made a Captain in Military Intelligence. He was 53 years old. Ethel, Taylor and their three children traveled for four years and lived in Quito, Ecuador; Miami Beach, Florida; McAllen, Texas; Fort Sam Houston, Texas; New Orleans, Louisiana; and Little Rock, Arkansas.

When the war was over, Taylor went back to his work with Canal Zone Customs. The Compton family lived in a duplex on Colon Beach right next to Christ Church by the Sea, House 422B, which afforded them a perfect view of all the ships entering and leaving the Canal. Taylor had a break in Canal Zone service between 1921 and 1926, when he worked in a bank in New York City. In 1951 he completed 30 years of Canal Zone service and retired, and the whole family left for New York.

The Compton missionaries birthed six children, but only two survived. Taylor Compton's sister, who was a twin named Blossom, was born in 1896. Blossom graduated from Balboa High School in 1914 and from Ohio Weslyan in 1918. She returned to Panama and taught school. She met Enrique Benitez, a Lieutenant in the U.S. Army stationed in Panama. He graduated from Ohio State in 1914. They were married in 1920. They had three children: Henry, born 1921; Babs, born 1923; and Ben, born 1928. For many years they were stationed at Fort Amador. In 1932, Rebecca Compton wrote, "Well in the course of events all of my children are a part of Uncle Sam's family in Panama." Taylor and Blossom remained close until their deaths in Florida, Taylor in 1960 and Blossom in 1961.

Writing about Panama in the 1930's, Rebecca Compton wrote:

I experienced such a concern for the people of Panama that Panama has been one of my living prayers ever since. Oh! I said, if only all the churches can unite and send a few families to this forlorn place — no matter if they do not teach them anything about religion but to show them how to live. My prayers were sincere but fraught in great ignorance. God answered them just the same, but it took the genius of a "General Gorgas" to clean up the Isthmus and create a healthy Zone. It took the knowledge of Goethals to engineer the great work of digging and building the Canal. It took Uncle Sam's millions to finance the whole affair and the rebellion of Panama which made it a separate nation in 1903 so that it could deal with the U.S. when Colombia refused to consent to the plan. It took gambling France out of the contract and gave it to our United States so that a higher grade of civilization might be established for all concerned on the now well conditioned Canal Zone in Panama Republic.

Christ Church by the Sea, Colon. Oldest Episcopal Church in Central America, built by the Panama Railroad Company in 1864.

PHOTO BY CAPT. WILBUR VANTINE

The William H. Conley Family

AS TOLD BY ROGER CONLEY

In the year 1906, the *Atlanta Constitution* ran an advertisement for men to build a canal in the Central American country of Panama. It listed several types of workers, including railroad engineers.

William H. Conley (Bill) and his two brothers, Richard S. (Dick) and David F. Conley, signed up to work for the Isthmian Canal Commission (ICC). They soon learned that "running engine" in Panama was very different from the same job in the United States.

William H. Conley

I remember my grandfather telling me: "When the sun got so very hot, they would take off their overalls and put them on the firewall to dry out the sweat; and they would run engine in their underwear. If it appeared that they were about to wreck, I would tell the fireman to jump, and he would jump without saying a word or looking around. He would go right out the window. Once when I had to jump, I hit the ground on my face and cut my gums to pieces."

The Conley brothers were said to have wrecked more trains than anyone else.

Uncle Dick was the engineer who took President Teddy Roosevelt across the Isthmus. Roosevelt said, "I want to shake the hand of the man who took me from coast to coast in 90 minutes." He asked Uncle Dick what he could do for him, and he replied, "You can give me a raise." The president nodded favorably to the pay master.

After the completion of the Canal, Bill and Dick stayed to work on the Panama Railroad. One day the boss called them into the office and started a conversation on the topic of what they had accomplished in building this Canal. Through trickery he found out who had seniority. He looked at Dick and said, "Which one of you stepped off the ship first?" to which he replied, "My brother Bill."

The boss then said, "Well, Bill, you have the day shift and Dick, you have the night shift." Uncle Dick wouldn't speak to his brother for a long time, only relenting when my grandfather named his first son, Richard.

William Hodge Conley married Byrd Patton Keith, and she gave birth to Emily K., Richard F., William H., Jr., and Keith D. Conley. They grew up on the Prado where they could go to see the silent movies in their time and listen to the bands at Stevens Circle. The clubhouse did not have a theater in the early years.

Richard F. (Dick) Conley

My father, Dick Conley, married Harriet Rogers, who had just finished high school in St. Paul, Minnesota. Her father was a mechanic for Isthmian Airways. Dick was working for the Canal Zone Police Department at the time. They had a son, Roger (yours truly), at Gorgas Hospital.

Dick's brother, Keith, wrote this tribute to him, my father:

One of the Finest—a Hero.
Dick Conley, a Canal Zone cop in 1935, was more fleet-

of-foot than flat-of-feet. He walked the Balboa Docks on the graveyard shift. The time was midnight; the tide was low, when he heard a call for help.

He dove off the dock in the dark, scraped the buttons off his tunic, and lost his Stetson uniform hat. The man he saved was a tuna boat engineer. The governor put a commendation for heroism in Dick's file. John McGroarty wrote a poem about him as the youngest man on the force. And, best of all, the man he rescued sent Dick a Christmas card every year thereafter.

Roger R. Conley

I remember riding in his sidecar at night from the Thatcher Ferry Slip to the station through La Boca. I can still see the bright red fender of the motorcycle.

Later, he went to work on the Panama Railroad as a hostler in the Balboa roundhouse. At night we would go to pick him up after his work shift. Dad would put me on the steam engine with him and run it to the water and sand towers. He would let me run the engine. What a thrill this was for me. I would drive it up to the turntable and then he would take over and put it on the turntable. Then we would turn the engine to the correct stall.

After that came the diesels. They were more comfortable to run, but they were not as glamorous as the steam engines.

After living in the States for many years, Roger returned to the Canal Zone and worked at various jobs with the government, starting with Coco Solo Hospital, and then at Gorgas, where he was an X-ray technician. He then worked for the Transportation Branch, Grounds Maintenance in a management position, Sanitation and Customs and Immigration. He retired in 1983 and returned to the United States.

Roger and his late wife, Carol, have two sons, Richard F. and William R. Conley, who were both born at Gorgas Hospital. Richard is a safety and training coordinator for Delta Airlines, and William is a sea captain for West-Pac Express. Roger and his wife, Olga Yolanda Johnston Conley, live in Albuquerque, New Mexico.

David F. Conley
Roosevelt Medal No. 5170 with One Bar

Richard S. Conley
Roosevelt Medal No. 1614 with Three Bars

William H. Conley
Roosevelt Medal No. 1621 with Three Bars

See also the *William C. Caley Johnston family history.*

PHOTO COURTESY OF DENNIS WHITE

Panama Railroad Engine #299, Balboa Train Station.

The Theodore S. Corin Family

AS TOLD BY TED AND GEORGIA CORIN

In the Spirit of Adventure

Theodore (Ted) Stephen Corin was born in Detroit, Michigan, in 1928 to Russian immigrant parents. He served as an infantryman in the Korean War (1950-51) and was awarded the Purple Heart. While working toward his master's degree at the University of Miami under the GI Bill, he met his soul mate Georgia as they were crossing a street together; they married in 1960 and moved to Tallahassee, Florida, for Ted to pursue a Ph.D. in Higher Education and for Georgia to work toward a bachelor's degree in Art Education at Florida State University.

In 1965, while still living in Tallahassee, the Corins had their first opportunity to go to the Canal Zone when Ted was offered a job for the summer trimester as Acting Director of Florida State University Canal Zone Branch, at Ft. Davis on the Atlantic side with quarters in Margarita. They were fascinated by the beauty of the tropical jungle bordering the Gatun Locks, the ruins of Fort San Lorenzo, tarpon fishing at the mouth of the Chagres, dining on fresh-shucked raw oysters at the Cristobal Yacht Club, and best of all — transiting the Canal. They left with some of the best memories of their lives.

Then in 1968, Ted was offered the position as Director of FSU's Canal Zone Branch at Albrook AFB, and in the spirit of adventure they packed up and were sent first class on Pan American Airlines with their Siamese cat in a carry-on basket to take up residence in Curundu. Breezing through the friendly U.S. Customs, no one ever asked what was in the basket.

That September, to make the Corin family complete, our adopted daughter Stephanie was born in Miami and brought to the Canal Zone at three weeks of age. Ted's mother Anita joined them shortly thereafter to help care for her new granddaughter.

In the spring of 1970 Ted was asked to speak to the Canal Zone College graduating class held in the courtyard of the Tivoli Hotel. Soon thereafter Dean Glen Murphy invited him to join the college faculty as instructor in the Social Science Department, and they made the move to the Canal Zone Schools Division.

Dr. Theodore Corin (right), teacher of psychology and education at the Canal Zone College, receives his notice of promotion to Associate Professor from Frank Castles, Superintendent of the Canal Zone Schools. October 30, 1971.

Their first quarters under Canal Zone housing were in Gorgas Apartments — the remodeled part of the hospital featuring some unusual floor plans — theirs was the "bowling alley." The living room windows overlooked the mortuary where they could occasionally see smoke coming from the crematorium, and once they witnessed the ambulances delivering 42 bodies from a chiva bus that had plunged from the Bridge of the Americas, smashing into the tank farm below.

During their first two years in the Canal Zone, Georgia began researching the molas of the San Blas Indians for her master's degree form Florida State University which she completed in the summer of 1970; she presented a bound copy to the Canal Zone College library reference department. She also became a popular speaker on the mola at many community clubs and organizations.

Stephanie began kindergarten in Balboa Elementary School, and continued in the Zone schools through Balboa High, class of '86. She enjoyed all the activities available to Zone kids, including dance classes with locally famous Vera Bomford; but she had a special passion for horses, starting early on with riding lessons at Fort Clayton stables and lasting until their final horse show held the month they left in 1986.

At Canal Zone College Ted taught courses in psychology and study skills and was promoted to the rank of professor and Chairman of the Social Science Department. Georgia joined the part-time faculty to enjoy teaching art history for 10 years but that ended because of the nepotism rule which went into effect when Ted became Assistant Dean of the College in 1981. By then she was also busy being a horse-show mom, had fallen in love with horses herself, bought a thoroughbred from the race track, and decided to give riding and horse shows a go herself.

For weekend retreats, the Corins bought a beachfront lot, high on a bluff in the interior of Panama near the Rio Teta overlooking the Pacific Ocean, and began building their "Finca Sea-Esta." They started out with a bohio tipico and small camper trailer, hauling fresh water from town and using kerosene lanterns. Eventually they built a comfortable house that had electricity, a solar water heater, deepwater well, swimming pool, dozens of fruit and shade trees and a bush pony for kids to ride. Sea-Esta became well known to their friends who were always welcome to share its natural beauty and unique stairway to the beach built from 54 used tires. Occasionally Margot Fonteyn and her husband Tito Arias, who were their neighbors at the time, dropped over. Margot loved to swim, and Tito enjoyed swimming with the help of an assistant.

Community theater was also a large part of the Corins' life. They participated in many Canal Zone College, Surfside Theater and Amador Theater plays as actors, dancers and costumers. On one occasion in 1977, in the college production of *Feiffer's People*, the three Corins performed together — daughter Stephanie just age nine at the time. It was during rehearsals one night that someone came in to announce that the U.S. Congress had finally passed the Panama Canal Treaty by just one vote. It was so quiet you could hear a tear drop.

On October 1, 1979, the Carter-Torrijos Treaty went into effect, and the interim transition from the tiny Panama Canal Company to the giant Department of Defense began. The Canal Zone College was renamed the Panama Canal College where Ted would spend his last years as its Assistant Dean. In January 1986 the Corin family retired to Austin, Texas.

We feel privileged for our 18 years in the Canal Zone, to learn and teach in an "almost utopia" where everyone had a standard of living that was adequate for the health and wellbeing for himself and his family, to experience a unique American society amid a foreign setting. It may indeed have been "American style socialism," but it worked and served the world and America well.

PHOTO COURTESY OF PANAMA CANAL MUSEUM

Canal Zone College, La Boca, 1971. Bridge of the Americas in background.

The John Paul Corrigan, Sr., Family

As told by the Corrigans

Our Corrigan family story began in 1833 when Peter Corrigan married Jane Corrigan in Clogherhead, County Louth, Ireland. Their homestead was built in Parsonstown, where their six children were born. Son, John, our branch of the Corrigan family, married Anne Tiernan in 1872 in Drogheda, County Louth, Ireland, and during the great potato famine in Ireland moved to England, where their six oldest children were born: Peter Francis; John Paul, Sr.; twins Patrick and Owen; Jane F.; and Joseph Aloysius. The family immigrated to the United States aboard the *SS Germanic* arriving in the Port of New York on July 4, 1885. John became a U.S. citizen in 1893 in New Jersey. The family settled in New Jersey, where the seventh child, May, was born. From 1913 to 1921, John worked as a carpenter for the Panama Canal during Canal construction; however, he did not perform the two-year continuous service required to be eligible for the Roosevelt Medal. John and Anne left the Canal Zone on May 20, 1921, retiring to the east shore of Maryland where he died and was buried.

John Paul, Sr., was born on June 24, 1875. He married Mary Cecelia O'Connor in St. Aloysius Roman Catholic Church in Newark, New Jersey, where he became a naturalized U.S. citizen. John Paul, Sr., and brother Joseph Aloysius served in the Battery M. Fifth US Heavy Artillery Unit as sanitary inspectors during the Spanish-American War (1897-1900). As sanitary inspectors, they worked on eradicating mosquitoes with Dr. Walter Reed and Dr. Carlos Finlay in Cuba.

Having fought with Theodore Roosevelt and the Rough Riders in Cuba, Joseph Aloysius and John Paul, Sr., moved to the Canal Zone at the request of President Roosevelt. Joseph Aloysius moved there in 1905 to work as a sanitary inspector in Pedro Miguel and later worked on the Atlantic side. Two years later in 1907, John Paul, Sr., moved to the Canal Zone to work as a sanitary inspector in Balboa. A third brother Peter Francis moved to the Canal Zone in 1907 and worked for the Panama Canal as a carpenter; he left the Canal Zone in 1921 and returned to Tom's River, New Jersey.

John Paul Corrigan, Sr., 1875-1955.

All three brothers earned the Roosevelt Medal which was awarded to Panama Canal employees who had at least two years of consecutive service during the construction era (1904-1914), and each bar earned signified an additional two years of continuous service. Joseph Aloysius earned Medal No. 260 with three bars; John Paul, Sr., earned Medal No. 2283 with two bars; and Peter Francis earned Medal No. 2348 with two bars.

John Paul, Sr., was responsible for a major contribution in the eradication and control of mosquitoes. In 1917, John Paul, Sr., invented the sectional concrete ditch bottom which allowed water to be moved from ponding bodies of water in which mosquitoes would breed. The ditches were 30" long by 10" inside diameter with interlocking ends. They were easy to transport, assemble, and install.

The work that the sanitary inspectors did with eradicating mosquitoes in the Canal Zone and Panama directly and significantly impacted the successful building of the Panama Canal.

John Paul, Sr., and wife Mary had eight children who are the third generation Zonians: Margret Elizabeth "Peg"; Mary Theresa; Kathleen Marvoureen "Kay"; John Paul, Jr.; Owen Joseph; Gerald O'Connor; Peter Tiernan, Sr.; and Joseph Steven. Unfortunately, Gerald O'Connor died at a young age due to a falling accident on a ship. Mary and Kathleen left the Canal as young adults. Margaret "Peg" resided in the Canal Zone, where she married Alwyn DeLeon. An interesting note is that the famous Zonian artist, Alwyn "Al" Sprague, was named after him. The boys in the family all had long careers in the Canal Zone on both the Atlantic and Pacific sides, and all retired from the Panama Canal Company. John Paul, Jr., "Jack" worked on the piers on both sides. Owen worked as an anglesmith at Mt. Hope Industrial Division. Peter Tiernan, Sr., worked as a plumber and housing maintenance supervisor.

These third generation Corrigans had many children which constituted a fourth generation of Zonians. They are John Paul III; "Jackie"; Juanita "Tita"; Terry; Larry; Margaret "Peggy"; Robert; Pete, Jr.; Eddie; Tim; Mike; Collin; and Irene. These fourth generation Corrigans had many family trips and experiences together.

For instance, a typical car trip in the early 1950's to one of the beaches in the interior of Panama would go something like this: The dad usually drove the car, and the mom usually sat in the passenger seat with the youngest child sitting on the car's front middle bench seat between mom and dad. The rear seat would have any number of children three, four, five, or more. The older kids got the rear windows, no matter who "dibs" the window because of the simple fact that they were bigger and stronger. They made younger kids sit in the middle of the back seat. Sometimes there were so many kids on the trips that they stood behind the front seat and held on to the strap attached to the rear of the front seat. The family dog always sat on someone's lap, next to a window so he could stick his head out. There always seemed to be enough room in the car to fit another friend or cousin. The more the merrier. Although the cars had no air conditioning, the family was still thrilled to be going.

John Paul Corrigan family, 1916, Ancon. Back row L-R: John Paul, Sr.; Margaret; Mary holding Peter, Sr. Front row L-R: Mary; Owen; Gerald; John Paul, Jr.; Kathleen.

> *Dad would either drive up to the Ferry or the Miraflores swinging bridge, neither of which was quick. There would be a quick stop in Chorerra to buy a block of ice, Balboa beer for the parents, and a soda or raspado for the kids.*

Dad would either drive up to the Ferry or the Miraflores swinging bridge, neither of which was quick. There would be a quick stop in Chorerra to buy a block of ice, Balboa beer for the parents, and a soda or raspado for the kids. This was a good sugar high for the remainder of the trip. Families picked up the block of ice for refrigeration of perishable foods because none of the houses in the interior had refrigeration; they only had an ice box.

Once arriving at the beach, the kids were involved

in many fun activities such as swimming in the ocean, riding horses, or maybe just climbing a mango tree. Mom and Dad would put baby oil or Noxzema on the kids and themselves in an effort to help prevent sunburn. We all know now that didn't work because we came home with sunburn and were sometimes sunburned so badly, once we even blistered. The Corrigan family, alone, has made a number of dermatologists rich. Following a couple days at the beach the return trip was always much less fun. Everybody was tired and cranky because they had gotten too much sun. The chances of anyone being in school on Monday were very slim.

After this generation of Corrigans grew up in the Canal Zone, they had the fifth generation of Corrigan Zonians. These include John Paul IV, "Johnny"; Jennifer; Jessica; Jolene; Gilbert; David; Lisa; Tiernan; Gerald; Brian; Edward, Jr.; Eddylynn; Cassie Lynn; Timothy Brian "Tim"; Michael Peter "Mike"; Christopher "Chris"; and Colleen.

In August 2007 a Corrigan family gathering was held in our Irish homeland, Drogheda, County Louth, Ireland, and 126 cousins attended from Spain, France, Aruba, Canada, England, the U.S. and Ireland. Representing the John Corrigan branch were: Charles Heimbold, Ann "Flynn" and Jim Bryan, Susan Corrigan (Mrs. John Paul) and her daughter, Maureen Heim, all from the States; Aline Corrigan and her children, Patrick and Elvira Corrigan and daughter, Caroline (Corrigan) and Oliver Klementieff and their families from France; Mary Lou (Jones) Ramey and her sons and their families, Kay (Jones) and Jack Lee and two of their children, Pete, Jr. (a fourth generation Zonian), and Rosie Corrigan, from the States; and Mai Enjuto Quinn and husband, Avelero Hernando, and twin sons from Spain.

Some Corrigans worked during the Isthmian Canal Commission construction era; some of them worked for the Panama Canal Company and the Panama Canal Commission. Collin Corrigan worked as the chief of the Mt. Hope Industrial Division for the Panama Canal Company, then for the Panama Canal Commission, until noon on December 31, 1999, when he retired from the U.S. Government. On January 1, 2000, Collin began his service for the Panama Canal Authority (ACP) performing the same duties for about four more years.

Our latest family member is Didi Bremer Rogers Corrigan who married Larry Corrigan in 2007.

We Corrigans are proud of our forefathers and our accomplishments and services to the Panama Canal from 1905 to December 31, 1999, and after. At family gatherings like weddings, funerals, reunions, parties, we always sing the "Corrigan Song," you know, the song the Harrigans try to claim.

There were two kinds of Zonians — Corrigans and those that wished they were. That's our story and we're sticking to it!

John Paul Corrigan, Sr.
Roosevelt Medal No. 2283 with Two Bars

See also the Peter Tiernan "Pete" Corrigan, Sr.; the Charles Clarence Huber; the John J. and Grace N. Jackson; and the Patrick Joseph and Jane "Jennie" Quinn family histories.

Mt. Hope Industrial Division.

PHOTO COURTESY OF PANAMA CANAL MUSEUM

The Peter Tiernan "Pete" Corrigan, Sr., Family

As told by the Corrigans

Peter Tiernan "Pete" Corrigan, Sr., was born on March 26, 1915, at Ancon Hospital, Canal Zone. He married Helen Jean Nash, who was also born at Ancon Hospital, on September 18, 1919. They were married on December 1, 1939, at St Mary's Catholic Church in Balboa, Canal Zone. Helen was the daughter of Edward Hugo Armstrong Nash and Florence Emma Huber, both long time residents of the Canal Zone. Helen was a third generation Zonian by virtue of her grandfather, Charles Clarence Huber.

Pete, Sr., graduated from Balboa High School in 1934. His first job in the Panama Canal Company began in 1935 as a deli boy in the Ancon Commissary making $62.50 per month. Later in 1935, he entered the Plumbing Apprenticeship Program, completed it in 1938 to earn his plumbing license, and then worked as plumber for the Panama Canal Company on the Atlantic side. Pete, Sr., was promoted to the position of Plumbing Foreman in Balboa. In 1953 he was in charge of preventive maintenance at the Balboa Heights and Ancon District. In 1960 Pete, Sr., transferred to the Balboa Housing Office as the District Maintenance Construction Supervisor. Pete, Sr., retired from the Panama Canal Company Housing Division in 1970 and moved to Sarasota, Florida. From 1973 to 1988, Pete, Sr., worked for Sarasota County as the Chief Plumbing Inspector. In 1988 at the age of 70 he retired from his second career.

As a young man in the Canal Zone, Pete, Sr., was well respected for his baseball ability as a pitcher and for his vast knowledge of the game. He became an outstanding baseball umpire, having umpired in the Canal Zone Twilight League for many years. He umpired in the Latin American Olympics in 1946 held in Colombia and again in Venezuela in 1951. The complete umpiring crew for those two events included Pete, Sr., his brother Jack, and the Williams brothers (Roger and Willy Williams). All four were considered the best umpires in the Canal Zone.

Pete, Sr., managed many Pacific Little League baseball teams and Fastlich League teams. He also managed VFW and American Legion Teenage Baseball League teams, as well as the Twilight League Men's baseball teams. He enjoyed great success because he was able to motivate players and employ his exceptional knowledge of the game. Over the years, many people considered him the best manager they ever played for.

Helen Nash (1919-1997) and Peter T. Corrigan, Sr. (1915-1998).

Pete, Sr., and Helen had four sons: Peter Tiernan Corrigan "Pete," Jr., Brian Edward "Eddie," Timothy James "Tim," and Michael Joseph "Mike" or "Dink." All of them were very active in Canal Zone activities and were good all around athletes as kids.

Peter "Pete" Tiernan Corrigan, Jr., married Rose Marie Seefried on September 9, 1961, in Lockport, New York. They had two daughters: Teresa Marie and Suzanne Marie. Teresa was married to David Greenwald, and they had two boys: Andrew Theodore Greenwald and Daniel Leo Greenwald. Pete, Jr., worked as a yacht broker in Sarasota, Florida, for 32 years retiring in 1996. He and Rosie raised Paint horses on the Corrigan 4-Cs Ranch in Sarasota, including some national champions. Daughter Suzie was a great handler and rider of Paint horses and won a national championship in 1989 at Ft. Worth, Texas, with Mr. Exclusive, their main stud horse. Unfortunately, Suzie passed away in 2002.

Brian Edward "Eddie" Corrigan married Lynn Egger, and they had three children (all born in Coco Solo, Canal Zone): Brian Edward, Jr. (who died several days after birth); Eddylynn; and Cassie Lynn. Eddylynn is married to Eric Thomas Kledzik. They have one son: Colton. Cassie Lynn married Thomas Green, and they have a daughter, Victoria. Eddie Corrigan later re-married Susan Marie Buss. They had twin boys: Timothy Brian "Tim" and Michael Peter "Mike," born in Gorgas Hospital, Ancon, and two stepsons: Glen and Robert Walker, Jr. Son Tim married Cody Wigenton and they had a daughter: Kelsi Alexa. Mike married Kristen "Kristy" Bryant and had the newest member of the Corrigan family, their daughter Mikayla Elizabeth, who was born on January 28, 2008. Eddie served in the U.S. Navy for nine years and then worked for the Panama Canal Water and Labs Division. Eddie died in 1976 in Sarasota, Florida.

Timothy James "Tim" Corrigan married Taffy Grace Koepke on July 26, 1969, at Sacred Heart Chapel, Ancon, Canal Zone. Tim graduated from Balboa High School in 1964. He was voted "Most Athletic" in his class. Tim attended Canal Zone College and Florida State University in Albrook AFB, Canal Zone, and in 1980, he earned his Bachelor of Science degree from Nova University (Davey, Florida), with Magna Cum Laude honors, majoring in Criminal Justice and Corrections. In 1982, Tim earned his Master of Science degree from Nova University in Human Resources Management. He began his 32-year career with the U.S. Government in 1966, as a Federal Police Officer with the Canal Zone Police Division. He served in a variety of leadership and managerial positions during most of his service with the Panama Canal Government and subsequent Panama Canal Commission. Most of his 16 years as a Police Officer was served with the Canal Zone Penitentiary. With the reorganization resulting from the termination of the Canal Zone Police Division, Tim worked as a Contract Specialist for the Storehouse Division, then as an admeasurer with the Marine Bureau. He was the Assistant to the Division Chief of Sanitation at Corozal, Canal Zone. He culminated his career with the Panama Canal Commission as the Chief Printing Officer

The Corrigan boys, L-R: Pete Jr., Eddie, Tim, and Mike, 1952.

for 11 years. On June 20, 1998, Tim retired, and he and his family relocated to Centreville, Virginia.

Tim devoted many hours to youth activities coaching his two children's sports teams, along with many neighborhood children, who affectionately called him "Coach." His coaching included such sports as football, cheerleading, softball, baseball, golf and "Cayuco," a Sea Scouts sponsored Ocean-to-Ocean race unique to the Canal Zone. Tim reflects, "There is no greater privilege than to work with kids, gain their respect, earn their trust and most of all become their "friend," which is an honor that few adults are lucky enough to achieve."

Tim and Taffy have two children: Christopher David "Chris" and Colleen Grace. Chris was born on October 13, 1976, in Gorgas Hospital, Ancon, Canal Zone; and on December 9, 1983, Colleen was born in Paitilla Hospital in Panama City, Republic of Panama.

Chris graduated from Balboa High School in 1994. This was a memorable occasion for his father Tim and Grandfather Pete, Sr., who were both present because grandfather Pete graduated 60 years prior (1934) and father Tim graduated 30 years prior to Chris (1964). Chris attended the Georgia Institute of Technology, where he met and married the former Alejandra del Carmen "Alex" Ramirez. Chris and Alex are both graduates of Georgia Tech, he with high honors in civil engineering and she in management. Chris continued his studies at Georgia Tech and earned his Master's Degree in Geotechnical Engineering. Subsequently, Chris and Alex moved to Florida, where they are raising their two daughters: Nichole Grace and Lauren Elizabeth. Chris and Alex work for Bolivar Trading, the Ramirez family business.

Colleen attended her freshman year of high school at Balboa High School, relocating with her parents to the U.S. as the curtain fell on the historic U.S. Canal Zone. She graduated from Fairfax High School in 2001 and attended Radford University, Radford, Virginia, where she earned a degree in elementary education in 2006, following in the footsteps of her mother. She is currently a kindergarten teacher for the Fairfax County Public Schools. Colleen is engaged to her fiancé, Joseph Wayne "Joey" Wilkinson from Bluefield, Virginia. Joey also attended Radford University where he earned his B.A. degree in Marketing in 2006.

Michael Joseph "Mike" Corrigan married Stephanie Lynn Zweig and had one daughter: Carey Lorraine. He later married Marjorie Bowen Hughes. He has two stepdaughters: Sheryl Lynn Broom and Leigh Ann Mahoney. He also has step-grandchildren Britney Boyd and Carina Boyd. Mike was in the U.S. Navy for four years, serving on the aircraft carrier *Bon Homme Richard*. He has been in the furniture business for his entire adult life. Mike currently lives in Chattanooga, Tennessee. His wife, Marge, works at the Tennessee Aquarium.

The Corrigan family is proud to have five generations of "Zonians": (1) John; (2) John Paul, Sr.; (3) Peter, Sr.; (4) Pete, Jr., Eddie, Tim and Mike; and (5) Brian Edward, Jr., Eddylynn, Cassie Lynn, Timothy Brian, Michael Peter, Chris, and Colleen. These seven people are also fifth generation Zonians under the Huber family tree. Thus, they are the only seven people in the history of the Canal Zone who are fifth generation Zonians on both sides of their families (Corrigan and Huber).

Family ski trip, L-R: Joey, Colleen, Larry, Alex, Lauren, Chris, Nicole, Taffy, Tim.

See also the John Paul Corrigan, Sr.; the Charles Clarence Huber; the John J. and Grace N. Jackson; and the Patrick Joseph and Jane "Jennie" Quinn family histories.

The Robert Crooks Family

By Bob Crooks

My mother's side of the family came to the Isthmus of Panama in 1906 in the person of great-grandfather William Frank Morrison, who signed on as a blacksmith for the Isthmian Canal Commission. His wife, Mattie J. Morrison (my "Meemaw"), joined him in July of that year and related these stories to four generations of our family.

Mattie left Texas with her five-year old daughter, Beast, and older married daughter, Cecil Lowe, and boarded a ship called the *Karen*, which was no bigger than a tugboat. She arrived on the isthmus July 12, 1906. Grandfather William met them in Colon and took them to Gorgona, House No. 86. That night, Meemaw discovered there were no screens on the windows, no plumbing, no electric lights, and only one pipe of running water — which you could not drink. Fortunately, distilled drinking water was brought to the family daily in big demijohns carried by West Indians. Ice came from Cristobal and was placed on the local train station platform, where each person had to salvage his own order. In those early years, Mattie became friends with Colonel Goethals and Admiral Rousseau and hosted them in her home. Eight years after she arrived — the year the Canal opened in 1914 — Meemaw took her first vacation back to the United States.

Grandfather Walter Brown came to the Canal Zone in 1907 with his wife Nannie Morrison Brown (daughter of William and Mattie) and their daughter, Mattilee. Nannie and Walter had four children born in the Canal Zone: Frances; Minnie; Walter Gay, Jr.; and Jack. Grandfather Brown was a machinist with the Isthmian Canal Commission until he retired and moved to Arraijan.

Meemaw loved to tell how Uncle Walter was the first baby to transit the Canal. On the day the Canal opened, August 15, 1914, the first ships through were the *SS Ancon* and the *SS Advance*. There was great demand for accommodations on the ships, so the invitations stipulated that no children were allowed. Because Uncle Walter was still nursing, Meemaw's daughter, Nannie, would have been denied the trip. But Meemaw's friend Colonel Goethals gave special permission for Nannie and the baby to go aboard for the transit, making my Uncle Walter the first baby to transit the Panama Canal.

Both my great-grandfather and grandfather were Roosevelt Medal recipients. The story of the Crooks family at the Panama Canal began with my father, Homer Crooks, who was born in Island City, Oregon. He was a third-class petty officer in the Navy when his ship came though the Canal. At that time, his brother, Mike Crooks, was working for the Canal at the Motor Transportation Division. While visiting his brother, Homer met the Brown family and dated Mattilee, the oldest of the Brown girls. After leaving the Navy, he returned to the Canal and

The Morrison family

signed on to be a Canal Zone policeman, a job he held for 30 years. One day, Dad spotted a lovely young lady and inquired about her. He learned she was Miss Minnie Brown, youngest daughter of the Brown family. He lost his heart to Min-

Sgt. Homer Vincent Crooks

nie that day and set his cap for her. What Homer did not know is when he was dating her older sister, Mattilee, young Minnie was hoping he would become her brother-in-law. Minnie got her wish to have Dad in the family, but as her loving husband. They married on April 14, 1930, and had three children — Nancy, Judy and me (Bob) — all born in Panama City Hospital.

I have so many fond memories of growing up in the Canal Zone. As kids and teens, we had so much freedom that it was like living, as one of my classmate's husbands said, in "the Land of Oz." Life was very safe and uncomplicated. As a kid, you could wander all over town, and your parents never had cause to worry.

I attended Balboa High School (1955) and have fond memories of our sports programs and trips by train to watch the games at Cristobal. I played golf at Summit Hills with Sandy Hinkle, Jack Hammond, Bob Zumbado, and Reggie Hayden. I fished along the Amador Causeway and on tuna boats with Al Sprague and Bill Kommenich, Doug Schmidt, and Sam Beckley. And I sat in the crossbeams under the pier at the Amador Yacht Club at night and listened to *The Happy Wanderer* and *Ebb Tide* playing on the jukebox up at the Yacht Club.

Every time I get together with Al Sprague, we recall fishing off the tuna boats at Pier 6. The police didn't like us fishing down there and would run us off if they saw us.

On one particular occasion, Al was back by the bait wells on the boat, and I was up forward. I saw the "Green Hornet" police car coming down the pier, and I started aft to warn Al. Suddenly, one foot slipped out from under me, and I went sliding past Al on my back with my fishing rod up in the air, yelling "Cops! Cops!" Al nearly fell over the side, he was laughing so hard.

After graduating from BHS, I joined the Navy, attended the University of Arkansas, and then joined the Army for eighteen years, retiring as a Command Sergeant Major in 1978. After two tours of duty in Vietnam, I returned to Washington, D.C., where I met Nicola K. Bucher, my wife of 40 years. We married December 9, 1967, in Arlington, Virginia; and, shortly thereafter, I left

CSM Robert Glenn Crooks, 101st Airborne Division, Ft. Campbell, KY, 1977-1978.

for Fort Gulick in the Canal Zone. Nikki joined me upon her discharge from the Navy in March. Our first son, Jeffrey McKenzie, was born in Coco Solo Naval Hospital in 1970. Our second son, Brian Michael, was born in Birmingham in 1972; and our third son, Benjamin Gaylord, was born at Fort Bliss, Texas, in 1974.

This is our legacy from our beloved Canal Zone.

William Frank Morrison
Roosevelt Medal No. 1258 with Three Bars

Walter G. Brown
Roosevelt Medal No. 1734 with Three Bars

The John M. Davis and Michael Kenny Families

As told by Bonnie Davis Dolan

A Great Life!
Bonnie Davis Dolan

A Story of Generations of Proud, Dedicated Men and Women

In 1907, John M. Davis, Roosevelt Medal recipient, joined the Canal Zone Police (CZP). John was born in 1882 in Williamsburg, Kentucky, and was a U.S. Army veteran of the Spanish American War in Cuba. While in the service, John received a letter from Laura Agnes Smith (age 13), Harrogate, Tennessee, asking him to be her pen pal. They began writing; and when John returned, he saw Laura only a few times before he accepted a position in Panama with the CZP. He feared the position would be filled, so he left for Panama immediately without seeing Laura to say goodbye. John continued to write Laura. In 1910, John took leave for Harrogate, Tennessee, and married Laura returning with his pen pal and bride. John and Laura Davis were blessed with four children. Son Ralph was born in the old Colon Hospital in 1917. Daughter Margaret was born in the United States, and daughters Norma and Myra were born in the Canal Zone. During John's tenure with the CZP, he worked in every townsite in the Zone, including Gorgona and Empire, which were construction towns that no longer exist today. A joke backfired as John, reaching District Police Commander rank, was to inspect the gallows for a hanging at the Gamboa Penitentiary, and Captain Foley jokingly pulled the lever as John stood on the platform. Foley was the only one laughing.

In 1914, Michael Kenny, ornamental plasterer, was hired by the Isthmian Canal Commission. He was born in 1870 in Roscommon County, Ireland, came to the United States in 1883, and became a U.S. citizen in 1896. He was married to Molly Van Glin in Brooklyn, New York, and they had four daughters: Marione, Lillian, Beatrice and Norma, all born in New York. Some of Michael's work included the ceiling in the rotunda of the Administration Building. Michael and family returned to Brooklyn after he became ill with tuberculosis. Michael died in 1925, and his wife and three daughters returned to live in the Canal Zone with their daughter, Marione, who had remained in Panama. Marione was seen coming down the steps of the Ancon Post Office by Ewing C. Journey (born in Tennessee), employed with the U.S. Army Corps of Engineers. When he saw her, he said he had met his Waterloo. They were married in 1919 and had two children who were born in Gorgas Hospital: Marion Lucille (Cele) (1920) and Ewing (Bud) (1923). In 1928, Michael and Molly's fourth daughter, Norma, married Ernest Angermuller. They had two sons born in the Canal Zone. Ronald (U.S. Marines) married Linnea (U.S. Air Force), and they have two children: Larry (retired U.S. Navy) and daughter, Britta. Michael Angermuller (U.S. Navy) married Roberta (Bird) and had one son, Sean Angermuller. Ron retired as Director of Admeasurement and lives in North Carolina with his wife, Linnea.

In 1938, John and Laura's son, Ralph, while attending Canal Zone Junior College, traveled to the Amazon River as a photographer for the National Geographic Magazine with Professor Vinton. On his return, Ralph began an apprenticeship with the Panama Railroad. Ralph was introduced to Michael and Molly Kenny's granddaughter, Cele, by his sister Myra. Cele graduated from BHS (1938) and went to the Johns Hopkins School of Nursing, Baltimore, Maryland. While she was in school, they corresponded; and during a trip to the States, Ralph proposed. In 1940, Ralph Davis and Cele Journey were married. They had two daughters: Karen and Bonnie, born in Gorgas. Ralph's position was critical to the war effort; and during World War II he remained with the railroad transporting troops, ammunitions and supplies for the Canal and military. Cele volunteered for the USO and worked at the U.S. Army Hospital in Ft. Clayton. Bud Journey joined the Navy and was stationed on the battleship, USS California serving in the South Pacific (combat veteran). After the war, Bud went into the Panama shrimping business; married Louise (Bricky) Coleman; and had one son, Tom Journey Pattison. Later he married Harriet Johnson, teacher, Schools Division, and had three children: Hayes, Andy and Molly. They moved to the States in 1962. Harriet has 13 grandchildren. Grandson Sean Sullivan is a 1st LT., U.S. Army.

In 1947, John M. Davis (Dida) retired as Captain, Balboa District Police Commander, after serving as Police Chief, Balboa District. He moved to the Gamboa Ridge with his wife, Laura (Nana). Favorites were family picnics at Summit Gardens and grandchildren's visits to their home, which was off the ground four-family quarters overlooking the Gamboa Golf Course. Their home was filled with aromas of Nana's butterscotch bars baking in the oven, sweet lavender, or Dida's Cuban cigars.

John and Laura's daughter, Norma, married Joe Short, a pioneer aviator in South America. Later, Joe was employed with the Department of Defense in the Canal Zone. They had two sons: Milton who served as a U.S. Marine, combat veteran, Vietnam War. He married Cindy Klocek and has two daughters, Melissa and Mindy, and five grandchildren. Milton retired as a police sergeant from Miami Dade Sheriff's office. Norman served in the U.S. Army. Both sons (Curundu boys) live in Florida. The youngest daughter, Myra Davis, married Bill Walston, who was with the Canal Zone Police and the Dredging Division. They have two daughters: Laura Walston married Phil Sanders (Electrical Division) and had two children: Lisa and Randy. Leilani Walston works for the U.S. Air Force at Travis Air Force Base in California, married to Herb Collins, retired U.S. Army, combat veteran, Vietnam War. They have one daughter, Myra. Myra Walston has three grandchildren and two great-grandchildren.

Ralph and Cele Davis's daughters, Karen and Bonnie, have memories of family gatherings at their beach house in Gorgona every weekend. Their dad, Ralph, was swept out to sea in an undertow, but was found alive on the beach the next day by the U.S. Army. Summers were mostly spent in Gorgona, Volcan and El Valle. In 1956, Karen and Bonnie made their first trip to the States.

Ralph worked as conductor on the Panama Railroad for 36 years. Sunday's train left Ancon Station at 2 p.m. Conductor Davis often took his grandchildren on the

Davis family gathering in their home. Top row, L-R: Ralph Davis, Bill and Myra Davis Walston, Laura Walston, Cele Davis, Laura and John Davis with Norman Short (baby); Middle row, L-R: Joe Short, Karen (Kay) Davis, Margaret Davis; Front row, L-R: Leilani Walston, Milton Short and Bonnie Davis.

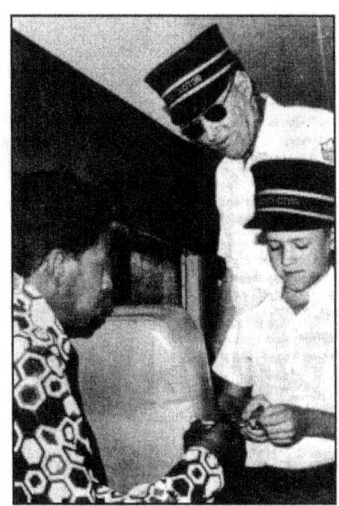

Conductor Davis makes a run to Colon with grandson, Eddy Dolan.

fastest train in the world traveling from ocean to ocean in one hour, give or take a few minutes for derailments. At several of the stops, Conductor Davis helped the natives that lived near the railroad tracks and often received conejos with their thanks. Cooking conejo was Ralph's specialty.

Karen (Kay) Davis graduated from Balboa High School (1959) and attended the University of Arkansas, where she met and married Ron Pyeatt. They had two sons: Ralph, born in Fayetteville, Arkansas, and Brian, born in Kansas City, Missouri. Ralph and Brian made many trips to Panama to visit their grandparents. Ralph is a federal agent with U.S. Department of Homeland Security and is married to Catherine Hwang, who is an attorney. Ralph is in the U.S. Navy Reserves and is a combat veteran, Iraq War. Brian is a business applications analyst for Vulcan, Inc. Karen married Steven Andrews and has a stepson, Matt Andrews. Steven and Karen retired from the Department of Defense, live in Florida and visit Panama often.

Bonnie Davis graduated from Balboa High School in 1962 with plans to go to college in Boston. She left Panama on the *SS Cristobal*, and as the ship pulled away from the pier, the breakwater and Colon fading from sight, her future was an unknown. But Boston was not for this Canal Zone girl, and she returned to Panama and began working for the military in Quarry Heights. In 1965, Bonnie married Edward V. Dolan, son of William and Cecile Dolan, and they had three children: Eddy, Jr.; Karen; and Tim; eight grandchildren: Morgan, Madison and Mackenzie Dolan; Jon Dolan, Kyle and Colin Castleton; and Michael and Angela Bapp; and a great-granddaughter: Kailey Dolan. In 1982, they left Panama when Ed's job was abolished by the Panama Canal Treaty of 1977.

Ralph Davis, age 57, died from skin cancer after a lifetime of exposure to the sun. Cele moved to El Valle and because of the Panama Canal Treaty later moved to the States. She passed away in 1998. John and Laura Davis passed away (1960's) and are buried at the U.S. National Cemetery, Corozal, Panama.

From 1907, John Davis, Michael Kenny and descendants are proud, dedicated men and women serving God and their country. Future generations always remember the sacrifice made, "The Land Divided, the World United."

John M. Davis
Roosevelt Medal No. 2926 with Two Bars

See also the Edward V. Dolan and the Bruce Gordon and Grace Aloise (Meister) Sanders family histories.

Davis and Kenny family descendants at Karen and Steven's wedding on October 13, 2007. Shown are (left rear) Tom Pattison, Ralph Pyeatt, Ed Dolan, Brian Pyeatt, Steve and Karen Bapp, Eddy Dolan. (Left front) Cindy and Milton Short, Catherine Pyeatt, Bonnie Dolan, Karen and Steve Andrews, Molly and Rick Ridder, Harriet Journey, Andy Journey and Linda Pattison.

The Fred B. Deakins Family

As told by JoElla Deakins

In 1905 Fred B. Deakins and Lita A. Goble met in St. Joseph, Missouri, where Fred was a streetcar conductor and Lita worked in a shoe factory. Lita resided in the "St. Charles Hotel for Women," which required a trolley ride to and from work. Fred lived with his widowed mother, Matilda, who wasn't happy with Fred and Lita's relationship as she didn't want to give up her son. So, Fred and Lita eloped and married in 1906. Fred then presented his "wife" to his mother. Lita spent the next fourteen miserable years in her mother-in-law's home. One day Lita noticed an advertisement for police officers in the Canal Zone. She gave Fred an ultimatum, that she'd take the children and leave him if he didn't get her out of his mother's house! So Fred went to Panama in 1919 and got a job as a police officer, returning in 1920 to bring the Deakins family to the Canal Zone.

Fred and Lita lived in Gatun, working and raising three children, Roger Lloyd, Allene, and Leta. Fred was a mounted police officer until his death in 1935. Lita stayed on in Panama by running the Cristobal Women's Club at the Gilbert House in Colon. With her son Roger's help, she bought a two-story house on Front Street which she turned into a boarding house for U.S. workers. This house was saved by Roger and friends helping the Bomberos during the great Colon fire in 1940. Lita left in 1942 to St. Petersburg, Florida, until her death in 1964.

Son Roger graduated from Cristobal High School in 1929. His first job was a banana checker at Gatun Landing, afterwards working for awhile with PanAm Airways. He then went on to work for the Electrical Division. Violet Sylvia Randall came to the Canal Zone with her parents, Bob and Daisy Randall, and her two sisters, Charlotte and Arlene, in 1929. Violet graduated from Cristobal High School in 1933. Roger and Violet married in 1936. As newlyweds, they moved to Madden Dam, which was very remote with jungle all around. While Roger was at work, Violet, being an excellent seamstress (She made her own wedding dress.), had a Panamanian woman teach her how to make the native dress of Panama, the "pollera." It was the start of her lifelong passion to learn all about these beautiful dresses. She conducted many a class, teaching many Zonians how to make them. The family still has a collection of Violet's polleras.

L-R: Fred and Lita Deakins, Roger, Allene and Leta in Gatun.

Roger and Violet's first son, Randall Lloyd, was born on July 10, 1940, in Colon Hospital, growing up in New Cristobal. Roger and Violet went through World War II in the Canal Zone with Roger working in the marine Electric Shop that did repair work on transiting warships and Violet active in the USO.

Roger and Violet's second son Terrell Carver was born on January 14, 1944, growing up in Margarita and Gatun. Roger and Violet passed on their love of Panama to Randy and Terry by taking trips to the interior, fishing at Gatun Spillway, and fishing in Panama Bay. They built a beach house in Santa Clara, enjoying one of the best beaches in Panama. Terry and Randy both graduated from Balboa High School and Canal Zone College, and then they both went into the military. Both returned for a career with the Canal. Roger and Violet retired in 1966 after 32 years of service with the Panama Canal Company, moving to Titusville, Florida.

Randy met JoElla Sue Jenkins in 1963 when he returned from the Navy. JoElla was the daughter of Joe and Elnora Jenkins, who were missionaries in the Republic of Panama. Since JoElla's parents were not government employees, she had never lived in the Canal Zone. She came to Panama with her parents and brothers, David and Dan, in 1953. Her family lived in Arraijan and paid tuition for her to attend Canal Zone schools. Because JoElla did not have Canal Zone privileges, she could not attend Balboa Theater. On several occasions when her friends sneaked her in, Terry Deakins, a theater usher, would kick her out. Needless to say, when Randy asked JoElla for a date, it was almost a "no go" when she found out who his brother was! Randy and JoElla were engaged when the riots of 1964 broke out. Since her parents lived in Panama City at that time, Randy's cousin, Fred Robinson, offered them the use of his bachelor apartment in Balboa. They were married on June 6, 1964, in the Curundu Protestant Chapel. They moved to Gamboa as Randy was working for Dredging Division as an electrician. He went on to become a chief engineer on both tugs and dredges.

Randy and JoElla's three children were Rick Lloyd Deakins, born June 24, 1965, at Gorgas Hospital, and Tim Cash Deakins and Tom Carver Deakins, born February 24, 1969 — twins! What a wonderful place Gamboa was for raising a family. So many memories: attending Gamboa Elementary School, fishing on Pipeline Road, exploring jungles, learning how to swim at Gamboa Pool to get a"B" badge, jumping off Gamboa Bridge, skateboarding, seeing movies at Gamboa Theater with bats flying overhead, playing kick-the-can at the "big square," eating mangos and rose apples, and, best of all, enjoying a whole town full of wonderful friends. Randy and JoElla retired to Florida in 1984 due to the Panama Canal Treaty. Randy worked ten more years at Cape Kennedy for the Space Program. Son Rick married Tracie Wyss, and they have two children, Ricky L. II and Meghan Mae. Rick works for NASA.

Son Tom married Cheryl Sherritt and works as an electrical engineer for Harris Corp. And son Tim is married to Alice DiBiase, and they have a daughter, Nikki. Tim is a Lab Supervisor at a hospital in Rockledge, Florida.

Randy has built a 100-year time capsule pertaining to the four generations of Deakins working and living in the Canal Zone. The capsule was sealed in 2005.

PHOTO BY CAPT. WILBUR VANTINE

Front Street, Colon.

The Sylvester P. Dennis, Arthur E. Baker, and William C. Bain Families
As told by Bonnie Bain Dunbar

The History of Three Families Joined Beneath the Southern Cross

To this day, my memory of the Canal Zone brings to mind romance, buccaneering adventurism, gold, jungle, snakes, iguanas, the King's Bridge, Las Cruces Trail, Chagres River, La Boca, Diablo, Cristobal, Colon, Darien, parakeets, molas, Farfan Beach, and a myriad of terms that describe the long-standing history of the Panama Canal. These terms were bigger than life as I was only eight years old when my parents moved the family from the Canal Zone to the States; and my adult memories are those of tales told by grandparents, parents and other relatives. Our family has always been proud of our isthmian heritage, and we tend to consider our forefathers as pioneers — perhaps not of the American West, or even the moon, but a wild, unfamiliar land between the oceans.

Sylvester P. Dennis was born in Wisconsin in 1890 to parents who immigrated to the area from England. His mother died when he was very young, and his father married again. Sylvester was never told that the woman who raised him was his stepmother. Upon discovering this fact as a young man, though he loved his parents dearly, he became angry, fled his home and found himself in the Canal Zone. It would appear that it was his personality and adventurous spirit that led him to this place, about as far away from Wisconsin as he could get. His anger subsided and he made peace with his parents over the rest of his life.

He arrived in Gatun, Canal Zone, in December 1910, at the age of 20 and was employed in the construction of the Gatun Locks as an operator of locomotive cranes. His skill with a crane was uncanny and his speed phenomenal. It was here that he received the nickname of "Spillway Kid." On completion of the locks, he went to work on the hydraulic dredges as an operator and worked from one end of the Canal to the other. He was a charter member of Local No. 596, Steam Shovel and Dredge Men. For many years he was chairman of the committee for those seeking licensure necessary for working on the dredges. He was also a Master Mason. His greatest pastime was hunting, and he was a student and authority on the habits of wild animals.

Around the same time that Sylvester Dennis arrived in the Canal Zone, a gentleman from New England also arrived with his family. Arthur Elliott Baker was born in New York City in 1870 to a family long established in this country, arriving on ships soon after the *Mayflower*. He married Annie Loughery on December 23, 1888. Arthur was an engineer for the New Haven Railroad. He took the first electric engine from New York to New Haven. Sometime in 1910, he was fired from the railroad for falling asleep at the controls. Subsequently, he was hired by the Panama Canal Company as an engineer. He retired from this engineering post in 1932 and returned to the United States.

Arthur and Annie Baker went to the Canal Zone with two daughters — Josephine Raymond Baker, age 17, and her younger sister, Melba, age 12. As fate would have it, Josephine Baker and Sylvester Dennis met and fell in love. They married on June 16, 1912. Arthur and Sylvester were already acquainted, and he was accepted immediately as a son-in-law.

The only street name I remember my grandmother citing was Barnebey Street. I do know that she said she lived in a boxcar temporarily, either awaiting quarters with her parents or after marrying.

The Dennis family grew over the years, first with four daughters — Dorothy, Josephine, Jean and Marjorie — and then one son, Robert (Dink). Childhood was a busy time in the Canal Zone with swimming, hunting, sledding on palm fronds down Sosa Hill, fishing, horseback riding and all manner of outdoor activities. After graduating from Balboa High School, Josephine and Jean left the Canal Zone to train as registered nurses in New Jersey. There they married, and each had two daughters.

Marjorie and her husband, daughter and son, and Robert and his wife and son, left the Canal Zone in the early 1950's. Sylvester Dennis died on July 2, 1933, of complications of kidney disease. He was cremated. His last day of employment with the Panama Canal Company's Dredging Division was June 28, 1933.

During the Depression in the late 1930's, William Charles Bain, who was originally from Scotland, applied for work with the Panama Canal as a heavy metal blacksmith. Unable to find work in the United States during these times, he and his wife, Catherina Hill Bain, also from Scotland, elected to move the family to a new start. Their eldest son, William James Bain (Billy), a first generation American born in Bridgeport, Connecticut, was in his senior year at Brooklyn Technical High School when his parents moved him to Balboa High School, where he received his diploma.

Billy Bain entered employment with the Panama Canal Company as an apprentice and became a journeyman sheet metal worker. He moved from his parents' quarters into bachelor quarters and worked for the Mechanical Division. In the meantime, Billy met 15-year old Marjorie Dennis at Balboa High School in 1935. They married in August 1940 and set up housekeeping on Empire Street in Balboa. Billy and Marjorie had two children.

During World War II, Billy worked as a night superintendent on the Pacific side and was involved in maintenance and repair of warships from the Pacific theater. The ships came in under cover of darkness and had to be back at sea by daybreak. Marjorie worked days for the Navy during this time. The Canal Zone was blacked out during the war and heavily protected by curfew.

During this time, Billy was sent by the Panama Canal to Brooklyn Navy Yard to train in the repair and installation of the newly invented radar equipment just coming into use by the military. He and one other man were the only individuals trained in this knowledge during the war and supervised and taught others.

William Charles Bain died in 1947 of black water malaria. He was cremated. Sylvester Dennis and William Bain never knew one another in life. Their ashes were scattered together on the waters of the Panama Canal at Culebra Cut.

Billy and Marjorie were living on the Prado in Balboa, across from the Balboa Elementary School, when the family left the Canal Zone in 1953 to settle in New Jersey. Josephine Raymond Baker Dennis was retired from the Panama Canal Company and had been living with Billy and Marjorie. She left with the family to live in New Jersey. Her son Robert followed in 1954. Her oldest daughter, Dorothy Dennis Douglas, remained in the Canal Zone and worked until retirement in the 1970's at which time she and her husband moved to Florida.

Thus came to an end the Canal Zone history of three families forever joined beneath the Southern Cross.

William (Billy) James Bain died on October 9, 2006, at the age of 89 and was the last of the generation that worked for the Panama Canal Company. There remain 32 living sons, daughters, grandchildren, great-grandchildren and great-great-grandchildren of the Sylvester Dennis and William James Bain arm of the family, all born between 1943 and 2007.

There are memories and tales too numerous to print here that make the Panama Canal experience unique to those of us whose families contributed to the monumental accomplishment that is the "Big Ditch."

Sylvester Phillip Dennis
Roosevelt Medal No. 6468 with One Bar

Arthur Elliott Baker
Roosevelt Medal No. 6829

The Edward V. Dolan Family

As told by Bonnie Davis Dolan

The Irish Came To Panama. . . . Their Legacy Still Remains

In 1942, Edward V. Dolan's father, William G. (Bill), and his uncle Joseph F. (Joe), of Mamaroneck, New York (born of Irish immigrants), were hired by the Canal Zone Fire Division, Cristobal Central Station. In 1944, Bill's wife, Cecile (McCarthy), and two sons born in New York, William E., and Edward V., flew to Panama on Pan American Airline's China Clipper. They finally arrived after four attempts caused by mechanical problems and lack of spare aviation parts due to the war. When the war ended, Bill and Joe remained with the Fire Division.

Bill and Cecile's family grew to nine children with seven more born in Panama: George (died in 1967, age 20); Jimmy, civilian contractor, U.S. Air Force, Vietnam War; Madeleine; Robert; Richard, Fire Chief, Florida (following in father's footsteps); Patrick, retired Staff Sergeant, U.S. Air Force; and last but not least, Celie. In 1972, after an exciting career, Bill retired as Chief, Fire Division, and moved with Cecile to Jacksonville, Florida. They enjoyed family get-togethers celebrating "ping dings," remembering their life in the Canal Zone. They have 61 grandchildren and great-grandchildren. Bill died in 1991, and Cecile died in 1993. Bill and Cecile's oldest son, William E., married Melissa Downing and had seven children: Sean, Maureen, Deidre, Kevin, Brendan, Terrance and Bridgid. William E. died in 2002.

Joe married Ann Chase of the Canal Zone. They raised eight children: Kathy, Joseph (U.S. Navy), Michael, (U.S. Air Force), Eileen, Paul, Ginny, Bridget, and Suzanne. Joe retired in 1979 as Chief, Canal Zone Customs, and moved to Texas with Ann. Both Joe and Ann died in the late 1990's.

Edward V. (Ed) graduated from Balboa High School (1960), attended Canal Zone Junior College, and joined the U.S. Navy in 1961. He traveled to many countries always remembering a girl with a blond pony tail back in Panama. Upon returning from the Navy, Ed and Bonnie Davis were married on July 3, 1965.

Ed joined the Canal Zone Police and worked through the ranks to Lieutenant. Ed was on the board that wrote the Joint Patrol Manual and Training Program for the Canal Zone Police and the Panama National Guard.

Bonnie Davis and Ed Dolan Reception, Roosevelt Room, Tivoli Hotel, July 3, 1965. Back row, L-R: William E., George, Jim, William G., Ed, Bonnie, Cele and Ralph Davis; front row, L-R: Melissa, Richard, Cecile, Celie, Madeleine, Patrick and Robert.

In 1982, the Canal Zone Police was abolished and Ed retired. Ed, Bonnie and their three children, Eddy, Jr., Karen, and Tim, left Panama that year.

Life in "Estados Unidos" was an adjustment for the Dolan family, but they always continued on with old traditions, family recipes, and dancing to Lucho Azcarraga (famed Panamanian musician and organist) keeping memories of Panama alive. At Panama Canal Society reunions in Orlando, they reunite with family and friends. Ed makes a dynamite ceviche handed down by his father-in-law, Ralph.

Bonnie, Karen, Tim, Eddy receive U.S. citizen certificates (witnesses Ed and INS official), Canal Zone, 1976.

Ed was hired by U.S. Customs shortly after arriving in the States, served as vessel commander on "go-fast" boats, and retired as a special agent in 1996. Bonnie retired in 2006 after 34 years of federal service both with the Panama Canal Commission (most years with the Transportation Branch) and as a Mission Support Specialist with the U.S. Department of Homeland Security. She also had a brief assignment with U.S. Customs at the U.S. Embassy in Panama assisting in the restoration after the U.S. invasion of Panama, the arrest of Manuel Noriega, and the removal of the Panama National Guard.

Dreams do come true. Son Eddy and family returned to Panama, where Eddy served as Special Agent, U.S. Department of Homeland Security, American Embassy, Panama. He is a major in the U.S. Army Reserves and was activated during the Iraq War. After a tour of six years, they recently transferred to the Washington, D.C., area. Eddy is married to Carson Miller Dolan and they have three children: Morgan, Madison and Mackenzie. They have a vacation home in Coronado Beach, Panama.

Bonnie and Ed's daughter, Karen L. Dolan, returned to Panama in 1994 and worked at Balboa High School and for the U.S. Southern Command. In 2000, at the end of the U.S. military era in the Canal Zone, Karen continued working for the U.S. Southern Command in Miami until 2007. Karen is married to LTC Steven E. Bapp, U.S. Army. Steven is a combat veteran of four wars. They have five children: Jon Dolan, Michael and Angela Bapp, Kyle (born in Panama, 1996), and Colin Castleton, and one granddaughter, Kailey Dolan.

Bonnie and Ed's son, Tim J. Dolan, is a staff sergeant, active U.S. Army National Guard, Florida, C Troop, 1st-153rd Calvary Infantry Scouts, and is also a combat veteran of the Iraq War. Tim continues to travel to Panama visiting family and friends. Tim, Eddy and Ed are active members of the BPOE Lodge 1414 in Panama, which gives us opportunity to visit Eddy and Carson's home in Coronado Beach.

Ed and Bonnie have eight grandchildren and one great-grandchild, and live in Deland, Florida.

Passed on to six generations, four generations born in Panama, are our love and unity of family. Memories are still being made with Dolan-Davis gatherings, visits to Panama, and stories of old.

To our future generations of family and friends, know that the waters flowing through the Panama Canal carry your ancestry. Forever we will hold a place in this land we called home.

Ed and Bonnie Dolan's family. Top: Eddy, Carson, Madison, Mackenzie and Morgan; Center: Ed, Bonnie and Tim; Bottom: Angela, Steve, Karen, Jon, Michael; Front: Kyle and Colin.

See also the John M. Davis and Michael Kenny family history.

THE THOMAS EDWARDS FAMILY

AS TOLD BY PEGGY A. HUFF

Thomas Edwards, son of Morgan (1855-1937) and Sarah Williams Edwards (1854-1926), was born January 30, 1878, in Wales. The coal mining family immigrated to the United States after the inundation of the Tynewydd Pit, Cymmer Rhonda Valley, on April 11, 1877. He married Bess Apgar, born February 1875; they had two children — Kenneth and Glenora.

Thomas went to Panama in the spring of 1913 and worked as a foreman for the Municipal Engineering Division. He built roads; laid pipe lines; and cleared jungle for airfields, sometimes having as many as 300 West Indian, Hindu, Chinese and U.S. labor force workers under his command.

Bess and the children joined him the following year. They left New York on April 25, 1914, on the *SS Advance*, an overgrown launch carrying 63 people fully loaded; the voyage took ten days and ended at Dock 11 in Cristobal. The family's first stay in Panama was very short. The day after they arrived, the powder magazine in Panama City blew up and broke all the windows in the Corozal quarters where they were staying. Cinco de Mayo Plaza was named in honor of the firemen they picked up after the blast. In the later part of May an earthquake occurred. There were cracks in the roads and most all of the family's dishes were broken; that and the powder magazine explosion finished the Zone for Bess. After seeing the *SS Ancon* transit the Canal in August 1914, Bess and the children returned to Pennsylvania. In January 1915 Thomas was transferred to the Atlantic side and assigned quarters in Gatun. Bess decided to give Panama another try, and she and the children returned. In the spring of 1916 the family moved to French quarters in Old Cristobal and then to 8th Street in New Cristobal. During World War I Thomas carried a metal check known as a watch fob with stars designating that he had unlimited entry to all Army posts. He terminated his employment with the Panama Canal Company in 1922 and returned to the family farm near Round Top, Pennsylvania. They lived for many years in Antrim, Pennsylvania. Thomas died June 15, 1956, and Bess December 3, 1965. They are buried in Williamsport, Pennsylvania. Glenora Edwards, born September 25, 1902, in Landrus, Pennsylvania, was pregnant when she died in 1925.

Thomas and Bess Edwards with great-granddaughter, Peggy Hale Huff.

Bess was a descendant of Johannes Apgar, who arrived in the American colonies in 1734 from Germany. Johannes's family name in Germany was Epgert, from the village where they lived. Because he could not write his name, the captain of the ship on which he sailed may have written it as Apgard, which later became Apgar. The resulting name became Johannes's family name in the new world, and anyone with the surname Apgar is related to the original immigrant Johannes. More than 145 members of the Apgar family served in the Union Army during the Civil War, participating in every major campaign. Dr. Virginia Apgar, the first woman at Columbia University College of Physicians and Surgeons to be named a full professor, designed the Apgar Score, the first standardized method for evaluating a newborn's transition to life outside the womb.

Thomas and Bess's son, Kenneth Morris Edwards, was born September 25, 1900, in Landrus, Pennsylvania. Ken attended Balboa High School his first year, graduating from Cristobal High School in June 1918; there were four graduates in the class. He went to work for the Electrical Division, moving to Los Angeles in August 1921 to finish his electrical apprenticeship. Times were rough in California; Ken joined the "hobo jungle" and spent time in jail for jumping trains. He returned to Panama in September 1922, and the following month the family returned to Pennsylvania. In 1928 Ken moved to Virginia, where he met Ellen Doris Ashton, daughter of William F. Ashton, for whom his father Thomas had worked in Gatun. Kenneth and Ellen married on July 17, 1929, in Petersburg, Virginia. They lived there through the Depression, and in November 1936 Kenneth returned to the isthmus to work as an electrician on the locks. Ellen and their two daughters, Anne Elizabeth and Helen Virginia, joined him the following August and lived for many years in Pedro Miguel on Incubator Row and Frog Alley. Ken was an avid hunter, fisherman, and jungle expert. Ellen never knew what she would find in the burlap bag that Ken would leave in the kitchen sink after a night of hunting. Instead of the traditional dog for the family pet, they had a wild boar named Ferocious, complete with long curled tusks. Three-year-old granddaughter Peggy didn't know she was supposed to be afraid of him and fed him dog food out of the can with a spoon. Ferocious lived in the basement and was exiled when he started climbing the back steps to wait outside the kitchen door for Ellen. Ken was also very handy with a pen knife and carved monkeys from peach seeds; only about ten are known to have survived. The Edwards moved from Pedro Miguel to the Gavilan area in Balboa and then to Carr Street in the Flats. Ken retired in 1960 as a senior lockmaster on Pedro Miguel Locks. They

ID badge, Kenneth M. Edwards, Lockmaster.

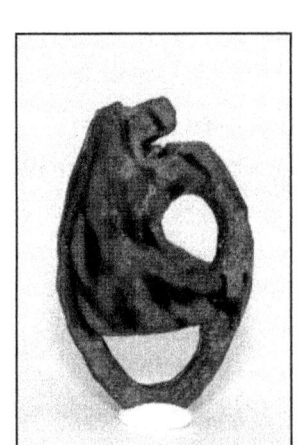

Peach seed monkey, carved by Ken Edwards.

moved to Fort Lauderdale, Florida. Ellen passed away January 24, 1972, and Ken February 4, 1974. Their remains are interred in the columbarium at St. Luke's Cathedral in Ancon, Panama.

Daughter Helen was born February 11, 1934, in Hopewell, Virginia. She attended grade school in Pedro Miguel and learned to swim in kindergarten. She was an avid swimmer, holding many records in her teens. Her hair was normally green from spending so much time in chlorinated pools. She graduated from Balboa High school in 1952. The next year Helen married Raymond Henry Magan, an airman at Albrook Air Force Station. Ray went into the active Air Force Reserves, and they moved to Pueblo, Colorado, where they lived their entire marriage. Ray was a member of the Pueblo police department from 1956 until his retirement in 1985; he retired from the Air Force in 1979. Helen worked in a savings and loan for 15 years. They both enjoyed photography, had their own dark room, and processed and printed color photographs. They owned a Piper Comanche airplane and made several cross-country trips. On one memorable trip to Fort Lauderdale to visit Helen's parents, they got lost in inclement weather; and when they were finally able to land and refuel, the 50-gallon tank took 49.4 gallons. On the return trip Helen started a small fire in the cockpit when she did not fully extinguish a cigarette. They both held amateur radio licenses, rode motorcycles and enjoyed gully riding and road trips. Ray enjoyed listening to music and had elaborate music systems. He learned to play the organ by ear, progressing from a spinet to a three manual organ. Ray could ski barefoot, and Helen could shoe ski. They had no children.

See also the William Francis Ashton family history.

THE HENRY EDWIN FALK, SR., FAMILY

AS TOLD BY MURRAY FALK

"ERII? What does that stand for?" I asked my Dad. With his chest puffed out and smiling from ear to ear, he told me that the stick pin I was holding was handed to him by the Queen. It was her way of showing her appreciation for his professionalism as the Panama Canal pilot responsible for her safe passage through the Canal while aboard the HMS Gothic.

"ERII stands for Elizabeth Regina the Second, Queen of England," he told me. What took place before this event to put Henry in this position as lead pilot?

He was the second born to Harry and Wilhelmina Hulnagel Falk. He was born August 1, 1899, on a Lighter ship in New York Harbor. His two sisters were Elsie and Grace. Personnel documents reflect that at the early age of 15, Henry worked for the Panama Canal from July 11, 1915, to December 8, 1915. He returned to New York to continue his education. In 1917 he enrolled in the New York State Maritime College and in 1919 graduated with honors. Part of his seagoing training was on the New York State Nautical School Training ship *Newport*. Henry worked on many ships for many steamship lines during which time he earned his Mates and his Masters Licenses. In 1923 he returned to Panama; personnel documents show that on August 1, 1923, he was hired as a pilot and thought to be the youngest. Henry's father, Harry, had already become a pilot in 1917, and they became the first father-son pilot team.

Wedding bells sounded for Henry on May 19, 1930, when he married Margaret (Peggy) Chisholm Murray, daughter of David and Lillian Chisholm Murray. David worked for the Canal as a machinist, and Margaret worked as a bank teller. Henry Edwin Falk, Jr., was born on April 9, 1931, and Murray Falk was born on July 24, 1934, both at Gorgas Hospital.

The family lived on the Pacific side during the time they were there, but visited the Atlantic side. This was done primarily by taking the world's fastest trans-continental train. It took less than an hour to cross the continent. The trans-isthmian highway was available to them, but it was too slow.

When we lived in Balboa on Empire Street, the oil tank farm was behind our house. One day Henry, Jr., and Murray went out to see the old caretaker who raised goats. Remembering their Dad saying how good goat's milk was, they brought some home for him to have with his dinner. Well, he told the boys it had to be chilled, so the kids didn't get to see him drink it the next morning with his breakfast. And it did disappear.

Floating dry dock on its side. Henry Falk was responsible for moving the dry docks through the Canal.

Henry was proud of his accomplishments, particularly when he was responsible for getting the floating dry dock through the Canal to the Atlantic side. Because of its size, it couldn't pass through the locks. So it was determined that the dry dock had to be floated through on its side. With the help of the crane *Hercules* and 50 gallon drums filled with sand strapped on to one side, they gradually flipped the dry dock. The sand was then released from the drums which gave the dock buoyancy. Henry was the pilot in charge during the first leg in Balboa Harbor through Pedro Miguel Locks. Tugs were used to get it through the Canal with the exception of when it went through the locks. That was when the "mules" were used to guide it.

It was difficult to guide ships at night during World War II because there was limited lighting in the Canal. As a result of ascending Jacob's ladder to the ship, then groping in the darkness for the ladder to the bridge, Henry banged his knee on a piece of equipment. In the meantime, he stood for relentless hours, while continuing to pilot the ship through the Canal. Henry suffered water on the knee and was hospitalized in Gorgas Hospital. Upon discharge from the hospital, Henry commented that he knew he never wanted to get "house maid's knee" again. On another occasion, he was coasting a war ship towards the east chamber at Miraflores Locks. The ship had no power and was at a "dead stop." Erroneously, water was spilled from Miraflores Lake at that moment which hit the bow of the ship and caused it to swing into the west chamber. Through my dad's skills, he backed the ship out of the west chamber and entered the east chamber as ordered.

Henry had just completed a northbound transit and was on the launch heading towards Cristobal, where he would pick up the "jitney" back home, when he saw the sky was red and billowing with smoke. The City of Colon was on fire. He gave up his ride home and stayed in Colon doing whatever he could to help fight the fires and aid the less fortunate. It was two days later when he returned home, tired, dirty, and with clothes reeking of smoke.

The pilots were a proud bunch. Whenever called upon, they always went that extra mile. And when World War II came, Dad was one of the first to offer his services to his country to return to sea as a Reserve Naval Officer. He retired a Commander, USNR, or Merchant Marine Captain. The War Department declined the offer, saying his job as a pilot was more critical. As a token, the pilots were allowed to purchase khakis at ships' stores or exchanges (PXs) and wear them when working.

My dad was very family oriented. He took us to the United States about every six years, picked up a new car, and spent roughly three months traveling across country, visiting relatives and points of interest. Some of the places he took us were Coney Island, the Petrified Forest, Carlsbad Caverns, Hoover Dam, and an Indian Village in New Mexico where Murray "fell in love" at the age of six with a little Indian girl. That was also the year that Henry, Jr., and I got to swim in Salt Lake. The boys ran across the salt and dove into the water, not knowing its density. Ouch! Both boys thought they had hit a wall. When they came out of the water, they were instantly covered with salt from head to toe.

We would visit Peggy's mom and dad, David and Lillian, and her brother Jim and his family, all living in San Francisco. Many hours were spent in Golden State Park seeing all it had to offer. Back home, Dad would take the family to see the ruins of Old Panama, picnic at Las Cruces Trail, and go to Far Fan Beach for a swim. We also went to Summit Gardens to enjoy the local fauna and to El Valle for a week at the AAA rental house to enjoy the local area. There were also Sunday dinners at the Hotel Tivoli. The adventures never stopped.

Living at 132 Ridge Road in Balboa Heights was an adventure for Henry, Jr., and me. There was a peach mango tree on one side of the house and an alligator pear tree on the other. The house was situated on the side of a hill overlooking the Orchid Garden. To the rear was another alligator pear tree, only shaped like a baseball. There were also a banana tree and a lime tree. A mud path led down from the house through clumps of bamboo to the garden. Both of us tried to chop down a black palm tree to no avail. The machete only bounced off, not scarring the tree at all. But we were successful in taking bamboo, making planters, and selling them — just so long as a tarantula didn't scare us when harvesting the bamboo. In the garden there were several ponds teaming with pollywogs, frogs, and water lilies. Imaginations ran wild, and we had lots of fun and made some good memories. Oh, the imagination of two adolescent boys. Even the fact that we might have been living over an old French cemetery didn't deter us.

On August 1, 1957, Henry, Sr., retired as the senior Panama Canal pilot and was honored by the Governor with a presentation of "The Key to The Panama Canal." The publisher of one of the local newspapers came out with a "special edition" about his retirement, exalting his achievements and how much he would be missed. Henry and Peggy returned to the United States, residing in St. Petersburg, Florida. From 1970 to 1972, Henry, Sr., was President of the Panama Canal Society of Florida, a society of former Zonians created "to preserve American Ideals and Canal Zone Friendships."

The Thomas Paul and Mary Denn Foley Family

AS TOLD BY KAREN HUSUM CLARY AND LORRIE HUSUM ALLEN

Thomas Paul Foley was born in Butlerstown, Waterford County, Ireland, on November 17, 1883. He emigrated to the United States via New York in 1906. Soon after, he met his future wife, Mary Winifred Denn (born in New York, October 30, 1887), whose sister was married to his uncle John Cunningham. He was naturalized as an American citizen on January 11, 1910. Their courtship was interrupted by Tom's military service at Ft. Monroe, Virginia, until 1911, and then by his move to Panama in 1912. According to Panama Canal Company employment records, Tom started work as a policeman at Cristobal Police Station on October 6, 1912. By 1930, he had risen to the rank of lieutenant at Balboa Police Station with a salary of $281 per month.

By the time Tom moved to Panama, Mary Denn had become a New York City schoolteacher, a job she loved and was reluctant to give up for either marriage or life in Panama. Tom, an articulate and faithful correspondent, courted her by mail for ten years. Some of his letters to her still survive. From Cristobal in August 1914, he wrote, "When we are joined together by the ties of holy matrimony, are we still going to live apart? Are you coming to Panama or am I going to New York? I think that with the assistance of those with more experience, we will be able to arrive at some definite conclusions and one more American girl will come to Panama to waste her sweetness on the tropical air." Dreaming of their children to come he wrote, "I would be the father of these children, for whom you would suffer all the pain and agony to bring into this world. One of these little ones, the outcome of a spontaneous and united love, would be as dear to me as the making of the Panama Canal was in lives and money."

Foley family circa 1921, L-R: Thomas Paul, Mary Agnes, Mary Denn, Genevieve Kathleen, Thomas Francis.

On August 18, 1916, Tom arrived in New York on the *SS Cristobal* to fetch Mary. They married on August 28, 1916; returned to Panama; and had five children, Mary Agnes (born 1917, Colon), Thomas Francis (born 1918, Colon), Genevieve Kathleen (born 1920, Colon), John Michael (born 1922, Colon), and Ellen (Eleanor) Matilda (born 1928, New York).

After transfers to Pedro Miguel in 1916 and Old Cristobal in 1917, the Foleys moved to Balboa in 1927. Tom took to Canal Zone life with gusto. Mary gave up teaching for motherhood. She maintained a strong interest in piano and composed songs that were played at gatherings.

Tom was an accomplished amateur actor who loved the limelight. His antics, both on and off stage, made interesting news copy for the *Panama American* and *Star & Herald*. In July 1922, in "Officer 666," a melodramatic farce, he played the starring role. The play won rave reviews on the isthmus and was performed at the National Theater in Panama City on July 22, 1922. James Zetek

(*Star & Herald*) wrote, "Perhaps the most important role in the whole play is Officer 666, and there is but one character on the whole Isthmus that could fill the boots of Officer 666, and his acting in Colon during the two performances there last month has been pronounced 'inimitable.' — Lieut. Tom Foley of our Cristobal Police Force is the genuine Officer 666."

Among his chief accomplishments on the police force was naming the thousand plus islands in Gatun Lake. In a November 2, 1930, newspaper article, it was reported that the police team, once the lake was filled, was to visit every island. The natives were consulted to pick appropriate names, such as Barro Colorado. "If it developed that the natives had named their island, the name was retained, but otherwise, some outstanding feature of the tiny bit of land was added as the name. All the names are in Spanish and none of them glorify American officialdom."

Police Lt. Thomas Foley, Balboa, 1940's.

The lighter side of Tom's Canal Zone police work is remembered by family and newspaper accounts. Daughter-in-law Eileen McGuire Foley recalls a time in Ancon when a dead horse was found in the street. Tom was told to go investigate and handle the matter. The dead horse was on a street named Frangipani. Tom didn't know how to spell "Frangipani," so he had the horse moved to a street nearby with a name that he knew how to spell, so that he could complete his report.

On June 10, 1940, in response to Italy's entrance into World War II, an Italian liner, the *Comte Biancamano*, was seized at the Canal's Balboa entrance. Tom and Bob Hull were assigned by U.S. marshals to stand guard on ship until the matter was settled by Canal Zone District Courts. The ship transited the Canal and anchored in Limon Bay. The assignment lasted six weeks. Lt. Foley was housed in the "luxury of the ship's bridal suite, enjoying movies in Italian, twice per week, and serenaded by the band and orchestra with concerts by the ship's musicians at frequent intervals" and "ate miles of spaghetti." Feeling the need for exercise one day, Tom (then 57) climbed down the ship's ladder and swam a mile and a half to Shimmy Beach at Fort Sherman, whereupon, wearing nothing but swimming trunks, he was apprehended by a sentry guarding the beach. He convinced him that he was neither an escaped Italian deserter nor an Irish lunatic, but instead Lt. Foley of the Canal Zone Police, upon which he was returned to the ship to continue his watch.

The Foley children attended Canal Zone schools, all graduating from Balboa High School. Much of Foley family life centered around swimming. They belonged to Henry Grieser's Red, White, and Blue Swim Troupe (100 strong) at Balboa pool. The Troupe perfected its swimming and diving, performing regularly for the community in "Water Spectacle" and "Aquacade" fundraisers, swim meets, and exhibitions at Madison Square Garden in the 1920's. In one of the water spectacles, Tom officiated in the pageant as King Neptune,

Thomas Foley as King Neptune, Balboa Swimming Pool, 1940's.

along with "his dancing and singing mermaids, including the most fantastic girls in swimming." Daughter Genevieve showed extraordinary talent as a springboard diver. In her teen years, she performed with the Troupe in diving exhibitions off the Pedro Miguel Locks and the *Atlas* crane in Gamboa. When Grantland Rice, the famed American sports journalist, visited Panama in 1936, he filmed the Troupe diving off the locks and swimming at Galliard Cut and included Genevieve in his film.

In 1924, son John, then age 2, was struck by polio. Although crippled in one leg and with a noticeable limp, John became an accomplished swimmer and diver and was one of two children who greeted President Franklin Roosevelt during his 1934 Panama visit. Daughter Mary remembers the Governor giving John the honor of standing next to the girl who presented FDR with flowers. When FDR shook John's hand, he said, "You and I are alike. We both have polio, and we both like to swim to stay strong, and we both know how to 'lick' polio." John's life was cut short by kidney disease. He died January 2, 1950, at Gorgas Hospital and is buried at Corozal Cemetery.

All of the surviving Foley children met and married their future spouses in the Canal Zone. Mary married William J. Cronan. Their children are John, Celia, James, Joseph, William, and Mary. William Cronan retired as a Police Sergeant, Cristobal Police Station, and the family moved to Santa Maria, California. Thomas married Eileen M. McGuire. Their children are Irene, Rita, Pauline, Joseph, and Thomas. They left Panama in 1950, and Thomas retired from civil service in 1973 in Orlando, Florida. Genevieve married Cyrus W. Field. Their children are Melvynn, Cyrus, Marianne, and Kenneth. Cyrus Field retired as the Production Superintendent of the Industrial Division in Mt. Hope, and they moved to New Port Richey, Florida. Ellen married Edward J. Husum. Their children are Janet, Karen, Edward, Maureen, Lorraine, John, Gregory, Raymond, George, Michael, and Mary. Edward Husum retired as a Police Lieutenant, Balboa Police Station, and the family moved to Tallahassee, Florida. In all, descendants of Tom and Mary Foley number 26 grandchildren, 48 great-grandchildren, and 17 great-great-grandchildren.

Tom was a shameless promoter of all things Irish and prided himself in being a son of the Irish Free State. He was a member of the Knights of Columbus and achieved the rank of Grand Knight of both Isthmian (Atlantic and Pacific) Councils. He was a regular contributor to its newsletter, *Discoverer*.

Tom Foley retired with the rank of Lieutenant in 1945. Three hundred guests attended a party given for him by the Atlas National Brewery in Panama City. After retirement, the family moved to Cocoli. He died on August 22, 1953, at the Old Soldier's Home and is buried in the Old Soldier's Home National Cemetery, Washington, D.C. Mary died in her sleep at her daughter Ellen's house in Diablo on Good Friday, April 1955, and is buried at Corozal Cemetery.

Thomas Paul Foley
Roosevelt Medal No. 7310

General Inspection, Balboa Central Police Station, June 19, 1943.

PHOTO COURTESY OF PANAMA CANAL MUSEUM

Ralph Edward Furlong, Jr.

As Told by Brenda Furlong

He was born March 21, 1925, in Limerick, Maine, population 850. His family consisted of his father, Ralph Furlong, Sr.; mother, Alice (Smith) Furlong; and an older brother, Harold. They lived in a large two-story farm house which was occupied by his extended family — his grandparents, his uncles and their wives and children. Altogether, at any time, there were 20 people living in the house. The house had a large wagon house attached, which burned down when he was three years old. Then they had to rent a house in town until a new, smaller farmhouse was built. There was also a large barn which was two stories, 82 feet long, and three bays wide for cattle, pigs, hay storage, and other miscellaneous items. This property was one mile from town, which was without public transportation, street lights, electric power, or water facilities. His home was exactly one mile from the town center, which had one doctor, one hotel, and two grocery stores. This was Uptown Limerick, named so because it was on top of a hill. Downtown Limerick was, of course, down the hill. Downtown held the main and only industry, the textile mill, a three-story building employing roughly 600 people. This mill was Ralph's first paying job at 30 cents an hour. Ralph's formal education was through ninth grade, which was mandatory, and he quit afterwards at 16 to work in the mill.

Ralph's children from his former wife are Ralph David (April 15, 1949, D: 2005), Carol Anne (Dusty) (January 5, 1951), Gail Marie (March 26, 1952), Darlene Joy (November 4, 1956), Valerie May (May 10, 1959), and Daniel Edward (April 12, 1961).

Ralph married his current wife, Brenda Argelis Grajales Guerra, on March 9, 1965, and they have two children: Brenda Alice (February 6, 1968) and Ralph Edward III (July 20, 1969). Ralph and Brenda stressed education heavily, especially because Brenda was an educator and principal, and because Ralph did not want his children doing back breaking labor, knowing the intensity of it.

The textile factory life proved to be very calm and simple, until the start of World War II. Ralph, Jr., joined the Merchant Marines at age 17 years, 8 months. He did his training at Sheepshead Bay Naval Station in New York City. He started out as a mess boy for one year, and then he changed over to the deck department, where he started as seaman. He maintained everything outside, such as chipping rust, painting, splicing cable, and cared for the ropes. He was also deck maintenance and boatswain, depending on which job was available when he was ready to ship out. He would sign up at the union hall when he was discharged from a trip and go home for a month of leave. He stayed in the Merchant Marines through World War II, and he saw action one time sailing with a convoy which came under attack from a German submarine. The submarine sank one ship in front of him. For this time served, he achieved military veteran status, due to the Merchant Marines falling under the Coast Guard's jurisdiction, and was honorably discharged from the military on August 15, 1945.

Ralph E. Furlong, Jr., Merchant Marine, 1944.

When he transited the Panama Canal many times, Ralph would talk with several Canal workers. It was from these conversations that he decided to apply for work at the Panama Canal Locks Division, and he was hired for a one-year period during a locks overhaul. He was part of a team which moved and cleaned the gates and then put them back in place. Also, he worked as a locomotive operator to pull ships through the Canal. This work completed his first-year contract.

After the locks overhaul, Ralph was hired as a construction inspector to check work performed by contractors. As a general inspector, he would check earth work, steel work, carpentry, doors, windows, cabinets, tile, and debris haul away. As the workforce was being reduced, and more contracts were being subcontracted, the Construction Division was created. Ralph worked as a construction inspector, supervisory inspector, and finally retired as chief inspector. During this time he received two outstanding performance certificates (May, 29, 1974, and June 8, 1976) for outstanding record of performance achieved and a special achievement award for superior job performance (October 6, 1977). While he was chief inspector, he had over 16 subordinates. The town of Paraiso was built under his watch, along with the utilities, roads, and everything else.

Ralph and Brenda lived in the town of Balboa under a year and then moved to Los Rios, where they remained until 1979. Their children and their friends fondly remember playing in the jungle across the street from their house, and the honey bears coming up to their front door.

Afterwards, the family moved to La Boca until Ralph retired in 1988 from the Panama Canal. Ralph affectionately recalls the fact that everybody knew everybody else in the Canal. This helped keep children safe and neighbors friendly, and the Canal became a paradise.

All of his life, Ralph was an avid fisherman. He loved fishing in Panama because he could fish in both oceans on the same day. Around 1980, Ralph started picking up the hobby of woodworking; and to this date, he has a garage full of modern woodworking machines and tools, from which many cabinets, armoires, tables, chairs, desks, bookshelves, beds/bunkbeds, sheds, and other items have been created. He is well known in Tampa, Florida, for his woodworking skills, and many houses hold some of his creations.

Ralph E. Furlong, Jr., 2008.

PHOTO BY DON GOODE

Miraflores locks overhaul, dry chambers.

The Geddes Brothers

As Told by the Geddes Sisters

Our Claim to Fame Is a Shovel Full of Dirt

Brothers Albert, Joseph and Thomas Geddes went to Panama to work on the construction of the Panama Canal at ages 26, 22 and 20, respectively. They were all employed in construction jobs in New York City as crane men before embarking on this new adventure. All three were employed as firemen by the Panama Canal and were shortly promoted to steam shovel engineers. Al was in Panama from 1905 to 1913, Joe from 1907 to 1910, and Tom from 1906 to 1913. Their mother Jane and sister Jennie came to Panama in 1907.

Al was a trained machinist and quick to learn. Al watched the crane men laboriously yanking the "dipper trip" which releases the bottom of the big dipper and lets the mouthful of earth and rock drop into the railroad car. He perfected a steam device that accomplished the same purpose by turning on steam. Canal engineers tested this method and highly praised its effectiveness. Al's invention was adopted on all 95 ton-shovels. Shovel output was increased by 100 cubic yards per day for a savings of $67 per day.

Tom was the only one of the brothers that married during the time in Panama and brought Loretta Ellis from New York to Panama in 1910. His marriage proposal was on the back of a postcard with a picture of the SS Cris-

Albert K. Geddes

Joseph C. Geddes

Thomas K. Geddes

tobal and said "For better or worse, let's take a chance." Loretta was very happy living in Bohio in spite of the heat, humidity, mosquitoes and other inconveniences. One of their favorite pastimes on Sunday afternoon was to walk on the railroad tracks. Their first two children, Margie and Al, were born while they lived in Bohio. Loretta was expecting their third child, Robert, when the family moved back to New York in 1913. Tom died in 1943 and Loretta returned to the Canal Zone after World War II and lived with her daughters, Carolyn and Gloria.

Their sister, Jennie Geddes, married Guy Johannes in Bohio in 1909. Guy eventually became Chief of Police in Balboa, and they lived there until Guy's retirement.

Steam shovel operations in Culebra Cut were permanently suspended on September 10, 1913. The last steam shovels to stop working in the bottom excavation were Nos. 204 and 226. No. 226 was operated by Al Geddes. The last train of spoil was hauled out of Culebra Cut on September 13, 1913; and the last shovel of dirt was put on the train by Steam Shovel Engineer Albert Geddes. A large number of amateur and professional photographers took advantage of the occasion to snap pictures of this historic event.

One interesting story about working conditions during construction was depicted in the *Star & Herald* newspaper edition dated May 10, 1907, where it was reported that the steam shovel men had received notification that their demand for an increase in pay from $210 to $300 per month could not be entertained. At a meeting some men were in favor of suspending work; but the majority was opposed, and work was resumed.

The brothers did not escape the health issues well noted during those years. Al purchased a steerage passage to Panama in 1905 and arrived in Panama with $5 in his pocket. Two weeks later he lay in Gorgas Hospital with yellow fever. Al, Joe and Tom all suffered recurring bouts of malaria and took annual trips to Hot Springs, Arkansas, to "sweat out" remaining vestiges of the disease. At one time Tom was near death with what was called "Blackwater Fever." Their personnel files bear out the annual trips on various sailing vessels, always at a cost of $20 each way. Their employment contracts stipulated an annual vacation.

Unfortunately, none of the three brothers was there to see the first Canal transit in 1914. All three were awarded the Roosevelt Medal to honor their service. Al had three bars while Joe and Tom each had one. Unfortunately, no trace of the medals was ever found by any descendants. Their contribution to the construction of the Canal is noted in *The Makers of the Panama Canal* and *Representative Men of the Panama Republic* book dated 1911.

The Geddes family's history in Panama was further written when Al established a construction contracting business in later years. One of his biggest projects was building Albrook Field. In 1930 brother Tom was back in Panama working for Al on this project.

In 1940 Tom and his four sons, Al, Robbie, Bill and Frank, left New York and returned to Panama to find work. Initially they all got jobs as heavy equipment operators with Uncle Al, then with the Canal and/or the U.S. Navy. Tom and sons, Al, Bill and Frank, returned to New York in two or three years. Robbie and his wife Florence remained and raised children Patricia, Dianne, Bob, Bill and Barbara before they retired to Florida following 33 years service to the Panama Canal. Patricia and Bob made careers out of Canal employment and retired under the beneficial terms of the Panama Canal Treaty. Barbara worked for the Canal for ten years before moving to Colorado. Patricia's two daughters, Wendy and Lori Flores, grew up in the Canal Zone as well. Pat and her husband, Robert J. (Bud) Risberg, retired to Florida. Bob also raised children, Christine and Michael, in the Canal Zone. Bob retired to Anacortes, WA.

After World War II Tom's two youngest daughters, Carolyn and Gloria, returned to the Canal Zone, where their husbands had prewar jobs. Carolyn married Joe Coffin, a member of the Fire Department. They had nine children: Lynne, Joe, Tom, Jon, Marjorie, Jerry, Marcia, Kathy and Nannette. Jon retired as a tugboat captain and still lives in Panama and continues to enjoy the Panama lifestyle. Gloria also raised her children, Cheryl and Brian DeRaps, in the Canal Zone. She and her husband "Pos" Parker retired to Florida. All the third and fourth generation Geddes descendants were educated through high school almost entirely in the Canal Zone schools.

Albert Geddes
Roosevelt Medal No. 388 with Three Bars

Joseph Geddes
Roosevelt Medal No. 2928 with One Bar

Thomas Geddes
Roosevelt Medal No. 4739 with One Bar

See also the Joseph W. Coffin family history.

The William B. Godfrey Family

As told by Dorothy (Godfrey) and Ira Brandt

Born in 1886, William Belding Godfrey (Will) was the third generation civil engineer in his family. After attending the University of Washington he took a summer trip to see for himself the famous Panama Canal then under construction. He was hired in August 1912 by the Municipal Engineering Division of the Panama Canal Company and stayed for 36 years, interrupted only by two years of military service in World War I.

One of his first assignments was to establish a survey camp on the Chagres River and then to cut through the jungle to lay the benchmarks for the border between the Canal Zone and Panama. Later, he worked on designing and building water and sewer systems, streets, and sidewalks for the Canal Zone and Panama. These are still in use today. As Canal traffic increased, he served on a commission to make preliminary plans for locating a third set of locks.

William B. Godfrey, Canal Zone, 1942.

In 1917 he joined the U.S. Army Corps of Engineers, went to France, and fought in the battles of St. Michel and Argonne Meuse.

On his return trip in 1919 Will stopped in Washington, D.C., to visit his brother whose wife introduced him to her sister, Grace Elizabeth Bell, then working for the U.S. War Department. They were married in 1920 and went to the Canal Zone. She traveled by cayuco (Indian canoe) to join him at the survey camp until residential quarters in Balboa became available. Their daughter, Dorothy, was born in 1923.

Will was fluent in Spanish and had good friends on both sides of the border. He was president of a Rotary Club in Panama, joined the American Legion, and was a Mason and a Shriner. An avid golfer, he played on courses in Gatun and Ft. Amador. He collected rare tropical hardwoods (such as black palm, coco bolo, lignum vitae, mahogany and nazarine) from which he made furniture and small items, including a black palm military swagger stick which was presented by the American Legion to General Eisenhower on one of his visits to the Canal Zone.

Grace Elizabeth Bell Godfrey, Survey Camp on the Chagres River, 1920.

Although it no longer exists as "Zonians" knew it, the Canal Zone was an extraordinary place to live. It was orderly, clean and secure, populated only by employees of the Panama Canal Company, their families, and U.S. military personnel. No one was rich or poor and all had a strong sense of community and shared purpose. Basic needs and services were provided at low cost or free. Canal Zone schools from kindergarten through junior college were excellent. Teachers were highly qualified, and the curriculum was about a year ahead of grade-level standards in the United States. Because of the heat the workday began at 7 a.m. and ended at 3 p.m., leaving plenty of time for recreation. Swimming and world-class deep sea fishing were popular. "Panama" means "abundance of fish," and the delicious corbina was a favorite catch.

Except for a short dry season it rained at about the same time every day, not in drops but in vertical sheets so clearly defined that a walker could often just cross the street to keep dry. Light bulbs were used to heat "dry closets" as protection from mold and corrosion.

During the torrential rainy season one part of the Canal, the Culebra (Gaillard) Cut, was subject to landslides that blocked Canal transit. After such slides, Will sometimes took Dorothy to watch the dredging operation as thousands of tons of earth were removed and carted away in dump trucks to be spread on low-lying land. Entire towns were built on such "fill" from the "cut."

Families took vacation trips to the States on company ships, sailing to New York.

One way to buy a new car was to order it in Panama, take delivery at the factory in Detroit, and drive it until time for the return voyage, car and all, to Cristobal.

A railway and eventually the Trans-Isthmian Highway connected Cristobal on the Atlantic and Balboa on the Pacific sides of the Canal. It can be disorienting to discover that, in Panama, the sun rises over the Pacific Ocean and sets in the Atlantic. The Southern Cross is visible in the clear night sky. When the moon is full, it shines with equatorial brilliance, pouring like rich cream over rooftops, trees and bushes, bright enough to read by.

There was no malaria in the Canal Zone because of a vigorous program of mosquito control. Oil was spread on the surface of all standing water that could be found, on puddles, and even in flower pots. The program did not extend into remote jungle areas, however, so when Will worked there in the early years he did get a recurrent form of malaria from which he never fully recovered.

A landmark event occurred in 1929 when everyone turned out to watch as Charles Lindbergh landed his plane in Cristobal and delivered the first airmail between the U.S. and the Canal Zone.

In 1934 during Will's assignment as Superintendent of the Northern (Atlantic) District of the Municipal Engineering Division in Cristobal, tragedy struck the family. Grace became seriously ill, died in Colon Hospital, and was buried in Mount Hope Cemetery.

Will and Dorothy returned to Balboa. Dorothy became a Girl Scout Mariner and attended the Girl Scout Summer Camp in Panama, as well as Youth Group retreats on the picturesque offshore island of Taboga.

In 1937 Will married Irena Ewing (Rena) who took an enthusiastic interest in Panama. She joined the Interamerican Women's Club, studied Spanish and the local cuisine, and learned to make a Pollera, the national costume of Panama. She explored Panama City and visited friends in the interior of Panama, with its higher elevation, mild climate, spectacular beaches and interesting towns, one of which, Boquete, was the No. 2 choice of an AARP list of the best places in the world for retirement living.

> *A landmark event occurred in 1929 when everyone turned out to watch as Charles Lindbergh landed his plane in Cristobal and delivered the first airmail between the U.S. and the Canal Zone.*

After graduation from Balboa High School in 1939 and from Washington State University in 1943, Dorothy came home and found that there was a wartime civilian personnel shortage so she went to work in the Commandant's office, 15th Naval District in Ft. Amador. She left her family home and moved with three other girls to Navy quarters on the shore of the Pacific entrance to the Canal.

When the war ended in Europe, she witnessed the dramatic redeployment of the U.S. fleet to the Pacific theater of war. Occasionally a battle-scarred ship returned for repairs in the Canal dry docks. The skippers of all these ships reported to the Commandant's office for new orders. When in port, no shore leave was allowed and seamen were confined to the pier. The Red Cross brought coffee and baked goods, to which Rena contributed more than 300 dozen home-made cookies. Dorothy joined a USO group to help with mail and conversation. There was entertainment and music for swing-dancing (which was popular at the time).

In 1946 Captain Ira Brandt, U.S. Army Medical Corps, was ordered to duty first in Quarry Heights and then in Fort Clayton. He and Dorothy met when each had a roommate who played in Panama's National Symphony Orchestra, and all four went to a concert together. They were married at the Tivoli Hotel in 1947 and sailed to New York aboard the SS *Ancon*. Ira pursued an academic-medical career on the faculties in sequence at the University of Miami, Yale University, University of California at San Francisco, and Indiana University. Their four children were Elizabeth, Laura, William and Rena.

In 1948 Will retired to Santa Paula, California, where he built his own house, became the City Engineer and worked as a Consultant for the California State Water District. He and Rena traveled and hosted many family visits. The conclusion of Will's well-lived life came with his death in 1976 at the age of 90.

In 1998 Dorothy and Ira took a nostalgic tour of the Canal Zone, visiting familiar landmarks and noting the many changes under Panamanian ownership. There had been much commercial development: tourist attractions, elegant hotels, residential property for sale on the formerly protected watershed shoreline of Gatun Lake, and the hiring of a Chinese company to operate the Canal ports.

But Panama's decision to build the third set of locks to meet the demands of a changing world demonstrates its commitment to the challenge and promise implicit in the century-old Panama Canal motto: *The Land Divided – The World United*.

Dorothy and Ira Brandt, Tivoli Hotel, 1947.

Tivoli USO. The USO building was located below the hill from the Hotel Tivoli. After WWII, it was used as a bus terminal in what is known as Shaler Triangle.

The Thomas I. Grimison Family

AS TOLD BY RICHARD GRIMISON, TOM GRIMISON, AND JANICE SCOTT

Thomas Irvin Grimison and his wife Jessie Gertrude Shaver were in their early 20's when they arrived on the Isthmus of Panama from Easton, Pennsylvania, in 1907. Tom was employed by the Isthmian Canal Commission as a conductor on a dirt train during the Canal construction. He was awarded the Roosevelt Medal with two bars. Tom continued his career working for the Panama Railroad and retired in 1940. He was reemployed during World War II as a marine traffic controller and later with TACA Airline at Albrook Field.

Tom was born on June 9, 1887, in Huntingdon, Pennsylvania, to Thomas Whitney Grimison and Ada Irvin. His father owned a cracker and confectionary business and a fireworks factory in Easton. Tom was known for selling 4th of July fireworks from a kiosk near the Balboa Police Station that were sent from the Easton factory. Jessie was born in Drifton, Pennsylvania, on September 12, 1886, to Andrew Benton Shaver, a Cox Brothers Coal Mine manager, and Sophia Elizabeth Miller. She was one of ten daughters and four sons. Jessie met Tom walking up College Hill in Easton, Pennsylvania, when he was returning from a cruise on a U.S. Merchant Marine training vessel.

Tom and Jessie raised three children in the Canal Zone—Helene (born 1907), Janice (born 1909) and Thomas Richard (Dick) (born 1911). The family lived in several Canal Zone neighborhoods including Ancon, Balboa (Barnebey Street), Summit, Lacona, and Cocoli. Tom was an avid baseball fan and was an officer of Balboa Elks Lodge 1414. Jessie always loved the Tivoli Hotel dances and said that the construction days were the best time of her life. The family enjoyed spending summers at the Hotel Aspinwall on Taboga Island, where son "Dick" played in a band. Jessie worked on the Atlantic side for many years with the Supply Division commissary. Jessie Grimison's employees reported that her inventory always balanced to the needle.

The Grimisons, 1926.

Tom and Jessie retired to their beach-front home in Gorgona, Panama. They bought the property in 1935, and Tom used obsolete Canal equipment which he purchased from Section "I" while building their beach house. The Grimison grandchildren have many fond memories of spending time at the Gorgona beach house with "Pappy" and "Nana," and many family members and friends from the States spent memorable vacations there. During World War II family members were eye witnesses to a crash on the beach of a P-39 Bel Air Cobra flying out of Chame Field. The plane flew right over the house prior to crashing.

Tom and Jessie enjoyed a social life playing poker, canasta and bridge at Santa Clara beach

The Grimison beach house, Gorgona, 1952.

on the weekends.

Tom died July 24, 1958, in Gorgona, Panama, and Jessie died in Miami, Florida, in 1972. They are buried at Corozal Cemetery in Panama.

Daughter Helene Shaver Grimison was born July 7, 1907, in Phillipsburg, New Jersey, and accompanied her parents to the Canal Zone that year. She was known as the "Belle of the Ball," and her charm and charisma matched her good looks. After graduating in 1925 from Balboa High School, she enrolled in Cedar Crest College, Allentown, Pennsylvania, and returned to the Canal Zone to recuperate following a serious automobile accident. In 1926 she married Austin S. Keeth, a U.S. Naval officer, and they were transferred to Shanghai, China, where he served on the Yangtze River patrol until the Boxer Rebellion. They were later transferred to Lisbon, Portugal, where Helene died in 1940 of spinal meningitis just months prior to penicillin becoming readily available.

Tom and Jessie Grimison, 50th wedding anniversary, Gorgona, 1957.

Daughter Janice Gertrude Grimison was born in Ancon Hospital on May 10, 1909. She was a 'tom boy' and played sports until her graduation from Balboa High School in 1927. She was working for the *Panama American* newspaper when she met and married a New Zealand boxer on tour, Edward W. "Ted" Scott, who later became the well-known editor of the *Panama American*. His popular daily column, *Interesting If True* By Eduardo or Edward the Unready, covered a period of 40 years about the Canal operation and the Zonians who ran it, as well as Panama and its political figures. Janice's career included working for the United Fruit Company and in various positions with the Panama Canal Company/Canal Zone Government including the Health Department, and as an accounting technician with the Maintenance Division. During World War II she was sent to Costa Rica to recruit labor for the Canal operation. She intercepted secret coded messages for the OSS operating in Latin America, where Ted was an OSS operative and Vice-President of TACA Airlines.

Janice volunteered for many years as secretary for the Pacific Civic Council, and was the scorekeeper and news reporter for the Fastlich Baseball League from its beginning in 1954 until 1962. In 1965 she retired to Florida, and she and Ted eventually settled in Cocoa Beach. Janice and Ted raised three children in the Canal Zone: Edward William III, BHS '56; Richard Alan McMaster, BHS '60; and Janice Grimison Scott, BHS '62. Ted died in 1989, and Janice died in 1990.

Son Thomas Richard (Dick) Grimison was born in Ancon Hospital January 4, 1911, and graduated from Balboa High School in 1928. After going to sea in the Merchant Marine for about a year, he worked for the Canal Zone Government as a customs officer. In 1935 he married Helen Alice Cawl, daughter of Canal tugboat Captain William and Anne McGrath Cawl, originally of Carbonear, Newfoundland. They had one son, Thomas William, BHS '53. Dick died in 1939 of cancer.

In 1948 Helen married widower Rex E. Beck, who had a son Rex E. Beck, Jr., BHS '62. Helen retired from the 15th Naval District, Canal Zone, in 1969. Rex Beck retired from the Canal Zone Government, Magistrate's Court, in 1972, and they moved to Merritt Island, Florida. Helen died in 1990 and Rex in 1992.

She intercepted secret coded messages for the OSS operating in Latin America, where Ted was an OSS operative and Vice-President of TACA Airlines.

As Tom and Jessie's four grandchildren pursued their careers, many of their sixteen great-grandchildren were born and raised in the Canal Zone.

Grandson Thomas William Grimison and wife Anne "Nancy" Burns raised eight children in the Canal Zone: Elizabeth, BHS '77; Christina, BHS '79; Thomas Richard, BHS '80; Patrick, BHS '82; Rebecca, BHS '84; Melinda, BHS '86; Eric; and Matthew. Tom was an architect with the Panama Canal Company and retired in 1986 after 25 years of service. He continued his career in architecture in Florida working for space industry contractors at the Kennedy Space Center.

Granddaughter Janice Grimison Scott raised four children in the Canal Zone with husband John Herring. The four Herring siblings are Katrina, BHS '81; Ian, BHS '84; Geoffrey; and Brian. Janice held various positions with the Panama Canal and Department of Defense during her career in the Canal Zone. She moved to the States in 1985 and served as the Mayor of Cocoa Beach, Florida, from 1999 to 2002.

Grandson Edward William Scott III and first wife Jeannine Larkin have two children, Edward IV and Heather. They were raised in the Canal Zone as young children while their father worked for the Panama Canal Company in personnel operations. The family moved to the States as Ed's government career progressed. He served as an Assistant Secretary at the Department of Transportation before leaving government service and embarking on a career in the technology industry. He worked at several companies including Sun Microsystems and was a founder of BEA Systems. Ed and second wife Cheryl Gilliland have a son, Reece.

Grandson Richard Alan McCaster Scott and wife Vanna Ketterman (deceased) have a son, Sean. Richard was a Peace Corps volunteer in Peru from 1966 to 1968. He is retired from his career as a social worker and psychotherapist in San Jose, California, where he specialized in work with the homeless and medically indigent. He also worked as an educational filmmaker.

Thomas I. Grimison
Roosevelt Medal No. 3619 with Two Bars

Taboga Island.

PHOTO BY KEVIN JENKINS

The Peter A. Hall Family

As told by Jane Kaufer Cochrane and Nancy Kaufer Lanfranco

Peter Adelbert Hall was born in Winchester, Wisconsin, on September 7, 1863, the only son of English immigrants. He and his sister were raised in a house built on a piece of homesteaded farmland.

Peter married a German lady, Emma Weiher, whom he met through her brother, the owner of the hotel where Peter worked in the bar. Peter and Emma's first daughter Norine was born in 1901, followed by a second daughter Jane in 1906.

Peter had always had an adventurous spirit, so when he heard that the United States was building a canal in Panama he applied for a job and was hired as a policeman on the recommendation of Colonel George Washington Goethals. Senator Stevens of Wisconsin, who was a former West Point classmate of Goethals, had written a letter of recommendation for Peter in which he said, "The only bad thing I know about this man is that he voted for me."

L-R: Peter, Jane, Emma, and Norine Hall at Empire.

Peter headed for Panama in 1909, a few months before his family, to get things organized before they arrived. They lived in a small cottage until 1910, when Peter was offered the much coveted job of caretaker of Camacho Reservoir, one mile from the town of Empire. They moved to a large two-story house overlooking the lake that furnished the town of Empire with water. Like a private estate, the 20-acre grounds were filled with orange, coffee and avocado trees.

Peter's duties included making a report on the daily rainfall and the elevation of the lake. He also kept records of the hours worked by the West Indian helpers who maintained the grounds with their machetes. Colonel Gaillard would sometimes pay a surprise visit to the area to check on the maintenance of the grounds and lake. He walked all the way from Culebra, two miles each way, dressed in his Army uniform, always with his swagger stick.

Peter took Norine and Jennie to the train station by buggy, pulled by their reliable horse Sandy. On the way to the station, Sandy would stop abruptly at the pump station where Peter often stopped to talk to the pump operator, and this almost threw the riders out of their seats. The girls rode to school by train to Corozal and later Balboa High School. For the next five years the family enjoyed living on the estate until tragedy disrupted their lives. Emma, their beloved wife and mother, passed away on February 22, 1914, after contracting tuberculosis. It was a devastating time for the three members of the family, but they gradually adjusted to life without Emma.

That same year the event that everyone in the Canal Zone had been anticipating came about — the opening of the Panama Canal, where "East meets West." Peter stood at the edge of the Canal watching the ceremony with his two daughters and felt pride and joy to have been part of this marvelous engineering feat.

In 1917 Peter married Mary McAuliffe, a woman he met while he and the girls were vacationing and visiting relatives in the United States. They met her on a train trip between Denver and Salt Lake City.

Norine remembered her high school years with fondness. There were only 12 in her class so they were a tight-knit group. They had several parties and picnics at Camacho Reservoir, borrowing Peter's boat to row out on the lake. The class of 1919 had their graduation exercises at the National Theater in Panama City. Each student was given a box seat for family and friends. They felt very special walking down the aisle to the "Triumphal March" from *Aida*.

After graduation, the family moved to Gatun Lake, where Peter had a new job as custodian of Agua Clara Reservoir. Norine worked for the Army for a short while, but was offered a better job in the Mechanical Division of the Mount Hope dry dock. Her desk was next to Louis A. Kaufer, who worked as a comptroller. As fate would have it, he was the man Norine would marry three years later in 1922.

After being married by Father Burns at the Miraculous Medal Churchside Chapel in Colon, Norine and Louis

Reservoir at Empire.

Hall residence and path at Empire.

rode by car to Old Cristobal to be married in the Canal Zone. Following a reception that afternoon at Louis's mother's house, the newlyweds boarded a ship and sailed to New Orleans, where they lived for the next two years. Their first child, Jane Bernice, was born there in January 1924.

In June 1924 they returned to Panama and lived in Balboa, where Louis worked for the Mechanical Division. Their second child, Theodore (Ted) Louis, was born at Gorgas Hospital in October 1925. Louis was transferred from Balboa to the Cristobal shops in June 1930. Their third child, Nancy, was born in May 1933 in Colon Hospital. All three children attended Canal Zone grade schools and graduated from Cristobal High School.

Louis was active in the American Legion, the Veterans of Foreign Wars, the Camera Club and was a Boy Scout leader. He took the scouts on a trip to Maine and to the scout camp in El Volcan. He loved gardening and raised orchids and all kinds of fruit in his backyard. Norine was active in the American Legion Auxiliary and the Woman's Club. She also loved to travel.

Louis retired in 1959 after 38 years of government service in the Canal Zone. He and Norine moved to California to be near their two daughters. Ted remained in the Canal Zone and raised his family there before retiring to Tampa, Florida. Norine held the record in the family for living in Panama the longest period of time — from 1909 to 1959.

Peter Hall, the original member of the family to settle in Panama, died in 1934 in Seattle, Washington. Louis Kaufer died in 1971 and Norine in 1986, both in California. Ted died in November 2004 in Florida.

Peter A. Hall
Roosevelt Medal No. 5053 with One Bar

The William G. and Bernice A. (Sanders) Hill Family

AS TOLD BY ROBERT HILL

Fortiter In Re, Suaviter In Modo: Resolute In Action, Gentle In Manner

Bill arrived in Panama in 1919 as a four-year-old. He and his mother left Puerto Limon, Costa Rica, for a better life in Panama. As their passage was on deck and the seas were rough, his mother lashed him to the mast.

He attended the Brothers Catholic School in Colon until his mother married Charles Hill, an American Panama Canal employee, at which time he was inserted into the Canal Zone schools. He spoke no English so he had to absorb the language and his studies together. He graduated from Cristobal High School in 1936 with high honors.

Bill married Bernice Aloise Sanders, the second child of Bruce and Grace Sanders. The United Fruit Company was his employer from 1936 to 1944. The burning desire to work for the Panama Canal organization required a U.S. citizenship, which through much effort he obtained in 1943. The Canal Zone Government hired him in 1944 to replace an employee who went into the Navy for the war effort. Any and all dealings with health and sanitation fell into his job requirements, such as ship/box car fumigation, mosquito/sand fly management (DDT spray trucks and the drainage ditch/Canal systems), animal quarantines, sanitation inspections of clubhouses, commissaries, barber/beauty shops, school cafeterias, swimming pools, hospitals, social clubs (Elks Club, Knights of Columbus Club, Tarpon Club, yacht clubs, American Legion Club, and golf/gun clubs). Also, he was charged to test Gatun Lake for radioactivity after nuclear powered ships were in the lake.

Bernice Aloise (Sanders) Hill and William G. Hill.

Being transferred was to be expected and accepted as he moved from New Cristobal, to Pedro Miguel, to Balboa, and to Margarita. In 1955, while in Balboa, he was told the still returning service men from World War II were requesting their jobs back which impacted him. He left the Canal Zone for employment in southern California for two years. The Panama Canal Hydrographic and Meteorological Division hired him back in 1957. He was with that service for only three months because at that time the Health Department heard of his return to the Zone and immediately had him transferred back to their organization. This meant being transferred from Balboa to Margarita, where he remained until his retirement in 1976 as Acting Chief of the Northern Sanitation District.

Bernice was employed by the Panama Canal Government from 1939 to 1944 except for a pregnancy service interruption during 1941. She was a Cristobal Clubhouse cashier; but having the least seniority, she was sent to the Gatun Clubhouse and even the Balboa Clubhouse for a day's work, if needed.

She remembers that following the great Colon fire, residents of Colon were allowed to eat meals at the Cristobal Clubhouse but not allowed to purchase anything else. She was assigned as a clubhouse cashier to transit the Panama Canal aboard the Panama Line's ship *Panama* during its Canal transit as part of the 25th anniversary celebration on August 15, 1939. Although she was born and raised in the Canal Zone, she had never transited the Canal, and this was to be her first transit. Her excitement was quelled quickly as she was assigned below deck taking money, making change and consequently not getting to see anything except the guests and the interior of the ship.

Bill and Bernice had three children: William G. Hill, Jr.; Robert Sanders Hill; and Bernice A. Hill, Jr. All three graduated from Cristobal High School. Bill, Jr., returned to the Canal Zone after many years and worked in the refrigeration/air conditioning service of the Electrical Division for about five years and then returned to the States. Bob landed a four-year cable splicing apprenticeship with the Panama Canal Company's Electrical Division. He migrated to the States upon completion of his apprenticeship to attend college and eventually to become a physician. Bernice, Jr., also went to the States to attend college and then went to work with the Cincinnati Bell Telephone Company, where she was accepted into training for and became the first female cable splicer in their employ ultimately becoming an outside plant engineer for the Northern District until she retired.

Bill and Bernice were married 71 years until his death at age 89 in Fort Myers Beach, Florida, where they had moved from their long time retirement home in Aiken, South Carolina.

See also the Bruce Gordon and Grace Aloise (Meister) Sanders and the Hermanus Kleefkens family histories.

PHOTO COURTESY OF WWW.CZIMAGES.COM

***DDT in the Canal Zone.** Taken in 1954, this photo captures DDT being dispensed from a truck on its twice-weekly rounds in a Canal Zone townsite. For years, DDT was used to control insects and mosquitoes before its danger to humans was determined. A favorite pastime of Canal Zone children was to gleefully chase after the trucks, inhaling the DDT to their hearts' content.*

CHARLES SANFORD HOLLANDER

AS TOLD BY DAUGHTERS ROBERTA WILLIAMSON-MUSCO AND ROSEMARY MCCORKLE

Founder of the Canal Zone Credit Union

Charles Sanford Hollander was a gentleman in every sense of the word. His two daughters never heard him utter the word "damn" throughout their lives, and he probably never did during his 93 years on earth.

His parents came to New York from Austria-Hungary, and Posen, Germany, by way of England in the 1870's. He was one of nine children (His birth certificate says "Sollie" because his immigrant mother did not speak English too clearly after the home birth on September 26, 1892.): Samuel, Joseph, Rebecca, Sarah (Sadie), Rose, Mary, Harriet, Charles and Dorothy. He was the only one to do two things: leave the United States for the Canal Zone and father children (Roberta and Rosemary), both born in the Canal Zone. He attended school in Newark plus business college and joined the New Jersey National Guard at age 17.

While working a postal route in New York City, he became acquainted with a man active in the budding Boy Scout movement, served part-time at $30 a month as a Deputy Scout Commissioner, and later became one of its first professionals. This included organizing scout troops throughout the New Jersey Scout Council as Field Scout Executive at $1,200 yearly. He worked with W.J.B. Houseman and James E. West, the first Chief Executive of the Boy Scouts of America. Troops he organized included one at a school for the deaf, one of delinquent boys and what was said to be the first troop of Catholics in Belleville, New Jersey. At one summer encampment for the Robert Treat Council as Assistant Camp Director, he was nicknamed "Girtchimik," Indian for "Mighty Beaver."

Charles S. Hollander ready to leave for Canal Zone, February 12, 1918, Newark, New Jersey.

In 1917, our father learned of an opportunity to work in the Canal Zone as a steno-typist at $103.00 a month and set sail February 1918 on the *SS Advance*. He had joined the U.S. Naval Reserves in New Jersey and was called up as a yeoman in the 15th Naval District from July 1918 to January 1919. This led to lifelong membership in the American Legion, including Adjutant of Post No. 1, Balboa, and Departmental Adjutant for the Republic of Panama. Later he joined the Payroll Division as roll keeper.

Having served his two-year contract, our father returned to New York and established a successful Albany restaurant business with seven branches and a loaning group called the North East Credit Corporation. A serious automobile accident left him bedridden for months. He lost his business and returned to the Panama Canal in 1929 to the Municipal Engineers Division, which was in the midst of constructing Madden Road. During the later construction of the Madden Dam Project, he rounded up tardy workers, was known as "Jefe," and also supervised the mess for the workers. During this time he partnered with Harry Bissell in a cayuco race against two young engineers. Was this the start of the traditional annual race?

Our father married Eleanor Freund, a Gorgas Hospital nurse, on August 15, 1931. After the birth of Roberta in 1932, the family moved from Pedro Miguel back to Madden Dam. It was a small group of homes, carved out of the jungle for government inspection forces. Once the dam was completed, there was another family move to Mindi Street in Balboa Heights, and Rosemary arrived in 1936. Dad had lived in the jungle for over six years, including a grass hut.

Eleanor's "prince of a fellow," 1931.

A change from the Maintenance Division to the Municipal Division found him doing accounting and payroll with the Grade 5 salary of $208.33 a month.

Along with his government job, our father gathered a group of friends who founded the Canal Zone Credit Union in 1936, with an initial savings limit of $2,500. A friend, Dr. Sampson of Pedro Miguel Dispensary, was concerned about employees, both local and U.S. rates, who had no insurance. Employees could not rent or buy a home in the Canal Zone. Burial was done generally by "passing the hat." As the Canal Zone Credit Union's first president with a strong box in his quarters, our father had many after-hour calls from Canal Zone residents with family emergencies in the States and would "open the safe" so they could have funds to leave. In his lifetime, he was involved with seven credit unions, including Jacksonville Naval Air Station and Mayport Carrier Base.

A friend lured him to the States in 1937 to start a coffee import business in Jacksonville, but unavailability of beans made an offer to return to the Zone enticing. In 1939, with World War II threatening, his family remained stateside. Returning from vacation with them, he was on a ship in the Caribbean when Pearl Harbor was bombed. War years found him maintaining storehouses of materials such as cement, lumber, reinforcing steel and plate steel at different lock sites; supervising mess halls; serving as Credit Union Treasurer and living in bachelor quarters. In the summer, 1944, his vacation was as camp counselor called Echochote at Scout Camp Immokale in Orange Park, Florida, where he was initiated into the Order of the Arrow.

We rejoined him in 1945; the quarters on Las Cruces were close to the Credit Union located upstairs at the Balboa Clubhouse. Our mother's death in 1947 left a void filled with his marriage to Ruth Goldstein in 1949. In 1952, the Credit Union moved to a newly constructed building. Charles worked on establishing five credit unions in the Canal Zone from Balboa to Cristobal, reorganizing the Air Corps Credit Union, and forming the Navy Credit Union. By the time he retired in 1953 after over 24 years of government service, he had served as Chief Administrative Assistant to the Chief Engineer in charge of building the Madden Dam, Chief Administrative Assistant to the Municipal Engineer, and Chief Administrative Assistant to the Maintenance Engineer.

In Jacksonville he continued working and consulting with local credit unions and volunteering with the Boy Scouts and the American Legion.

Patriotic, honest, hardworking, always an inspiration to his daughters and numerous grandchildren, Charles Hollander was proudest of his role in founding the Canal Zone Credit Union. He died June 15, 1985.

C.S. Hollander, Founder. Photo taken February 1953.

See also the Eleanor Freund Hollander history.

Eleanor Freund Hollander

GORGAS NURSE CHERISHED BY DAUGHTERS ROSEMARY MCCORKLE AND ROBERTA WILLIAMSON-MUSCO

Roberta Hollander Williamson-Musco

Eleanor Hollander's life reads like a storybook — some of it wonderful, some of it sad, and most of all, memorable.

Ella Hermine Maria Freund was born Christmas Eve, 1896. Her mother, Hermine Freund-Laux, was a spoiled descendant of German nobility dating back to 1424 with Count Kreller Laux, whose castle overlooked the Rhine River. Her family was in the jewelry business. Her father, Franz Anton Freund, was a watchmaker who also ran Prinzen Inn in Kappelrodeck, Germany, still in operation today. Frank suspected Hermine of running around with other men when she took off to a spa, leaving their daughter with a nurse. She denied it, but he stubbornly left with a chest full of family jewels and four-year old Ella. The child grew up believing her mother was dead. This resulted in a search for 40 years by the devastated Hermine covering three continents!

Ella Marie Freund, age 10.

Crossing the Atlantic Ocean by steerage in 1901 on the *St. Paul* so he could not be traced, Franz settled in Elmira, New York, near an older lady who had cared for Ella shipboard. He learned of the hoped-for Rhineland of the New World being established by a German named Cullman near Birmingham, Alabama. The climate was beneficial to growing grapes as well as iron in the mountains like the Ruhr Valley. He claimed to be a widower, managed an inn in St. Clair Springs, became "the catch of the town," married Kate, a local girl, also of German descent, and moved to Cullman. Ella, meanwhile, was being raised by nuns at the nearby Sacred Heart Academy/convent, learning fine needlework and piano. Franz bought an acre lot from Kate's family and personally constructed a two-story home. After sister Hildergarde's birth, Ella became her caretaker while Kate utilized her seamstress skills and "Frank" loaned money to locals. He was regarded with much respect and always referred to as "Herr Freund." But he had to cope with a rebellious teenager!

"Cinder Ella" believed she was just that with "a wicked stepmother" and ran away from home, finding employment with an older couple caring for the invalid wife. This led to Saint Vincent's Hospital, training to be a nurse while scrubbing floors to work her way through and graduating as a registered nurse with highest honors the week Armistice was signed in 1918. Ten years of nursing financed a trip to the West Coast, nurses' tour to Italy for an audience with Pope Pius XI, and, finally, reconciliation with her father and adoring sister. While doing private-duty nursing for Senator Lister Hill of Alabama, she heard of nursing opportunities in the Canal Zone and arrived October 12, 1928.

Long hours of work on her feet (sharing shifts with 100 other nurses at Gorgas) ranged from nursery detail to eye, ear, nose and throat wards. Letters written in June 1930 revealed she served as assistant to a dentist for men who had been at the South Pole when they returned to civilization after two years

Eleanor Freund, R.N., on a day off from Gorgas Hospital duties, 1930.

and needed extensive dental work. An unhappy suitor left behind was soon replaced by a number of bachelor admirers during her off-duty hours. Eleanor, as she now called herself, participated in horseback races, trips to beaches, a seaplane excursion to the San Blas Islands, and dancing at the El Rancho and the Union Club. It was "paradise," and she "should have come years earlier!"

A vacation trip to New York found her heart leaning towards Charles Hollander, "a prince of a fellow," and upon her return, Eleanor agreed to marry him on August 15, 1931. By 1932, she had learned to drive and soon was visiting her fellow nurse friends with daughter Roberta in a laundry basket in the back seat. They lived a relatively isolated existence at Madden Dam with marmoset (Jocko) and Cappy Boy, their dog. A States' vacation allowed the doting Hollander clan a chance to see the only grandchild and was followed by a move back into town and Rosemary's 1936 birth.

Sue Core wrote about the next event in her *Chuckles, Sniffs and Sighs* column in the *Panama American* newspaper. Mother's Day Eve, 1937, the only time Germany and the United States observed the holiday on the same day, a letter arrived asking if Eleanor Hollander was the same "Ella Marie Freund" whose father lived in Alabama. Her mother was alive, not dead, and Ella's place of birth was Germany, not Switzerland! Eleanor and her little girls crossed the Atlantic Ocean and celebrated her mother's 65th birthday with Roberta's fifth in the last days before Germany started World War II. A German agent watched from across the street the entire three months of their stay. They visited ruins of the Lauxberg Castle and met

L-R: Roberta, Eleanor, and Rosemary Hollander; Jacksonville, Florida, September 1942.

relatives Eleanor never knew existed. Although invited, Hermine refused to leave "her" Germany.

Return to Panama was brief, as a business venture for Charles moved the family to Jacksonville; then he returned to the Canal Zone until family quarters were available in 1940. In 1941, the family vacationed in Jacksonville, and the decision was made to stay "safe" in the States while he returned to work. A virtual "war bride," Eleanor passed the next four years volunteering as an air raid warden; serving as war ration board member; teaching home nursing and Girl Scout classes; organizing doll exhibits benefiting the formation of the Jacksonville Children's Museum; assisting the PTA; driving her girls to music and dancing lessons; raising a victory garden and carefully allotting ration stamps for sugar, gas and shoes. The summer of 1944 was spent with daughters as camp nurse at Girl Scout Camp Chowenwau near Green Cove Springs in Florida. Charles (on vacation) was a Boy Scout counselor across the river, joining her on weekends. Her spirit of volunteerism and service to the community was instilled in her daughters throughout these years.

With the celebration of VE Day, 1945, the trio flew home to Balboa, happily reunited with Charles. Cancer struck as they celebrated their 15th anniversary in August 1946; and on February 9, 1947, just fifty, Eleanor died. The fairy tale was over. Her grave is in Corozal Cemetery, her peaceful, beautiful "paradise."

See also the Charles Sanford Hollander history.

The David Cooper Hollowell Family

AS TOLD BY KARL D. ALLEN AND DAVID HOLLOWELL

Born in Drenchman's Bayou, Arkansas, in November 1882, Cooper was his given name, but he was known as "Reddy" for his flaming red hair.

The man-about-town bachelor was not happy with his life as a farmer or with the mercantile run by his father in Joiner, Arkansas. A friend who was working on the Canal encouraged him to "come on down." He took a ship to Panama from New Orleans named *Ellis*. He arrived in Colon on March 5, 1908, and began work as a cable splicer helper at $0.32 an hour; thus started a career that ended in 1945 after 35 years of service.

His future bride, Ida May Jordan, was born in June 1894 in Dubois, Pennsylvania. She arrived on the isthmus in September 1907, when her mother took her five children to meet her husband, William B. Jordan (Roosevelt Medal No. 2555), who had been recruited in 1907 to work as a Panama Railroad conductor. With his railroad experience, his starting salary was $170 a month.

The Jordan family arrived in Colon and took the train across the isthmus where passengers were dropped off at the various towns where they would live. When they arrived in Pedro Miguel, they transferred to a baggage car for the short trip to Paraiso which was their new home. They were quartered in a three-room house and quickly noticed that the legs of the beds were in cans of oil and did not touch the walls. The feet of the ice box and stove were also in cans of oil to protect them from insects. The family soon grew used to life based on the commissary system; ice and fresh meat were brought by train from Colon. Before their arrival, there was no school. With four school-age children, an 18-year old teacher was hired to teach classes in Pedro Miguel.

Ida, as a teenager, worked at the Administration Building in Balboa Heights. Her diary is filled with stories of work, dancing, trips to Taboga, living in Pedro Miguel, and meeting Reddy, the love of her life.

Cooper lived in a crew camp at the Canal's edge near Miraflores. When the couple courted, he would walk her to Pedro Miguel through the railroad tunnel. When it was time for Reddy to go back to camp, she would escort him back through the tunnel. Then he would have to escort her back. Thus the family legend of "the courting tunnel" was born.

Ida and Cooper Hollowell, June 1913.

The two married in 1913 and had a brief honeymoon at the Tivoli Hotel. They had four children: Victoria H. Allen, born in the old French Hospital in Ancon; William "Bill" Hollowell; Hope Hirons; and David Hollowell. The family moved to Gatun in 1925 when David was six months old. All four children graduated from Cristobal High School. The Jordan grandparents lived across from the high school.

Cooper was a fun loving father and enjoyed a good joke. He was very active in Gatun and helped to build the gun club. He loved to garden, and his side of the duplex was filled with plants and flowers. He was also active in the Masons and Shrine as was his father-in-law. The Hollowells attended the Union Church.

The family experienced a wonderful life in the Canal Zone with activities that included swimming, baseball, fishing, bowling and exploring the rain forest and old Spanish installations. In the 1920's and 1930's there were no restrictions to accessing or crossing over the locks; at times it seemed the locks were just an extension of their playground. During World War II, access to the locks was restricted. This stopped the use of Gatun golf course which had a real water safety hazard — the Gatun Spillway — and also stopped Bill and his friends from retrieving and selling golf balls that did not make it across the Spillway. You could still get to the Tarpon Club for meals and fish on the Chagres River. Many a tarpon was caught and turned into a commissary book by selling the fish at the Colon market. David caught an 85-pounder but it was not near the size of the one his father caught.

Gatun life was always interesting, with many events. The commissary was a source for the youngsters to pick up extra spending money by pulling groceries home in their wagons. July 4th was always a big event with tents set up in the open field across from the upper locks chamber; games, rides, and sales were held. A soap box derby was held on School House Road, which had a sharp curve at the bottom. After the real soap box derby was completed, a couple of dads decided they would compete. What a pile up at the bottom of that sharp curve! The hill behind the fire station was a won-

Hollowell family. Seated L-R: Ida May, David, Cooper. Standing L-R: Victoria (Pete), William (Bill), and Hope.

Cooper Hollowell with tarpon caught below Gatun Dam Spillway, Chagres River, early 1940's.

derful place to slide down on palm fronds, especially after a heavy rain. In Old Town, there was a bandstand in front of the clubhouse which was a source of good listening on weekends. The play sheds were great fun for volleyball, baseball, and basketball games; dances; and Halloween parties.

In the 1930's, Franklin Delano Roosevelt visited the Canal Zone. The President toured the Canal and then boarded a Navy ship at Gatun Locks. For easier access, the ship was raised in the chamber. He boarded at high noon, but at 6 a.m. soldiers lined the road that he was to travel. Travel to the Pacific side was only by train until 1943, when the highway across the isthmus was constructed. All learned to drive on the left hand side of the road.

The Jordans retired to the United States in 1933 and the Cooper Hollowells, in 1945. All four of their children held jobs in the Canal Zone. Two retired from there — Hope Hirons from Schools Division as a swimming pool manager at Gatun and Margarita and Bill Hollowell from the Water Works Division. All of the grandchildren lived for a time in the Canal Zone. The last Hollowell to leave was Cooper's grandson Cody, who retired just before the Canal was turned over to Panama.

David Cooper Hollowell
Roosevelt Medal No. 3830 with Two Bars

William B. Jordan
Roosevelt Medal No. 2555 with Two Bars

THE JACK B. HOOD AND PATRICIA C. PERRY FAMILY

AS TOLD BY JACK AND PATRICIA HOOD

Jack Brian Hood was born in 1948 in Clarkesville, Georgia. His parents were John M. Hood from Sautee-Nachoochee, Georgia, and Sarah G. Dyer from Choestoe, Georgia. Jack graduated in 1966 from Jordan Vocational High School in Columbus, Georgia. He received an A.B. degree from the University of Georgia (Phi Beta Kappa and Phi Kappa Phi) in 1969 and a J.D. degree from the University of Georgia Law School in 1971. In 1972 Jack earned a Diploma in International Law from the University of Cambridge in England while enrolled at Darwin College. Also in 1972 he became a Captain in the U.S. Air Force Judge Advocate General Corps and was assigned to Howard and Albrook Air Force Bases in the Canal Zone. In 1976, Jack opened a private law practice in Diablo Heights, Canal Zone.

In 1979 Jack became a full time law professor at the Cumberland School of Law, Samford University, in Birmingham, Alabama. He later became an Assistant U.S. Attorney in the Middle District of Georgia, a partner in an Alabama law firm, and an Assistant U.S. Attorney in the Northern District of Alabama. Jack is the author and co-author of numerous law books, five of which are also on the Westlaw database. He is a member of the Alabama, District of Columbia, Georgia, Tennessee, and former Canal Zone bars. Jack is a 5-string banjo player, and in 2004 he published a novel called *Banjo Lessons*, a part of which is set in the Canal Zone. He also wrote and performed the song, *Digging the Panama Canal*, which was produced on a 45 rpm record in 1977 by the Chiva Bus Bluegrass Band in Panama.

Patricia Christine Perry was born in Rio de Janeiro, Brazil, in 1951. Her parents were Randolph E. Perry of Concord, California, and Maria Luiza Trinidade of Lisbon, Portugal. Pat graduated from Balboa High School in 1969, and attended Canal Zone College for two years. She worked for the Gordon Dalton Travel Agency in Panama City, Panama, from 1972 to 1979. She then worked for several travel businesses after moving to the United States. Pat is an avid tennis player, having won many trophies and tournaments over the years. She is fluent in Portuguese, Spanish, English and French. She also played the washtub in the Chiva Bus Bluegrass Band.

In 1974 Jack married Pat in Ancon, Canal Zone. They are the parents of two daughters. Sara Marie Hood was born in 1979 in Birmingham, Alabama. She graduated from John Carroll Catholic High School there in 1997 and then graduated from the University of Alabama in 2001. She presently works for *Cooking Light* magazine as a Senior Marketing Coordinator. Laura Christine Hood was born in Macon, Georgia in 1986. Laura graduated from John Carroll Catholic High School in Birmingham, Alabama, in 2005. She married John C. Cleage in 2006, and they have a son, Walkin E. Cleage, who was born the same year. All of the immediate Hood family members live in Birmingham, Alabama.

The Hood family, 2007. L-R: John C. Cleage holding Walkin E. Cleage, Laura Hood Cleage, John M. Hood, Pat Hood, Jack Hood, and Sara Hood.

The Charles Clarence Huber Family

As Told by the Corrigans

Charles Clarence Huber was born on October 1, 1872, in Ohio and died June 1, 1939, in San Diego, California. He married Gertrude York on March 17, 1984, in Gallia, Ohio. Their six children, Henry Thomas, Theodora Julianna, Florence Emma, Clarence Phillip, Helen Louise, and Clara Ruth "Clae," were all born in Ohio. The family resided in Ohio and moved to Virginia in 1906, relocating to the Canal Zone in 1909.

Charles is the holder of Roosevelt Medal No. 5311 with one bar. He entered Panama Canal service in 1909 and was a machinist at the Gorgona shops working at Miraflores Locks. After the Canal opened in 1914, he worked as a tug boat captain.

Charles C. Huber, 1872-1939.

Gertrude lived with Pete and Helen Corrigan's family at one time. She is remembered for telling ghost stories and scaring the dickens out of the young children. Her stories always ended with "you better be good or the goblins will get you."

Our branch of the Huber family is Florence "Flo" Huber. Florence married Edward Hugo Armstrong Nash on March 7, 1917, in the Canal Zone. Their three daughters, Lois Ruth, Helen Jean, and Narvice Noreen "Rene," were born in the Canal Zone at Ancon Hospital. A cute story about Florence: She drove herself to the police station at Balboa to take her driving test. The officer questioned her on who drove her there. She replied, "I did!" She passed the test, got her drivers license, and drove herself home. Florence died in the Canal Zone in 1972 and is buried in Corozal Cemetery.

Hugo Nash was born in 1893 in Dry Harbor (now called Discovery Bay), Jamaica. He was educated in Canadian schools and at the age of twenty went to work for the then British American Tobacco Company, Tabaqueria Istmeña, S.A. (TISA), in Panama as the founder and president. Well known on the isthmus, Hugo was a past president of the Panama Rotary Club and a prominent member of the American Society in Panama. He was a past founding member and president of the original Old Panama Golf Club. Hugo was an avid golfer, swimmer, and fisherman. He held the Canal Zone underwater swimming distance record which was held at pier 18, Balboa. Hugo brought the first neon sign to Panama — a pack of Lucky Strike Cigarettes—which was placed on the roof top of the first building across from the Good Neighbor bar on Automobile Road. In 1949, Hugo retired to Lake George, Goldenrod, Florida, where he sold real estate.

Edward Hugo Armstrong Nash, 1893-1958, and Florence Emma Huber, 1899-1972, in Panama, 1917.

Hugo and Florence built one of the first three houses

at Bella Vista beach which was later turned into Swiss Chalet Restaurant. The other two families were Dr. Frank Raymond and the parents of "Chito" Arosemena. These houses were built when Bella Vista was still out in the country.

Helen and Chito became childhood buddies, riding horses, swimming at the Bella Vista beach, and helping cowboys herd cattle to the Abattoir. Chito became a renown pediatric physician in Panama.

Helen Jean Nash Corrigan was born September 18, 1919, in Ancon Hospital and graduated from BHS in 1938. She married Peter Tiernan Corrigan, Sr., in 1939. They raised their four boys, Peter Tiernan, Jr., Brian Edward, Timothy James, and Michael Joseph, in the Canal Zone.

As a teenager, Helen wanted to participate in a horse show being held in the Canal Zone. At the time they wouldn't let her participate in show as her family lived in Panama and not the Canal Zone. Helen, not settling for defeat, rode her horse, Tiny Tim, across the tarmac at Albrook Field, and participated in the race. She won the prize, but they wouldn't give it to her as she was not officially entered in the race. She really didn't care; she proved her point and just enjoyed the show.

Helen will be remembered as a mother to many children as well as to her own. She always found room to care for one more. If you wanted advice, you went to see Helen. The advice was never sugar coated and was usually pretty good.

Peter Tiernan, Jr., "Pete" graduated from Balboa High School in 1959, attended Canal Zone Junior College, and then moved to Florida, continuing his education at the University of Miami. He married Rose Marie Seefried in 1961 and had two daughters and two grandsons. Pete retired in 1996 following a 32-year career selling yachts in Sarasota, Florida. Pete and Rosie raised show horses, some of which were national champions.

The children of Charles and Gertrude Huber. Back, L-R: Florence, Theo, Helen. Front, L-R: Clarence, Clara (Clae), Henry, circa 1907.

Brian Edward, Sr., "Eddie" was in the BHS class of 1961. Eddie was an excellent baseball player. In 1957 he was a member of the VFW Canal Zone team that won the national championship in Hershey, Pennsylvania. He later joined the U.S. Navy. Eddie was married to Lynn Egger and had three children, Brian Edward, Jr.; Eddylynn; and Cassie Lynn. Eddie's second marriage was to Sue Buss. They had twin boys, Timothy Brian and Michael Peter. Timothy was married and has a daughter Kelsi. Michael is married to Kristy Bryant; and they have a daughter, Mikayla, who was born on January 28, 2008, and is our latest family member.

Timothy James "Tim" graduated from Balboa High School in 1964. He worked for the U.S. Government and the Panama Canal for 32 years. Tim retired from the Panama Canal Commission in 1998 and moved with his family to Centreville, Virginia. Tim married Taffy Grace Koepke on July 26, 1969. They have two children Christopher David and Colleen Grace. Tim has always had a love for golf. His grandfather Hugo Nash taught him how to play at age 9.

Michael Joseph "Mike" attended Balboa High School class of 1967. He served four years in the U.S. Navy. He served during the Vietnam War and was stationed aboard the *USS Bon Homme Richard* CVA-31. Mike has one daughter, Carey Lorraine. He has been in the furniture sales business most of his adult life. Mike and his wife

Marge live in Chattanooga, Tennessee.

The Huber family is proud to have five generations of "Zonians": (1) Charles Clarence; (2) Florence Emma; (3) Helen Jean Nash Corrigan; (4) Pete, Jr., Brian Edward "Eddie," Timothy James, and Michael Joseph Corrigan; and (5) consisting of the following: Brian Edward, Jr., Eddylynn Kledzik, Cassie Lynn Green, Timothy Brian, Michael Peter, Chris and Colleen Corrigan. These seven family members are the only ones in the history of the Canal Zone who are fifth generation Zonians on both sides of their family, Huber and Corrigan.

Charles Clarence Huber
Roosevelt Medal No. 5311 with One Bar

See also the Peter Tiernan "Pete" Corrigan, Sr.; John Paul Corrigan, Sr.; and John J. and Grace N. Jackson family histories.

PHOTO COURTESY OF THE CORRIGAN FAMILY

Panama Rotary Club members (Hugo Nash, far right) planting 100 caoba trees, August 14, 1938.

THE MERCER BLANCHARD HUFF FAMILY

AS TOLD BY EDITH LOUISE HUFF WILLOUGHBY

A Young Man Seeks Better Opportunities at the Panama Canal

The second generation to experience life during the American era of the Panama Canal was the Mercer Blanchard Huff generation, beginning with the arrival of a young man nicknamed Pete from Columbus, Georgia. In 1903 Pete left his family's home in Columbus at age 19 for New York to work in semi-professional and professional baseball leagues.

By 1906 Mercer, his brother James, and other Americans had read of the work available on the Isthmus of Panama; the Isthmian Canal Commission promised quarters in addition to a pay scale roughly double what they could acquire for similar jobs in the United States. The advertisements were telling young, able-bodied men to write, if interested, to "Chief of Office, The Panama Canal, Washington, D.C.," giving a brief statement of training and experience. Soon it would be evident that they would earn the good pay that was advertised. And so it was in 1906 that Mercer sailed to the Canal Zone to work for the ICC On February 8, 1906, Mercer Huff began work as a clerk in the branch of Labor, Quarters and Subsistence. He worked in this department for three and a half years in both the Atlantic and central divisions.

Mercer (Pete) Huff (seated to left of unknown man in white) and his brother, James B. Huff (seated to the right of man in white), on the ICC baseball team.

All of the quarters at this stage had not been prepared but some bachelor quarters were available in Panama City and would cost a mere $8.50 a month. Panama City was a town of 45,000 with no paving, sewers or water works. Therefore, his arrival was not a picture perfect one.

The Canal Zone originally had very minimal facilities for entertainment and relaxation for the Canal workers except for saloons. John F. Stevens began a program of improvements in 1906 as he believed that the kind of work that needed to be done could only be done by a labor force free of disease, well fed, and well housed. To avoid the saloons, a number of clubhouses were built and managed by the YMCA, which provided the men with billiard rooms, an assembly room, a reading room, bowling alleys, dark rooms for the camera clubs, gymnastic equipment, an ice cream parlor, a soda fountain and a library. Dues were only $10 a year; the remaining deficit of about $7,000 at the larger clubhouses was paid by the Commission.

When baseball grounds were built by the Commission, special trains were laid on to take people to matches, and a very competitive league soon developed. Mercer (Pete) and his brother, James B. Huff, were on the ICC baseball team. Pete was said to be "one of the stellar infield players on Canal Zone teams during the early construction period." During their bachelor days baseball reigned for the two brothers.

Mercer became one of five paymasters who rode in a railway pay car along the Canal route, meeting a very close schedule of stops at appointed locations along the railroad line. At the stops the pay car was met by lines of "silver roll" workers who boarded the car, under guard, to be paid monthly wages. Four of the paymasters, at special teller windows, paid the workmen in silver, issued to the paymasters in bags. A fifth teller paid in gold coins. Occasionally "gold roll" workers, who were paid higher wages and usually, but not always, were the white workers, received pay from the pay car, but ordinarily they would report to the paymaster's office.

Mr. Huff resigned in August 1909 to return to the States. It was at this same time that the girl he had been courting, Selma Maenner, the 19-year-old daughter of Ludwig Maenner, chief draftsman, was leaving the Isthmus upon the resignation of her father and the family's return to New York. Mercer left the Isthmus to marry Selma. He returned with his new wife in July of the following year and was re-employed in his old position. In 1911 he was transferred and promoted to Assistant Claims Officer in the Office of the Examiner of Accounts.

Selma reminisced about the picnics that she and Mercer took to the ruins of Old Panama from Culebra with friends. At other times during the early construction days, they adventured into the bottom excavation of Culebra Cut and photographed themselves around the bucket of a Bucyrus steam shovel or other earthmoving equipment.

One of the chief recreations for some young couples was a trip by train to Panama City from Culebra about once every two weeks. To add to the diversion, a dance hall had been constructed by the Commission at the Tivoli Hotel for employees and visitors. A group known as the Tivoli Club was given the privilege of holding a dance there the second and fourth Saturdays of each month. These dances continued to be a wonderful distraction from everyday living for Mercer and Selma. She was known for her love of ballroom dancing.

Selma gave birth to their first son, Maenner Blanchard, on January 18, 1914, in Ancon Hospital, a hospital that remained in use from the French construction days. The couple had been living in married quarters in Culebra until the birth of Maenner. After Maenner was born, they moved into the newly constructed cement, four-family quarters on the beautifully designed Prado in Balboa, next to where the high school gymnasium would be located years later.

In 1914 the ICC was renamed The Panama Canal, and the Panama Canal opened to maritime traffic. That same year Mercer B. Huff was awarded Roosevelt Medal No. 1223 and two bars in recognition of six years of ICC service.

Six months after the opening of the Canal, Mercer was transferred to the paymaster's office where he remained continuously employed. During his years there, a second son, Thomas Daniel, was born on July 29, 1918, at Ancon Hospital. The two brothers attended Balboa Elementary School and Balboa High School, which was in the Balboa Elementary School building at that time. They each left the Canal Zone upon graduation from high school to attend colleges in the United States in 1933 and 1937.

Selma Maenner and Pete Huff, upper right, holding hands, Culebra Cut, 1908.

In the paymaster's office, Mercer received numerous promotions. He was promoted to assistant paymaster and cashier on October 1, 1939. About four years later he was appointed paymaster of the Panama Canal by Governor Edgerton, effective September 1, 1942, to succeed C.L. Bryan upon his retirement.

Around the same time Mercer's older son, Maenner B. Huff, with his wife Sara Antoinette Baker Huff and young son, Donald Blanchard Huff, returned to the isthmus. Maenner had arrived to take part in the Panama City lumber business owned by his father-in-law, Donald T. Baker, who had been exporting tropical woods for about 25 years. Soon after Maenner's arrival, Donald Baker became very ill with cancer and died a week later. Mr. Baker's wife, Sara Weeks Baker, decided to sell the business after her husband's death; Maenner found himself without work and looking for a position with the Panama Canal. This would be the start of the third generation of the Maenner and Huff families in the Canal Zone.

Mercer and Donald shared about seven years working for the Panama Canal. During that time, Mercer and Selma's second grandchild, Christine Baker Huff, was born to Maenner and Antoinette on October 12, 1942, at Gorgas Hospital, the new hospital in Ancon.

Five years later, Mercer Huff retired as paymaster from the Panama Canal with 37 years, 7 months and 24 days of service. About 30 of these years were in the paymaster's office.

Upon retirement, Pete and Selma Huff temporarily made their home in St. Louis, Missouri. His annuity was $2,773.20 payable in monthly installments. Not long after retirement, Mercer Huff died at home, 906 N. Wayne, Arlington, Virginia, on Friday, September 26, 1947.

James B. Huff, Jr.
Roosevelt Medal No. 311 with One Bar

Mercer B. Huff
Roosevelt Medal No. 1223 with Two Bars

See also the Ludwig Theodore Maenner; the Donald Thompson Baker; and the Maenner Blanchard Huff (two) family histories.

Balboa Pay Train. Locomotive 215 (sister of 299), pulling a U.S. pay car along what later would become La Boca Road, in front of St. Mary's Catholic Church, with Canal workers lining up to receive their pay.

MAENNER BLANCHARD HUFF

AS TOLD BY EDITH LOUISE HUFF WILLOUGHBY AND CHRISTINE BAKER HUFF FEWELL

Circumstances Lead Third Generation to Follow in Father/Grandfather's Footsteps

Maenner Blanchard Huff lived a full and rich life that demonstrated many talents. His life had many aspects: growing up in the Canal Zone where he was born; finishing his education and early married life in the United States; and his return to the Canal Zone and subsequent 35 years of government service.

Maenner was the first son of Selma Maenner and Mercer B. Huff, who had met in the Canal Zone in 1906, married and returned to continue with employment. He was born on January 18, 1914, at Ancon Hospital and grew up on the Prado in Balboa. Maenner was awarded the rank of Eagle Scout when he was 14 years old, played clarinet and taught himself how to pole vault in high school after reading a book about it. Summer employment was with the War Department at Quarry Heights and later at Fort Clayton.

A few months after graduating from Balboa High School in 1931, he attended Washington University in St. Louis, Missouri, to study architectural engineering for two years. Next he went to Pace Institute in New York for two years of accounting and business. While riding on a bus one day he had a chance meeting with a girl he knew from BHS, S. Antoinette Baker, who was attending Pratt Institute in New York. The two dated for awhile and then eloped on November 1, 1935. Since Antoinette's mother highly disapproved of the manner in which they married, she planned another wedding at a church in Brooklyn on January 25, 1936.

Maenner's employment varied in New York as he worked as a bookkeeper with Universal Oven Company, New York City; a salesclerk at R. H. Macy (along with Antoinette in a different department); and a junior accountant and chief accountant at Fitzgibbons Boiler Company.

It was in 1939 that Maenner; Antoinette; and young son Donald, born in Brooklyn on February 17, 1937, went to Panama. Maenner planned to join his father-in-law, Donald Baker, in his lumber business exporting tropical woods. Unfortunately, Mr. Baker fell ill with sigmoid cancer and lived only a week beyond the diagnosis, dying on September 20, 1939. After his death, Mrs. Baker sold the business. Under these circumstances, Maenner found employment with the Signal Office, U.S. Army, Quarry Heights, in the corps's construction program. This provided him work for ten months. Then, with his father still working as a paymaster with the Panama Canal Commission, he was accepted as a general clerk in the accounting department with a starting basic pay of $187.50.

> *For the next eight years Maenner acted as business manager and board of directors member of the Balboa baseball team in the Isthmian Baseball League. He also managed and played on his own team in the Pacific Baseball League from 1940 to 1946.*

For the next eight years Maenner acted as business manager and board of directors member of the Balboa baseball team in the Isthmian Baseball League. He also managed and played on his own team in the Pacific Baseball League from 1940 to 1946. As with his father, Mercer Huff, baseball was an important part in his life at that time until he traded it for golf in later years.

Soon after Maenner was promoted to accountant in 1941, Christine Baker was born on October 12, 1942; and the couple and their two children continued living in a one-bedroom 12-family apartment in building #0767-A, Williamson Place. It was the location of many of the 12-family buildings; many people beginning work in the Zone started with an apartment there. New hires had no Panama Canal service and could not compete with other employees for quarters assignments. When Edith Louise was born on June 6, 1944, the family was able to move to family quarters, #1449 Owen Street, in the Balboa "flats." A fourth child, Linda Antoinette, arrived on March 13, 1947.

Maenner began working after hours with a number of organizations, serving as accountant for the Jewish Welfare Board under Rabbi Witkin until 1974; becoming a Mason and secretary of his Blue Lodge for two years; and auditing in the Zone and in Panama in connection with Army PXs and USO clubs. He also audited records for the BHS student association and the Scottish Rite and Shrine, and served as chairman of the supervisory committee and auditor of the Canal Zone Credit Union.

A local board of examiners selected Maenner to attend an intern program in Washington, D.C., designed to develop administrative abilities of employees. After five months of training, his title changed to administrative assistant, governmental accountant, and his service grade was changed to GS-10(a) with a raise in pay. He reached a grade of GS-12 with a salary raise and the title of systems accountant on January 19, 1959.

Maenner used his unusual talent for organization and could think through how things would fit together. This ability was used in his work as a systems analyst to manage the employee payroll and benefits. He designed and created an improved and workable pay statement in red, white and blue, sent to all Panama Canal Company and Canal Zone Government employees twice a month on payday. Maenner's sense of humor displayed itself outside work, where his family adored it, as well as at work. On one occasion he made wacky flag signaling motions to send silly messages to his children. He alone knew the meanings, but the children enjoyed his antics. The public was fair play for jokes as well. When hiking became his passion, he was thrilled to have planted small plastic mushrooms along the trail when a mushroom expert had joined the hike to inform everyone about them. He relished the moment the discovery was made and lived off the laughs of others.

Mooney Huff's habit of smoking would later bring him the title of "Keeper of the Stogies."

He was nicknamed "Mooney" for the rather rounded shape of his face at one point in his life and all who knew him used it. He was not Maenner, but Mooney Huff. Mooney was well known for his cigar smoking during most of his career. There was some doubt that others were as pleased with his smoking as it pleased him. Mooney's left shirt pocket, if not filled with pens, was filled with Dutch Master cigars. In the 1950's the Administration Building always had plenty of dark green spittoons supplied beside desks throughout the rooms.

Motivated by his son Donald's interest in learning to pole vault, Mooney volunteered as a BHS track coach specializing in instructing pole vaulting in the 1950's and '60's. He cherished this position and the time he spent at the track in Balboa. As a member of the Balboa Relays track and field judges, he also supervised the pole vault event for most of these years.

Before retiring in December 1974 Maenner bought a book with a title that read something like "Safest Places to Retire." He studied it thoroughly and being the very intelligent, organized and analytical type that he was, he figured the cost-of-living for each area and printed the information inside the front and back covers and anywhere he could find space for his calculations. Based on this work Maenner made a decision to retire to the small town of Brevard at the base of the Pisgah National Forest in the mountains of western North Carolina.

In December 1974, Antoinette and Maenner met with the governor of the Canal Zone at Balboa Heights. There, Governor Parker presented Maenner with a certificate which read, "The Panama Canal awards The Master Key to Maenner B. Huff in the grade of Keeper of the Stogies with all rights, privileges, and perquisites of a bearer of the Golden Key to the Locks of the Panama Canal." It was signed by David S. Parker, Governor of the Canal Zone and President, Panama Canal Company.

After the holidays, Maenner and Antoinette left for Brevard, N.C. After a very brief stay to select a home, they went to the Olympics in Munich, Germany. Following the games and additional travels in Europe, they returned to purchase their retirement home and began a new life. Maenner took up hiking with the Pisgah hikers and became a leader of the group. As he had at BHS, he adopted a student at Brevard High School for several years, instructing him in pole vaulting. He also became very involved in the Blue Devil Booster Club — bookkeeping for them, managing the food booth they supported at football games, and attending football and baseball games with friends.

Antoinette died in Brevard on January 5, 1990. While following the doctor's orders, and recognizing that he must give up hiking after a heart attack and begin walking at a mall outside Asheville, Maenner met and married Christine Leverette Cook on June 1, 1992. They lived in Maenner's house in Brevard for one loving year, until Maenner suffered a massive heart attack at the mall one morning and died on June 28, 1993.

Although all of Maenner's children left the Canal Zone after high school graduation, one daughter, Edith, returned to work as an elementary school teacher in 1967 after completing her college degree in the United States. She continued to work and live there until May 1999.

Maenner Huff receives Master Key Award from Governor Parker (pictured with Toi Baker Huff).

See also the Ludwig Theodore Maenner; the Mercer Blanchard Huff; the Donald Thompson Baker; and the Maenner Blanchard Huff (fourth generation) family histories.

The Maenner Blanchard Huff Family

AS TOLD BY EDITH LOUISE HUFF WILLOUGHBY AND CHRISTINE BAKER HUFF FEWELL

The Fourth Generation Completes the Family Circle (1906–1999)

Mr. and Mrs. Maenner Blanchard Huff had four children, Christine, Edith, Donald, and Linda. The first child, Donald Blanchard, was born on February 17, 1937, in Brooklyn, New York. Two years later the family of three moved to Panama, and Maenner found work with the Panama Canal Commission soon afterwards.

Donald attended schools in Balboa and graduated from Balboa High School in 1955. As a teenager he was a DeMolay and was very active in track and field as well as football. His father coached him in pole vaulting. After graduation Donald left the Canal Zone to attend Rensselaer Polytechnic Institute in Troy, New York. He left college and joined the U.S. Army, serving overseas in Germany. Following his discharge he took up residence in Detroit, Michigan, and eventually went to work for IBM and became an engineer for mainframe computers. He married Laureen Duquette and had two children, fraternal twins Kirsten Huff and James Huff, born on March 2, 1970.

Christine Baker Huff was born on October 12, 1942, in Gorgas Hospital. She attended schools in Balboa and graduated from BHS in 1960. Scouting was a favorite activity throughout her school years. One activity involved having the noted Panamanian artist Diana Chiari de Gruber teach the girls about native dyes and using them to create bateas. Red clay and other native plants were gathered; instructions given to make them into dyes; and then sanding, tracing pre-Columbian patterns, painting and shellacking followed to complete a batea. A small batea dated 1953 is a prize possession. Another memorable scouting activity was a three-day camping trip to complete requirements for the Pioneer Badge. The group of girls set out for a remote site on Canal Zone land near K-9, taking water and supplies. A latrine was constructed by digging a hole and arranging screening, a shower was devised with a watering can, tables were constructed for food preparation by lashing together sticks, food was suspended from trees to protect it from animals, and perishables were tied up in the nearby river for cooling. The girls slept in jungle hammocks pitched on the ground on a slight incline surrounded by a trench. This proved invaluable in the heavy rain that occurred during the night. For recreation during the day the girls made an excursion to a large waterfall to swim. Since the vegetation on the river banks was too thick to walk through, the waterfall had to be reached by wading in the stream.

L-R: Christine, 6; Edith, 4; Donald, 11; Linda (Toni), 20 months. Administration Building, Balboa, 1948.

In her junior year of high school, Christine attended Girls State and was elected governor, which gave her the opportunity to go to Washington, D.C., to attend Girls Nation as the representative of the Canal Zone. At Girls Nation, she was elected Secretary of the Navy, which gave her the privilege of being taken to the Pentagon to meet the real Secretary of the Navy. She was a member of the high school swim team and also participated in the youth group of St. Luke's Episcopal Church, where Grandmother Sarah Evangeline Weeks Baker was involved in many church activities. Mrs. Baker shook the hand of the Archbishop of Canterbury at St. Luke's Episcopal Church in Ancon in 1951. Sunday became a special day to spend time with Grandmother Baker. After church she took her granddaughters to lunch at the Tivoli Hotel (which later held a special place in our hearts) or the Ancon Clubhouse. It became a favorite routine. After church was lunch and then the sisters bused back to Balboa and walked home. Grandma Baker was also instrumental in starting Daughters of the American Revolution and Children of the American Revolution chapters, which was something in which all the girls became involved.

Sara Evangeline Weeks Baker, grandmother to Donald, Christine, Edie and Toni, shakes hand of Archbishop of Canterbury, St. Luke's Cathedral, January 7, 1951.

In 1960 Christine attended Antioch College in Ohio. She then attended the University of Chicago to obtain her master's degree in social work. Christine married Charles Kenneth Fewell, an attorney, in 1971 and has made her home in New York since that time. She has worked as a psychotherapist after completing training at the Institute for Psychoanalytic Training and Research in New York City. She also taught at New York University School of Social Work and obtained her Ph.D. there in 2006.

In the early hours of June 6, 1944, a third child, Edith Louise Huff, was born at 2 a.m. in Gorgas Hospital in Ancon. Like her older sister, scouting held many memories for Edith, and she took part in Brownies, Girl Scouts and Mariner Scouts through high school. Camping trips were new and educational experiences. Constructing a vagabond stove and buddy burner to cook food while camping were a favorite. Edie marveled at the success in cooking bacon and eggs for breakfast on what used to be a 10-ounce coffee can. Upon graduating from Balboa High School in 1962, Edith attended Canal Zone Junior College for two years and graduated from Barry College in Miami, Florida, in 1966 with a bachelor's degree in elementary education. After completing her undergraduate work, she returned to the Canal Zone and began teaching fifth grade at Howard Elementary School on Howard Air Force Base. She returned to Barry College during summer vacations and earned a master's degree in 1973.

Edie, a nickname from college, married John A. Willoughby (Jack) on April 12, 1973. He had returned to the Canal Zone in 1971 to teach industrial arts at Balboa High School. Edie and Jack enjoyed bowling on the teachers bowling league; canoeing and fishing on the Mandinga River, which emptied into the Canal near the Gamboa penitentiary on the banks of Culebra Cut; bottle digging among the ruins of the construction day town of Culebra, where her great-grandfather, grandfather and grandmother once lived; airboat rides and fishing on Gatun Lake; exploring and observing the sounds and beauty of the wildlife; and trips to El Valle, Santa Clara, the San Blas Islands, the Volcan and other areas of Panama to enjoy and partake of the mixed cultures Panama had to offer as well as to practice Spanish, the language in which her mother was fluent. Edie and Jack left Panama in May 1999 when the final year of school closed. With everything being prepared for the American turnover of properties to Panama and Jack retiring, they took residence in Merritt Island, Florida, with their four dogs.

Linda Antoinette Huff was born on March 13, 1947, at Gorgas Hospital. As a young girl she reveled in her dance classes with Dorothy Chase, a former Rockette. She took ballet, tap and jazz dancing and often practiced enthusiastically in the living room of our home. Christine, Edith and Linda were members of the Order of Rainbow for Girls. Christine left for college before becoming Worthy Advisor, but Edith and Linda would serve as Worthy Advisors before graduating. Membership in Rainbows provided opportunities to develop leadership skills and take on responsibilities. Linda invested much time and effort into her term as Worthy Advisor at a time when both of her sisters had left for college. After graduating from BHS in 1966, Linda attended Parsons College in Fairfield, Iowa, taking jobs in many States before she found residence in Minnesota and served in the Air Force Reserves as a journalist. At that time she began to use Antoinette as her first name and Toni for a nickname. San Diego, California, would become her home and soon she would own a business – Huff & Associates Recruitment Advertising. Toni was diagnosed with hepatitis C in 1993 and received a liver transplant in 1996. Her health improved for a while; then she was diagnosed with a new strain of hepatitis C and she died in her home September 8, 1999, at the age of 52.

The Panama Canal Society's annual reunions in Florida provide a special time for banding together and sharing memories of living and working in the Canal Zone. Some memories are shared again and again; other memories are kept in hearts to treasure and to share with new generations.

Perhaps those who lived in the Canal Zone might hold strong feelings as Christine did when she and Charles and her father, Maenner Huff, returned to the Isthmus in 1992 for the Christmas holidays with their two children — Anna, born July 23, 1975, and John, born January 30, 1980 — to visit her sister Edith and husband Jack Willoughby. Christine arose before the others and went out for a walk alone on the Prado, passing with nostalgia the BHS Class of 1960 Balboa Bulldogs wall (the style did not age well). The feel of the heavily laden humid air; the sight of the wonderful, vivid flowers and trees; the aura of the sunshine brought an overwhelming flood of sensory memories that shouted, "I'm home!"

See also the Ludwig Theodore Maenner; the Mercer Blanchard Huff; the Donald Thompson Baker; and the Maenner Blanchard Huff (third generation) family histories.

> *Christine arose before the others and went out for a walk alone on the Prado, passing with nostalgia the BHS Class of 1960 Balboa Bulldogs wall (the style did not age well). The feel of the heavily laden humid air; the sight of the wonderful, vivid flowers and trees; the aura of the sunshine brought an overwhelming flood of sensory memories that shouted, 'I'm home!'*

BHS Class of 1960 Balboa Bulldogs wall.

PHOTO COURTESY OF PANAMA CANAL MUSEUM

THE CHARLES D. HUMMER FAMILY

AS REMEMBERED BY JOANN HUMMER HAUGEN AND CHUCK HUMMER

A special place time and people
Charles W Hummer

100 Years of Hummers in Panama

The saga of the century of Hummers in Panama began when Charles "Charlie" DeRussey Hummer, the patriarch of the Hummer family, was born in New Brunwick, New Jersey, in 1876, son of German-Dutch parents. He grew up on a dairy farm in Franklin Township in Somerset County.

As a young man he served in the army in Cuba in the Spanish-American War, and after discharge he returned to Cuba in 1898 to serve as a civilian in the Quartermaster Department. Then the lure of travel took him to the Philippines for four years with the Quartermaster Department. His second taste of Panama was on his way home from the Philippines. It was not long after his return trek across the isthmus that he saw Panama Canal construction advertisements in the local newspapers in New Jersey. Still footloose and fancy free, his desire to see and be a part of front page events brought him to the Canal Zone in September 1904, one the earliest of the army of Americans that followed.

His brother James Louis Hummer joined him on this adventure. Louie, as he was known, was assigned as a steam shovel crane operator, and Charlie was first sent to Culebra shops where he assisted in assembling the first two Bucyrus and Marion steam shovels shipped to Panama. Charlie was promoted to locomotive engineer and was assigned as wreckmaster operating the 75-ton Bucyrus locomotive crane used for picking up overturned steam shovels in Culebra Cut.

Louis was killed on December 12, 1908, in one of the major accidents that occurred during construction when a premature explosion detonated killing 28 workers. Charlie continued throughout the construction as the wreckmaster, and he had plenty of work.

In 1909, Miss Helen "Nellie" Hunt went to the Canal Zone to visit her sister Ann and her husband, Patrick Maher, another construction worker. It was at one of the many social events that she met her "Charlie." They were married on August 15, 1909, in Lamont, Illinois, and returned to live in the town of Empire along Culebra Cut. Their first son, Joe, was born the following June, and their second son, Charles "Tuck," arrived, January 3, 1912. They lived in Empire until its closing and then lived on Morgan Avenue — second house up from the Balboa Elementary School, where Joe and Tuck were members of the Morgan Avenue Gang that included neighbor friends, Earl and Elmer Orr and Jimmy Marstrand.

Charles Hummer, sons Joe and Charles, and wife Nellie, 1913.

In the early thirties the Hummers moved to the last house on Amador Road where they were well known for the yearly display they hosted when their Night Blooming Cereus lilies bloomed. People came from all over the isthmus to see them. One night there were over 300 blossoms counted. When Charlie and Nellie retired, the plants were moved to a special bed at the foot of the Administration Building.

After thirty-four years, one month and twenty days, Charlie retired as the oldest of the old timers in 1938. He was awarded the Roosevelt Medal No. 144 and was one of only 41 American employees awarded four bars for his years of service during the ten-year construction period (1904-1914). The original Roosevelt Medal, Canal Completion Medal, Thatcher Ferry Bridge Medal and the Golden Anniversary Medal are all contained in a special table made by Tuck Hummer in 1966. The table, constructed entirely from rock specimens from the Panama Canal, depicts the Canal Zone Seal with the medals mounted in the four quarters. The table top is a part of the Panama Canal Museum collection as is Charlie's "protest" brass ashtray from the Society of the Chagres which says, "We did our damndest and now we pay rent." Of course the ashtrays were made in the Isthmian Canal Commission foundry.

Table made by Tuck Hummer from Panama Canal rock specimens.

Both the young Hummers went to Balboa High School and then on to the Canal's apprenticeship program where Joe became a journeyman machinist and Tuck a draftsman. Joe went to work for the Dredging Division on the suction dredge, *Las Cruces*, and then transferred to the Locks Division. When the *Las Cruces* was sold, Joe accompanied the dredge to Bolivia, where it was placed back in service. He returned to work for the PC Company Locks Division until retiring due to illness.

Joe, Charlie, Nellie, and Charles Hummer, on home leave, 1923.

Tuck served a short stint as a mounted policeman for the Canal Zone Police and then as a draftsman on the Third Locks Project for the Special Engineering Division. He continued his employment in the Canal Zone until 1968, when he retired from a job as construction inspector on the Bridge of Americas and the new Gorgas Hospital. Tuck and wife, Kathryn, retired to St. Petersburg, where they lived until their deaths.

Just before World War II the brothers bought beach property near San Carlos in Panama, where they both built vacation homes. Their homestead was dubbed affectionately as "Agewood Heights" after the locally distilled American Whiskey. Joe moved to Agewood Heights after his retirement. The beach homes provided years of enjoyment for the Hummer families; and the next generation of Hummers got their initiation into Panamanian life along a beautiful and pristine three mile almost private beach, rode horses, trekked through the jungles and learned to make the local raw sugar, raspadura.

Joe married Ann Szabo; and they had three children, Joann and twins, Bette and Bob. They lived first on Williams Place and then moved to Barnebey Street in Balboa. Memories of Christmas tree burns and raiding of the trees from others' stash are some things that are remembered with fondness. The family then moved to Gamboa, Pedro Miguel and back to Balboa. Joe married Edna Fluharty Henter in 1953, and Joseph Edward was born the next year.

Although Joann and Bette moved to the States after high school, Bob returned to the Canal Zone after his four years in the Air Force with his new wife, Wanda. Their three boys, Bob, Roy and Tony, were all born in the Canal Zone. Bob did his apprenticeship with the Electrical Division when he first went back and then transferred to the Industrial Division. They lived in Margarita for awhile and then moved to Balboa and later to Los Rios and then back to Margarita before retiring to Arkansas after 20 years of service.

After finishing school in Florida, Joseph Edward returned to the Canal Zone in 1972 and completed his apprenticeship with the Electrical Division. He went to work with the Panama Canal Commission in the Communications Branch and continued to work for the Panama Canal Authority until he retired in March 2004 with 32 years service. When the Treaty became final in 1999, Joe bought his house in Los Rios and continued to run the motorcycle shop he had opened. He married Paula Blanchette, and their son Russel was born.

Tuck married Kathryn Laurie, one of the five daughters of James and Charlotte Laurie. James was a radiologist at the Ancon Hospital. Tuck and Kathryn had one son, Charles, Jr. Charles, Jr., attended Canal Zone schools, graduating from Balboa High School in 1955. Upon his graduation from the University of Notre Dame in 1960, he and his bride Greta Navarro Chiari moved back to Panama where he worked for the Naval Research Laboratory. Charles transferred to the Panama Canal Company in 1970 where he was Assistant Chief of the Dredging Division in Gamboa on leaving Panama for Washington, D.C., in 1979. Charles and Greta had one son, Charles III, also born in Panama.

As this story is written, there is still a Hummer family along the Panama Canal, one hundred years after the first arrival in the tropics, a legacy that found some Americans making Panama their careers and their home. Joseph Edward and his wife are permanent residents of Panama.

Charles, Jr., along with Joe Wood and Betty Frassrand, was one of the founders of the Panama Canal Museum in 1998. He dedicated over ten years to the development of the museum in Seminole, Florida, where he and his second wife, Sandra Lary, lived after his retirement from the U.S. Army Corps of Engineers headquarters in Washington, D.C. After his retirement in 1989, and after spending four years in Holland where Sandra was posted to the American Embassy, he spent two years as an expert lecturer on the Panama Canal on cruise liners transiting the Canal and an author for an international dredging journal.

On September 24, 2005, the Panama Canal Museum Board of Trustees honored Charles, Jr., by naming the museum's library The Charles W. Hummer, Jr., Library for Historical Research.

Charles D. Hummer
Roosevelt Medal No. 144 with Four Bars

James Louis Hummer
Roosevelt Medal No. 1527 with One Bar

Patrick F. Maher
Roosevelt Medal No. 376 with One Bar

Floating cranes Ajax and Hercules frame the cutterhead dredge Mindi at the Dredging Division dock in Gamboa.

PHOTO COURTESY OF DENNIS WHITE

The Oscar R. Hunter Family

AS TOLD BY WILLIAM R. AND DOROTHY E. HUNTER

Oscar Ray Hunter, the elder of identical twins, was born on August 2, 1884. His parents, William Allen and Emmeline Hunter, were farmers in the area of Woodstock, Georgia.

OR, as he was known, left home at age 16 (about 1900) to work for a lumber company located in Plant City, Florida. I heard many stories about sawmills, steam-driven apparatus, and other related stories although I don't remember too much about that period in his life.

OR's connection with the Panama Canal began on November 23, 1907, when he reported for duty as a clerk at $125 a month. His position became permanent on April 25, 1908.

Oscar Ray Hunter, circa 1918.

Charlotte Georgina McCabe, circa 1918.

On May 26, 1911, he was assigned to the Washington Office — Browne's Division, which was a branch located in Wheeling, West Virginia, with a salary reduced to $83.33 a month. Apparently this position didn't suit him so he resigned on August 28, 1911. The history of his next two years is not known in detail; however, he spent close to two years working on the Madeira-Mamore railroad along the Madeira and Mamore Rivers in the Matto Grosso region of Brazil, now known as the Territoria do Guapore. I heard quite a few stories of life in the Matto Grosso, but was most impressed by the accounts of the many deaths of workers brought about by a disease called beriberi. It seemed to have great effect on those who ate polished rice and little effect on those who ate non-polished rice. I remember asking if he had seen many headhunters and the reply was that on occasion they would come across an Indian or two along the railroad line.

After the Brazilian episode he returned to work on the Panama Canal on July 5, 1913, at the Quartermaster's Office at $125 a month. He resigned from this position on June 23, 1914, returned to the States, and worked for the Shipping Board during World War I. It was during this period that he met my mother, Charlotte Georgina McCabe, known as Georgie to her friends. She was born in Parrsboro, Nova Scotia, and traveled to New York City to become a nurse at Bellevue Hospital, class of 1913.

Apparently the employment situation at the Panama Canal was better than that in the States after the war because OR resumed service there on November 29, 1918. He resigned on March 2, 1919, but returned on December 10, 1920.

Sometime early in 1921, he suggested to Georgie that she take the ship to Panama so that they could marry. The ceremony took place in the Colon office of a Justice of the Peace on April 2, 1921. About three weeks later on April 22 he resigned again and went to work for a private company, Hibbard, which was building roads in the interior of Panama, which at that time meant the side of the Canal closest to North America. He was their paymaster. At first they lived in David and then moved to Aguadulce as the road system became larger. I was born on October 22, 1924, while they were living in Aguadulce, although my mother Georgie went to the Canal Zone for my birth. At that time there was a medical doctor, Rafael Estevez, who had talked the Panamanian government into building him a hospital in Aguadulce. For most of the time we lived in that town, she was his chief operating room nurse. For years after I kept hearing all sorts of bloody stories about her experiences, enough to convince me not to become a physician.

The residences in David and Aguadulce were reasonable for North Americans. We lived in the house in Aguadulce for a year and a half when I was an infant. However, I saw it a number of times when I was a teenager on visits to Aguadulce, and it did not make a bad impression. The next town in which we lived was Anton. We spent about six months in an adobe house with a tin roof — definitely not appealing to North Americans. I remember stories about the wildlife that became pets — two deer, a buck and a doe, and a monkey. The buck became dangerous because he would rear up and try and get people under his front hooves, so they got rid of the deer. Another amusing story concerned ants coming out of a hole in the adobe wall which my mother tried to cure by pouring boiling water into the hole. The result was no reduction in the number of ants but a large hole in the adobe wall.

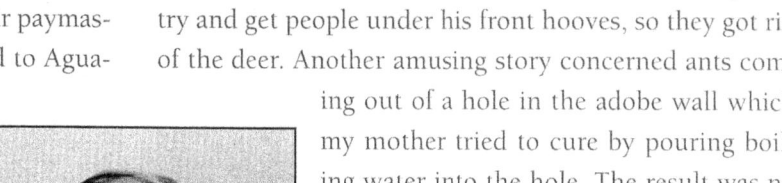

Oscar Ray Hunter, circa 1940.

A few years ago my wife and I found a number of letters my mother had written to her relatives in Canada in which she commented about cooking meals in Panama — one hand for cooking, the other for wiping away the sweat (no air conditioning).

On November 9, 1926, my father returned to work for the Panama Canal as a clerk in the Mechanical Division and worked for that division until voluntary retirement on November 30, 1945. He and my mother then took up residence in Ocala, Florida, until 1958 when she died. A few months later he moved to St. Petersburg, Florida, where he died on August 18, 1963.

Oscar R. Hunter
Roosevelt Medal No. 3526

Colon, Republic of Panama, December 1916.

PHOTO COURTESY OF PANAMA CANAL MUSEUM

The Hurst Family

As told by the Hurst siblings

Fond Memories of Our Time in the Canal Zone

The Hursts emigrated from the United Kingdom to North America in the late 1600's and settled in what would become the southern United States. Paul Hamilton (George) Hurst was born on April 25, 1916, in Ottawa, Canada. His parents were William Harris Hurst and Helen Deasy. Paul George arrived in Panama in 1934 while in the U.S. Army and fell in love with the country. On November 10, 1935, he married Serafina Quintero from Ocu, a small town in the province of Herrera. He left the Army shortly afterwards and started working for the Panama Canal Company as a clerk in the library of the Administration Building and later worked in the Supply Division. He lived with his family at 814-A Empire Street in Balboa. His first child was daughter Helen Hope Hurst born on January 8, 1937, at Gorgas Hospital. Paul Harris Hurst was born at Gorgas Hospital on January 14, 1939.

In 1941, with the start of World War II, Paul George left the Panama Canal Company, bought a cantina in Panama City, and became a successful businessman. He unfortunately died on May 9, 1945, at the age of 29, following complications from a ruptured appendix. He is buried at the Corozal Cemetery.

Paul Harris was six years old when his father died. He attended St. Francis Studios Grammar School in Balboa, Ancon Elementary School, Balboa Junior High School, and graduated from Balboa High School in 1958.

Paul Harris, Port Captain's Building, Balboa, 1971.

Paul Harris was ready to see the world, so he joined the U.S. Army and was stationed in Germany for two years. He then attended the University of Southwestern Louisiana in Lafayette, Louisiana, from 1960 to 1961.

On January 2, 1961, Paul married Delia Maritza Rodriguez from Ocu. They traveled to Louisiana together, but in August of that year, decided to return home to Panama. Paul Harris's career with the Panama Canal Company, Marine Bureau, Navigation Division, began on December 14, 1964, as a signalman. On November 19, 1967, he transferred to the Transit Operations Division as a marine traffic controller in Cristobal. In June 1969, Paul transferred to the Marine Traffic Control Center in La Boca, where he continued his employment until he retired on November 30, 1990. He served the Panama Canal Company for 30 years.

Paul Gabriel was born in Panama City on March 2, 1962, followed by Helen Marie on October 18, 1963. The Hursts moved to the Canal Zone in 1965, and their first home was on Tavernilla Street in Balboa. Harris William was born in Gorgas Hospital on June 7, 1967. The family then moved to Margarita when Paul Harris transferred to Cristobal. Paul Gabriel started school in Margarita where Helen also learned to swim and got her B badge. The family moved to Diablo for a short stay when Paul Harris was transferred to La Boca. They then moved to Balboa into a four-family home next to the post office. Paul Gabriel,

Helen, and Harris attended Balboa Elementary School. In 1972, the family moved to 0944 Amador Road. Paul William was born on October 19, 1973, at Gorgas Hospital. In 1979, the family moved to their last Canal Zone home in Los Rios. Paul William attended Los Rios Elementary School.

Parents Paul and Delia lived in Los Rios until 1992, when they moved to El Dorado in Panama City. They are enjoying their retirement living in Panama City, their beach house in Costa Esmeralda, their new house in Ocu, and visiting their children as often as possible.

Paul Gabriel graduated from Balboa High School in 1980 and attended Louisiana State University, where he obtained his degree in Industrial Engineering. He married Betty Gonzalez (Balboa High School, 1981) on August 31, 1985. They have three children: Taylor Elizabeth (December 13, 1989), Paul Joseph (October 4, 1991), and James Joseph (November 27, 1996). They live in Virginia.

Helen graduated from Balboa High School in 1981 and obtained her physical therapy degree from the University of Oklahoma. She then married Wally Loera on January 2, 1988. They have two children: Ashley Caitlin (November 5, 1990) and Gregory Harris (February 14, 1995). They also live in Virginia.

Harris William graduated from Balboa High School in 1985 and obtained his degree in mechanical engineering from Louisiana State University. He married Henrietta Wolf (Balboa High School, 1985) on January 2, 1992. They have two children: Madeline Marie (March 3, 1997) and Nicholas Harris (November 11, 1999). They live in Singapore.

Paul William graduated from Balboa High School in 1991. He attended medical school in Costa Rica and underwent residency in obstetrics and gynecology at the University of Missouri in Kansas City. On January 2, 2004, he married Jamie Backes, and they have one son, Paul Julian (May 18, 2006). They live in Texas.

Our memories are many:

From Paul: "My fondest memory is meeting my future wife Betty in the Canal Zone and sharing mutual memories. I remember playing baseball in Little League, Fastlich League, at Balboa High School, and ending my

L-R: Delia, Paul, Paul William, Harris, Helen, and Paul Harris, in Los Rios, 1985.

freshman year at Panama Canal College. I remember hours of practicing basketball at the upstairs gym at Balboa High School and with the soldiers at Fort Amador. And I remember, too, riding our bikes to the Balboa clubhouse and going swimming at the Balboa pool."

From Helen: "The most wonderful memories I have are from being a girl scout in the Canal Zone. I participated in many activities, such as primitive camping in San Blas over spring break, being a counselor at summer camp at Camp Carribean, hiking the Las Cruces trail. I also participated with Balboa Rams cheerleading, paddled in the cayuco race in the "Slave Galley," enjoyed New Year's Eve at the Amador Officers' Club, took trips to Costa Esmeralda to my parents' beach house and hung out in the hammocks, especially on our breaks home from college."

From Harris William: "My fondest memories in the Canal Zone are meeting my wife Henrietta and having a lifetime of shared memories and friends which we still avidly discuss. When we remember our youth in the Canal Zone, our thoughts always gravitate to sports and activities (horses, softball, baseball, cayuco, football, volleyball, ring-a-levio, kick the can, skim boarding, the slide, surfing, water skiing, beach trips, and Gatun Lake). How we had time and energy for it all was incredible."

From Paul William: "There are a couple of things I remember from living in the Canal Zone: being able to play in the neighborhood as a child on my bicycle or skateboard without worry, close friendships, and all the sports we played at all levels including the cayuco race. More fond memories include hanging out at the Amador Causeway, jumping off platforms in Gatun Lake, beach trips to Costa Esmeralda and, of course, the Balboa Yacht Club."

From Paul and Delia: "The memories we hold dear include the holidays with our children and family, especially the tradition of buying the Christmas tree. The tree was always beautiful regardless of when it arrived. We also have fond memories of entertaining in our home friends and family for Thanksgiving, Christmas, and many other occasions. We also enjoyed spending countless hours watching the activities of the children, especially the boys playing baseball. Shopping at the commissary, taking the family to Gorgas Hospital, and watching our four children grow up in a place that was a paradise are also things we will never forget."

L-R: Wally, Helen, Harris, Henrietta, Paul William, Jaime, Betty, and Paul in Panama, 2003.

Tavernilla Street, Balboa.

PHOTO COURTESY OF PANAMA CANAL MUSEUM

The John James and Grace Norris Jackson Family

As Told by the Koepkes

John James Jackson was born in Boston, Massachusetts, in the late 1800's. Having fought in the Boxer Rebellion, he was recruited by Theodore Roosevelt to head the Supply Division in the Canal Zone in 1905. Shortly after arriving there, John contracted malaria, and while in the hospital met nurse Grace Norris, who nursed him back to health. During his recovery they fell in love, and the couple married. As a result of John's service during the Canal construction period, 1904 to 1914, he received the Roosevelt Medal No. 683, with three bars, and Grace Norris earned Roosevelt Medal No. 5927. While in charge of the Supply Division, John conducted a survey of incendiary matches produced around the world, and selected the famous matches made for the Panama Canal Company by the Swedish firm Jonkopings Westra Tandsticksfabriks. After the first shipment of matches arrived in 1905, Zonians used the special Canal Zone matches from that date until the termination of the Canal Zone on October 1, 1979. John and Grace had one daughter, Gertrude Norris, born on July 20, 1914, in Culebra, Canal Zone.

During his tenure with the Supply Division, the Panama Canal Company terminated laundry services for the ships transiting the Canal. Opposing this decision, John went to the U.S. Congress in Washington, D.C., requesting permission to continue this service. Having rejected his request, one Congressman challenged John to open his own laundry if he felt so strongly about it. So, he did!

John returned to the Canal Zone, resigned from the Panama Canal Company and opened the Jackson Steam Laundry in Colon, Republic of Panama. He was very successful and his service eventually expanded to serve the local patrons.

John James and Grace Norris Jackson.

John and Grace's daughter Gertrude Grace Jackson grew up in Colon and graduated from Syracuse University in New York. Gertrude married Henry Bell Twohy, Sr., in 1937. Henry was a naval aviator who graduated from the U.S. Naval Academy in 1929. The couple had a son, Henry Bell Twohy, Jr., born on May 24, 1938, in Coco Solo, Canal Zone. Gertrude was widowed when Henry, Sr., died testing an aircraft for the Navy in 1938.

Gertrude returned to the Canal Zone and worked in the office of Censorship until she met William Robinson. Gertrude later married William Robinson in 1942 and had a son, William, who died as an infant in the Philippines. Unfortunately, her second husband, also a U.S. naval aviator, disappeared and never returned during a flight mission from the Philippines during World War II.

On March 17, 1945, Gertrude married Lyle Lawrence "Kip" Koepke from Michigan, also a graduate of the U.S. Naval Academy and a naval aviator. Kip had a great Navy career and retired with the rank of Captain. Kip was a natural athlete, and walked onto the Academy's football team tryouts, and made the team. In 1929, Kip was selected All American for his play as a Guard on the Navy football team. The following year, he was elected the 1930 Navy football team captain. Those achievements are more impressive since Kip never played high school football. At the Naval Academy, Kip also excelled on the Navy water polo team, and was a teammate of Gertrude's first husband, Henry Twohy, Sr., himself an All American in water polo.

Kip was the first Eagle Scout in the State of Michigan. He loved fishing, hunting, football and golf. When Captain Kip Koepke was the Commanding Officer of Coco Solo Naval Base in the Canal Zone, he played his most memorable round of golf at the nearby Brazos Brooks Country Club. Kip was playing in a foursome with Gil Morland, Frank Day, and Dr. Vern Prior, when he tied the course record of 68. His score recorded a 33 on the out nine, and 35 on the return nine. The course record he tied had been established previously by none other than his good friend "Doc" Mitten. During his tenure in Coco Solo and as a result of his work with Panama, he was awarded the Vasco Núñez de Balboa Medal from Panama.

While a test pilot during World War II, Kip was instrumental in preparing the Jimmy Doolittle Raid on Tokyo, helping pave the way to the U.S. success in bringing Japan to terms and ending the Pacific Campaign of World War II. While the Navy set out to ascertain if bombers could be launched from aircraft carriers, many of Kip's fellow test pilots conducted trial landings and takeoffs from similar length, land-based runways. Kip was the first pilot to actually achieve a successful landing and take off from an aircraft carrier with a B-17 bomber. Once Jimmy Doolittle knew he could operate his bombers from an aircraft carrier, he built on that information to plan his raid on Tokyo. The rest of the story is now history.

Gertrude had one son and three daughters: Henry Bell Twohy, Gertrude Kandi, Taffy Grace, and Kathryn Sherry Koepke.

Henry Bell Twohy, Jr., married Carolyn "Hogie" Holgerson on March 17, 1969, in Ancon, Canal Zone. They have two children: Kara Kathleen and Henry Bell "HB" III. Henry retired from the Panama Canal Commission in the early 1990's, having worked with the Canal Zone Police, Housing Division, Admeasurement Division, and Fire Division.

Kandi was born in Coco Solo, Canal Zone, on October 28, 1946. She married Stevin "Steve" C. Helin in 1967. They had two daughters: Tracey Ann and Stephanie Marie. As a classroom teacher and later a teacher for the gifted, she was much loved by her students. Kandi died on December 20, 1989, the first American casualty of Operation Just Cause, the assault of Panama by U.S. Forces to apprehend Panama's Dictator Manuel Noriega.

Taffy was born on January 21, 1950, at the National Naval Medical Center in Bethesda, Maryland. A typical military "brat," she spent the first years of her life moving around from Panama, Ohio, and Costa Rica, finally settling in the Canal Zone, where she graduated from Balboa High School in 1968. After receiving her Bachelor's degree from Florida State University and a Master's from University of Miami, she began her career in education as a classroom teacher in the Canal Zone. A year later, due to the transfer of function, a Panama Canal Treaty requirement, she transferred from the Canal Zone schools to the Department of Defense schools system. She has had a long career with the Department of Defense Education Activity (DoDEA) as an educator, school administrator, superintendent, and Director, Educational Partnership. She transferred to the DoDEA headquarters in Arlington, Virginia, in 1998. Her biggest achievement has been en-

L-R: Larry, Sherry, Gertrude, Taffy and Henry.

suring the quality of education for students around the world, especially in the Canal area. Taffy is most proud of the education provided for all the children, particularly in the former Canal Zone.

Taffy married Timothy James "Tim" Corrigan on July 26, 1969. They have two children: Christopher "Chris" David and Colleen Grace.

Katherine "Sherry" was born in Columbus, Ohio, on September 28, 1955. In 1960, she moved to Panama with her parents, where they lived for 12 years. Sherry graduated from high school at Balboa in 1973. She married Richard Allen Williams on May 5, 1988, in Cardenas. They have two children: Kathryn Georgiann "Kate" and Ryan Allen. Having moved many times throughout her life, she was attracted to the moving industry. Sherry worked for LACMA, Latin American and Caribbean Moving Association, until she moved to Centreville, Virginia, in 1999. She is currently living in San Juan, Puerto Rico, working with the moving industry.

Additionally, Kip had three sons: Ward, Larry, and Craig Koepke. Larry lives in Davison, Michigan, with his longtime companion, Patti. Larry has one daughter, Kathryn "Katie" Koepke.

It is interesting to note that Kandi, Taffy, Tracey, Stephanie, Kara, and Colleen have all been educators. Additionally, Henry, Steve, Tim, and Rick were all Sergeants with the Canal Zone Police Division.

The most memorable times for the family growing up were the weekends in Santa Clara. There always seemed to be lot of laughing and "partying," eating mangos and pivas, going to the beach, riding horses, and relaxing in hammocks.

John James Jackson
Roosevelt Medal No. 683 with Three Bars

Grace Norris Jackson
Roosevelt Medal No. 5927

See also the Charles Clarence Huber; the Peter Tiernan "Pete" Corrigan, Sr.; and the John Paul Corrigan, Sr., family histories.

Family Thanksgiving photo, L-R: Ryan, Cassie, Tim, Colleen, Patti, Larry, Henry, Kara, HB, Taffy, Chris, Nicole, Alex, Sherry and Kate.

The William C. Caley Johnston Family

As Told by Olga Johnston Conley

William Charles Caley Johnston was born November 10, 1870, in London, England. His father, William Walter Whitehall Johnston, was a military officer. William Caley's mother died when he was 15 years old. He received his early education in the Godolphin Grammar School, Beaconsfield, Buckinghamshire, England. He received his technical education in Anderlecht College, Brussels, Belgium. He also took a four-year study in mining engineering in the Royal School of Mines in Clausthal, Germany. He spoke German, French and Spanish fluently. After completing his technical training, he returned to England and worked in Wales and Cornwall.

Mr. Johnston came to Latin America to work for the welfare of the young republics and came to Panama in July 1893. He worked as assistant engineer for the Colombian Mining Co., manager of the Veraguas Mining Co., and manager of the Caribbean Co.

In 1899 he married Hermelina Grindale, daughter of William Grindale and Hermelina Castillo. Mr. Grindale, who also was from England, worked at the Tumaco mines in Colombia. Hermelina Grindale was a descendant of Edmund Grindale, Archbishop of Canterbury. There are still many Grindale descendants living in Spain. Mr. and Mrs. Johnston had three sons and three daughters: sons Arthur Percival and Eric (who both died in their 20's) and Francis Hope and daughters Enid, Gladys, and Irene Johnston.

After the formation of the Panama sovereign state, the young republic found there were many engineering problems to solve, so they turned to Mr. Johnston to avail themselves of his services. He was an engineer of sound learning who could discuss public affairs, and he was versed in the needs of Panama. In 1908 he became the official engineer for the government of Panama, and in 1909 he was made official engineer for the province of Coclé. In Coclé he found the fossil remains of an extinct megatherium (giant ground sloth). It was named Megatherium Johnstoni by the Smithsonian Institution. He also worked for Dr. Belisario Porras, President of Panama.

Mr. Johnston was chosen as Commissioner to represent the Republic of Panama in the settlement of the Canal Zone boundaries.

The following article appeared in an English Panamanian newspaper:

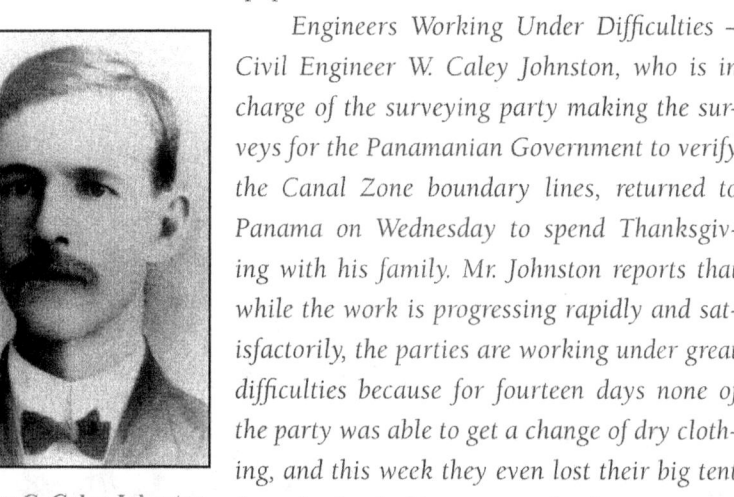

William C. Caley Johnston

Engineers Working Under Difficulties – Civil Engineer W. Caley Johnston, who is in charge of the surveying party making the surveys for the Panamanian Government to verify the Canal Zone boundary lines, returned to Panama on Wednesday to spend Thanksgiving with his family. Mr. Johnston reports that while the work is progressing rapidly and satisfactorily, the parties are working under great difficulties because for fourteen days none of the party was able to get a change of dry clothing, and this week they even lost their big tent from the floods. However, as the dry season approaches, he is in great hopes of resuming the work.

Another article appeared in a London newspaper saying:

Johnston in bad accident – W. C. Johnston, the Engineer in the Public Works department of the Republic of Panama, had a nasty accident when the coach in which he was riding was struck by a street car. Mr. Johnston was riding down Ancon Avenue, past the Tivoli Garage, and was horrified to see a street car approaching only a few feet away. The street car struck the coach and Mr. Johnston was thrown so hard that he rolled across the Tivoli road, bringing him up against the fence around De Lesseps Park. He said that the coachman did his best to get out of the way. Mr. Johnston went to his own physician.

Mr. Johnston was a member of the Mining and Metallurgy Association, Royal Society of Arts, National Society of Engineers, and British Club of the Isthmus of Panama.

His photo and biography appeared in the 1911 edition of "Makers of the Panama Canal," under a special section, "Founders of the Republic of Panama." He died of a heart attack at age 48 and is buried in Chiriqui, Republic of Panama. His widow, Hermelina Grindale Johnston, died of tetanus at age 75 and is also buried in Chiriqui.

Mr. Johnston's descendants are Francis Hope Johnston and wife Andrea Gonzalez (deceased), who gave birth to Maria, Isabel and Graciela Johnston. Maria and husband Arcadio Ruiz had five children and grandchildren. Isabel and husband Ludovico Cassatti had Ludovico, Silvana, and Italia (triplets), Roco and Blanca (who married Quinto Pan and have three children). Graciela and husband Aniceto Rodriguez had Yovanna and two grandchildren who live in Panama.

Francis Hope Johnston married Secundina "Nina" Cabrera in 1937. Francis H. died in 1985, at age 84; Nina died in 2004, at age 85. Their children were Olga Yolanda, Francisco, Emilsa, Melva, and Enrique. Olga Y. Johnston Conley (previously married to Dale Griffith) gave birth to Louis J. Griffith. During college vacations Louis worked for the Panama Canal Commission and the U.S. Army Supply and Services Division. He married Susan Ball, and their children are Abigail, Andrew, and Ashley Griffith. Louis is active in the church and is studying for a Master of Divinity. They live in Ft. Wayne, Indiana.

Olga and husband Roger Conley retired from the Panama Canal Commission and live in Albuquerque, New Mexico.

Francisco Johnston and wife Ernestina Cedeno gave birth to Francisco and Lorena Johnston. Francisco and wife Marlin had Catherine and Francis Johnston. Lorena and husband Efrain Batista's children are Stacy and Derek. They live in Chiriqui.

Emilsa Johnston married Guberto Grajales and had Guberto, Sandra, and Melissa Grajales. Guberto married Arianis Vannucci and had Guberto, Carlos Alfredo, and Manuel José; they live in Panama. Sandra Grajales Febles (previously married to Manuel Carvajal) gave birth to Aaron, Stephanie, and Alexander Carvajal. Sandra worked for U.S. Armed Forces Southern Command, Canal Zone, for several years. Sandra, husband Angel Febles, and their children live in Miami, Florida. Melissa, an architect and lawyer, lives in Panama.

Melva Johnston who died January 7, 2008, at age 62 years and G. Guerrero gave birth to James. James Johnston and wife Diana DeLeon are parents of James and Shirley. Enrique Johnston and wife Agustina Vigil had William and Linda Johnston; also Francisco and Enrique Johnston with wife Daisy Gomez. Linda and husband Julio De Gracia have one child, Juliany. They all live in Chiriqui.

Enid Johnston married Luis Prosperi (both deceased) and had Jorge Severino Prosperi and Luis Prosperi. Jorge and wife Margarita Ramirez (both deceased) had three sons. Luis Prosperi married Maria Elena Zarak and had Giovanna, Anna Lisa, Carlos Ivan Prosperi and seven grandchildren. Enid died in 1996 (90 years old). Gladys Johnston married James Purdy (both deceased) and had two sons. Gladys died in 1995 (85 years old).

Irene Johnston Fodell, the widow of Guillermo Fernandez of Costa Rica, remarried Folke Fodell in Panama and resided in the U.S. until her passing in 1992 at age 80. Her children were Elizabeth Fernandez, George and Charles Fodell. Elizabeth F. Grant (previously married to Steve Pedone) had Lisa, Michelle and Stephen Pedone. Lisa married David Tooke and gave birth to Kirsten, Ashley and David Tooke, and they live in Orlando, Florida. Michelle married James Cooper. Stephen married Kari Larsen and had Madeline Pedone, and they all live in New York. Elizabeth Grant worked for the Department of Defense, Canal Zone, for several years. She retired from the Voice of America, Washington, D.C., and now lives with her husband Dr. Michael Grant in Sanford, Florida.

George Fodell, senior officer to a security company, married Dianne Fraher and had sons Ryan and Connor. Charles Fodell, a telecommunications company engineer, married Cindi Hendren and their children are Matthew and Robert; they live in Raleigh, North Carolina.

We are proud to be Mr. Johnston's descendants; he was a beloved husband and father; these are cherished memories from his grandchildren — "for our grandchildren."

See also the William H. Conley family history.

THE KANE FAMILY

AS TOLD BY JOSEPH AND LUCILLE KANE

Leprechaun and Elephants

The newly wed Kanes left New York State for the Canal Zone in 1963. Joe was hired to start a new library for the Canal Zone College that recently moved from Balboa High School to the La Boca campus. He walked in the first day of school to an empty room 25 feet wide and 90 feet long. The only books were a set of encyclopedias donated to the college by Maurice Thatcher. At the first faculty meeting, Joe was told the books and library shelving were on order. The shelving arrived but two years later as we were headed home for free home leave, no books had arrived. That first home trip via the *SS Cristobal* was most exciting.

1965 saw the Middle States Accrediting Association place the Canal Zone College on probation. That action resulted in our obtaining a new library and Annex building. Do you know how exciting it is for a young librarian to have the opportunity to work with the Panama Canal engineers in the designing of a new library? It was a highlight of Joe's 25-year career as a librarian. The Canal Zone College library would continue to grow thanks to the college faculty, student assistants, and community.

Another major highlight in Joe and Lucille's life was the birth of our three children at Gorgas Hospital in Ancon. Sean, Nona and Devin would grow up in an environment that was free from violence and surrounded with multi-lingual families. It was a value-oriented society for them. It started with St. Mary's School, Curundu Junior High School, Balboa High School, and then the Canal Zone College. By the time they graduated, they could say they had visited the interior of Panama many times, been in two oceans in one day, and traveled to 49 states over the years when we were on home leave in the United States. They still enjoy traveling around the world and talking about their experiences in the Canal Zone.

The Kanes were blessed to have an ecumenical marriage with Lucille attending the Balboa Union Church and Joe and the children at Scared Heart and St. Mary's Church. Our relationships with the pastors, Vincentian Priests, Sisters of Mercy and the Maryknoll Sisters were wonderful. Who can forget the Balboa Union Church calendars or the beautiful Christmas Eve service where lighted candles were placed on the front lawn of the church every year? Lucille began her life- long Christian education ministry teaching Sunday School and serving on many committees. She was the Church Administrator when we retired. Joe was active in his church by singing in the Sacred Heart choir and lecturing. The boys were altar servers.

> *Panama was small enough that many of us were able to meet the President of Panama and ambassadors from other countries and attend functions with the Governor of the Canal Zone and others who were traveling through the country.*

In 1977 Lucille was asked by the Sisters of Mercy to become the principal of St. Mary's School. She served as Principal from 1978-1988. Under her leadership and the leading of the Holy Spirit she added a kindergarten, pre-kindergarten, and high school. The Vincentians turned the school over to Archbishop McGrath, and it continues to educate students at a new site in Albrook.

Entertainment played a major role in our lives.

Panama was small enough that many of us were able to meet the President of Panama and ambassadors from other countries and attend functions with the Governor of the Canal Zone and others who were traveling through the country. Joseph Patrick Kane was born on the 17th of March. His birthday celebrations introduced the Panamanians to Irish music. We would have a parade down our street with H. Loring White as drum major, Howie Atwell on the Scottish bagpipes and DeWitt Meyers on drums. These celebrations were held at the Tivoli Hotel, the Kanes' home, and the Elks Club. Of course, Joe dressed up with a green top hat and green suit. He was the tallest leprechaun in the country. He would tell Irish tales with an Irish accent to the children at St. Mary's school, and his picture made the *Panama Canal Spillway* several times. Lucille started a tradition on St. Lucia's Day, December 13 — she made Swedish cookies for the college faculty every year.

Joe Kane. Panama was our "Pot of Gold."

Our home was filled with elephants that were given to Lucille as a young child, and her collection increased to over a thousand. Students and faculty would bring her elephants from all over the world. The Guaymi Indians of Panama designed a dress for her with elephants on it. She found several molas made by the Kuna Indians with an elephant motif. The Indians would wonder why someone would collect elephants when they stayed in her home.

From our first night on the Isthmus of Panama at the Tivoli Hotel to the 12-family building in Williamson Place, the four-family dwelling on Tavernilla Street, the single home on Amador Road, and finally the home on Plank Street, we were ready to invite friends and strangers into our home and share a meal with them. In the end, it was the people of Panama that touched our hearts, and we were proud to have served the Panama Canal Company in the education field.

PHOTO COURTESY OF PANAMA CANAL AUTHORITY

Balboa Union Church.

The Andy Kapinos Family

As told by Carol Kapinos Smith

My first memory was living in Cocoli at the end of a street right next to the jungle: One night after returning home from a movie, a snake was coiled up on a step to our house which was built on concrete stilts. I also remember being under the house when my mother, Verna Fasiska Kapinos, who was washing clothes, took a tablecloth, wadded it up, and knocked a big black scorpion off my shirt that I was wearing. Moments before, I looked down and saw what she was going to knock off.

We moved to the Atlantic side before I started kindergarten. I remember scooping up tadpoles in a ditch and bringing them home in a large jar to watch them hatch into frogs. Some sprouted legs and their tails partially disappeared, but they never reached looking like baby frogs, probably from the lack of food. I also remember at night, along some stretch of road near Margarita, the thousands of crabs crossing the street with their pinchers held up high in the car's headlights. My father, Andy Kapinos, told me the smell of the dead crabs was, well, you can imagine. . . . Then we moved to Balboa and lived there for most of my elementary school days.

We lived on Akee Street in a four-family house that was also built on concrete stilts. You could park your car under the living quarters which had a maid's room that we used as storage. There were two concrete sinks on one side of the stairwell and two on the other along with a bathroom in between. Some days I would fill up the sink and loved to soak in it.

My mom taught me to cook even though she didn't know how when she was first married. She waited each night for my father to return home from work, and they would shop in the commissary for dinner. My father was stationed in the Canal Zone and was cook in the Army during World War II; when the war ended, he joined the police force.

My dad sewed a lot of my clothes. He bought my mother a Kenmore sewing machine, and she wanted no part of it, so my dad taught himself how to sew. I had dresses with circular skirts all through school. He used the same pattern. And then he started making my skirts. For my birthday I could pick out three yards of fabric and tell my dad if I wanted a pleated or gathered skirt, and that's what he made for my birthday. When I took sewing in eighth grade, I started making my own clothes and asked my father for sewing advice when needed.

> *My dad sewed a lot of my clothes. He bought my mother a Kenmore sewing machine, and she wanted no part of it, so my dad taught himself how to sew. I had dresses with circular skirts all through school. He used the same pattern.*

On Sundays after returning home from catechism, we would put our boat in the water at Gamboa. The government leased islands for $1 a year if you erected a bohio (four poles and a roof) and kept the grass cut. We had friends who had an island, and we would meet up with them and spend all day there. My mom would pack the usual lunch, ham sandwiches, potato sticks, canned beans, and sodas in glass bottles.

Once my mom forgot to pack our sandwiches, but we had enough extras to eat. Another boating time my father couldn't get the motor started when it was time to leave. He and my mother rowed all the way back to Gamboa. That was on a Sunday when we didn't see anyone else in a boat. The current in the Canal was going in the right direction so at least they weren't rowing against the current. Then it started to rain. We finally made it back to Gamboa about 7 p.m. We found a phone so my friend could tell her parents why we were late. I asked my mom the next day if her arms were hurting from all that rowing and, yes, they were. My dad also taught me to water ski since we owned a boat. Once when I was trying to slalom, he was so patient that afternoon waiting for me to drop one ski. We were going slowly so I wouldn't be afraid. It took all afternoon until I finally dropped the ski.

For school vacations, we would rent a house at Gorgona. I loved going to the beach. Once, after showering about 4 p.m. and having dressed for dinner, I looked down to where we all swam a half hour earlier to see a huge shark that got caught up in a wave and landed on the beach! It caught the next wave and swam out to sea. That really scared me since we were at that very same swimming spot 30 minutes earlier!

My father taught me to drive when I was 15, driving mostly on the road in the rain forest and also around the lake near Miraflores Locks. Once I stalled my father's 1953 Ford so much that my father couldn't get it started again. I don't remember how we got home because we were about a half hour from home.

The next summer in Virginia I got my driver's license when visiting my aunt and uncle because I was 16. I practiced for my driving test outside Washington, D.C. My dad and I headed to Washington, D.C., on Highway I-95. I remember my dad yelling for me to move over three lanes of traffic and turn around because we were nearly in the city and were quickly getting lost. When we returned to the Zone at the end of the summer, I wasn't allowed to drive until I was 17 years old.

I went to Canal Zone College where I met my husband, Robert Smith. We married in 1966 and have three children.

PHOTO COURTESY OF NELLREE BERGER

Madden Dam Road through the Forest Preserve on the Pacific side.

The William Henry (Harry) Keenan Family

As told by Charles Keenan

Harry was born in Willoughby, Northwest Territory, Canada, now Saskatchewan Province. His Irish father, John Arthur Keenan, married a Scottish lass, Agnes Stuart Cameron, whose parents had homesteaded in the area. He was active in sports and captain of the Prince Albert football association. He had worked several years for the Canadian railroad in Winnipeg when he received a letter from his uncle Dan Cameron, then living in Panama, that jobs were plentiful with the Isthmian Canal Commission. Eager to escape the harsh winters Harry sailed from Los Angeles to Balboa in January 1909, lived at the YMCA until hired as a boilermaker at 65 cents an hour, then moved to bachelor quarters in Gorgona. Harry soon learned that States hires got more benefits than locals and resigned in November 1910, expecting to rehire from San Francisco. On the trip north, he met his future wife, Harriet Underhill, crossing the isthmus by rail. She was traveling from Boston with her mother to visit Harriet's sister in San Francisco.

Harry got a job at the Risdon Iron Works in Oakland. On December 31, 1910, he applied for reemployment with the ICC. On May 4, 1911, he was rehired and arrived in Panama on the *Acapulco* on June 2nd. The next day he started to work at Gorgona as a boilermaker first class.

Apparently there were some love letters back and forth. Harry took leave in April 1912 to marry Harriet, now back in Boston. They had to elope because Harriet's mother did not think a boilermaker was a suitable husband for a proper Bostonian. They returned to Panama on separate trips so that Harriet could take time to appease her mother. The new bride got quite a shock at conditions in Gorgona. There were no streets or sidewalks, only mud. The first year was a nightmare. There was very little furniture, but Harry was handy with a saw and made crude chairs out of barrels and scrap wood.

In 1914 Harry was transferred to the Balboa shops and moved to quarters 118-B in Balboa. In 1917, with two small children, Howard and William, they moved to larger quarters, 1452 Balboa. Harriet's father shipped an old piano that belonged to her grandmother. She practiced several hours a day and soon was playing for the silent movies until the "talkies" arrived in the late 20's. She also played the organ for the different churches they attended. Harry was a Mason, and both were active in Eastern Star. The piano set the stage for a musical family. Howard became an accomplished pianist, William mastered clarinet and saxophone, and both played in orchestras. Marvin entertained family and friends with a fine baritone, and the girls, Virginia and Harriet, excelled in piano and voice. Athletics were encouraged for all the family. Harry played semipro baseball with the Panama clubs, Harriet was an ardent swimmer, the boys were active in Cristobal High School sports, and the girls became excellent golfers.

> *The new bride got quite a shock at conditions in Gorgona. There were no streets or sidewalks, only mud. The first year was a nightmare. There was very little furniture, but Harry was handy with a saw and made crude chairs out of barrels and scrap wood.*

Under pressure from Harriet's family, Harry resigned in 1918 and moved to Boston to operate a trucking business. After a severe winter, a flu pandemic that almost killed baby William and mechanical problems with the trucks, Harry decided the Chagres waters were calling and rehired as a boilermaker with the Panama Canal Company in the Balboa shops. In 1921 son Marvin arrived. In 1923 Harry transferred to Gatun Locks with quarters across from Gatun Elementary School. Harry's first daughter, Virginia, arrived in 1924. The older brothers suggested a middle name, Rose, for her bright red hair. In 1927 they moved to quarters 101, a duplex above the Gatun post office, the postmaster living in the other apartment. They had a spectacular view of Gatun Lake and the locks and looked down at the railroad station and commissary. In September 1929 daughter Harriet, the last of the children, arrived at Old Colon Hospital.

The kids spent a lot of time swimming in Gatun Lake. The attraction was a steel barge, 50 feet off shore, with a diving board on one end and a 20 foot tower on the other. To call them home for supper, Harriet would step outside on the porch and bang on an oriental brass gong. Another favorite was Gatun Locks, where they could walk about freely. Across the locks was the Gatun Golf Club, 18 holes built on the area dammed for the flooding of Gatun Lake. One of the three-par holes was across the lower end of the 110-yard wide spillway built for the release of water from Gatun Lake. There was always about one foot of water coming down the spillway from the power plant that furnished electricity for Gatun Locks and the Atlantic side. On weekends Marvin would put on spike shoes, chase golf balls from the duffers and sell them back for a dime.

Mah Jong was a favorite in the 20's and 30's, and they all learned to play. Monopoly was another favorite. A bunch of children would gather in the Gatun clubhouse and play until someone had to leave for meals; they left the game just as it was and came back at an appointed time. Some of these games went on for weeks.

In 1936 they moved to a new duplex, 123 Lighthouse Road, and in 1937 to a new cottage on a hill above the old post office, again with a nice view of Gatun Lake and the locks. By this time the two older boys had graduated from college and remained in the States.

Harry faced mandatory retirement at age 62 in 1945, and moved to Fort Davis to manage the golf club. In 1947 they moved to Santa Clara, where a number of ex-Zonians had located. He was fairly active until he died in his sleep eight years later. His ashes were placed in the columbarium in Saint Luke's Cathedral in Ancon. Harriet died in St. Petersburg, Florida, in 1968; her ashes are also in the Saint Luke's columbarium.

William Henry (Harry) Keenan
Roosevelt Medal No. 6681

PHOTO COURTESY OF PANAMA CANAL MUSEUM

New Quarters, Balboa, looking West, July 1917.

The Captain Dean and Joyce Kelly Family

BY ANNETTE KELLY MARSH

Our journey to the Panama Canal Zone started before my father and mother even met. My father's childhood friend, Sam Moses, had been stationed at Rodman Naval Base and decided to remain there after discharge. About that time, Dad's first wife was tragically killed in a car accident. Dad didn't have any insurance and was distraught at losing his young wife, mother of his six-month old daughter, Sharon. Sam convinced my father he should come to Panama and live with his wife and him so he could work as a civilian for the Navy. Dad stayed not quite two years working as a machinist. It was a God sent miracle for my dad.

He returned to Hornell, New York (rural western part of the state, south of Buffalo), where he met and married my mom. They had three children very close together: Annette, Raymond and Jonathan. Dad was working at Turbodyne as a machinist. They were on vacation near Detroit when he received a call offering him a temporary job with the Panama Canal Company. They came home on a Saturday; he picked up his tools on Sunday at the shop and flew to Panama on Monday. Mom and Dad had $50.00 which they split before he left for Panama. He lived in the bachelors' quarters in Balboa and worked in the shop. He became involved with the Sea Scouts as a leader.

Dad worked about nine months when he was given a permanent position. Now it was time for the family to make the move to Panama. My mom's best friend, Phyl, says she still can't believe what my mother did. Mom called Phyl and said, "Fred [our uncle] is going to take me and the kids to New York City, and we are getting on a freighter for Panama." So he packed us up, and off we went. I was 4, Raymond was 2½, and Jonathan was 1½. We spent 7 days on board the ship, sailing out past the Statue of Liberty. We did make one stop in Haiti. My father said it was the saddest sight you had ever seen, watching Mom coming down the gang plank. He said I came flying down, and my brothers, who were each in harnesses, came down dragging Mom. We collected the luggage, went through U.S. Customs and headed for Gamboa. That was the next shock to Mom's system; the four-family on the Gamboa Ridge with no windows, just screens. Dad used to say if Mom could have found a canoe, she would have headed back to New York.

Captain Dean L. Kelly, tugboat *Rodman*.

But we settled into family life. Mom came to love being in Panama. Dad became interested in pursuing life as a tug boat captain as he watched them from inside the hot shop. He would work all day in the shop and then would go to the tugs for training after his shift. So Dad wasn't around very often. He finally received his license and had his first job on the tugboat *Shammy*. His duty on the small tugs was as a dredge tender, which included moving pipeline and scows. I started kindergarten while we lived on the Ridge, and then we moved to the duplex, sharing with the Glass family. We used to slide down the hill behind the house on cardboard after it rained. Our baby sister, Theresa, was born while we were living there. Now we were eligible for an up and down style house. We moved to Williamson Avenue halfway down the street on the left. Eventually, the Duffus family moved in on the other side of the duplex. They became our neighbors from that time until my parents retired. We moved to a cottage up the street, and the Duffus family moved in next door.

Life was one big adventure. We were outside all the time exploring our world. My brothers were boy scouts. We loved being in Gamboa and being part of the Gamboa family. A lot of families started out in Gamboa; but as soon as housing was available in town, they moved. I don't remember ever wanting to move "to town." We spent a lot of time at the gym and pool. We had the movie theater, club house, tennis courts, ball field, the Chagres River, and the jungle — what more could you ask for? We attended the Gamboa Union Church. My brothers delivered the *Miami Herald* to neighbors' homes for years. Mom loved to sew; she made a lot of our clothes. She was involved with the church and taking care of her family. She worked harder than anyone I knew. Dad belonged to the Elks Club and the Gamboa Golf Club. The organization that played the biggest part in Dad's life was the Masonic Lodge. He was a 33½ degree Mason. He joined the Abou Saad Temple as a Shriner, enjoying his time with the Kitchen Crew. In 1973, he became a Jester. Not only was my father involved, my mother was a past Worthy Matron of Fern Leaf Chapter #4 of the Eastern Star in Ancon. My sister and I were past Worthy Advisors of the Ancon Chapter of the Rainbow Girls.

The four of us attended Gamboa Elementary School. I was in the first 7th grade class to attend Curundu Junior High School and the rest of my siblings followed. We graduated from Balboa High School in 1971, 1973, 1974, and 1978, respectively. Jonathan attended one year of Canal Zone Junior College before finishing his education in the U.S. Raymond, Theresa and I left right after high school graduation for college in the U.S. Nothing compared with the total education we received living in the Canal Zone, the diversity of people, the cultural differences, and the travel.

Kelly family, L-R: Jonathan (6), Raymond (7), Annette (9), Joyce holding Theresa (2), and Dean, 1962.

My father worked most of his career at the Dredging Division. He pushed scows working with the dredge and handled ships through the Galliard Cut. He continued to work his way up on the tugs until he was a senior master on the bigger tugs. His first senior master position was on the tug *Rodman*. He had to work out of Balboa until he could transfer back to Dredging Division in Gamboa. He worked on a variety of tugs, such as the *Culebra* and the *San Pablo*. He retired in 1979 as the senior master of the *Culebra*. He took his camera to work with him every day, and we are blessed to have many pictures of the things he would encounter during his day.

Mom and Dad returned to Hornell, New York, when they retired in 1979. Sadly, Mom passed away in 1982 after a brave fight with pancreatic cancer. She never really got to enjoy her retirement. Dad remarried in 1985. He passed away from heart disease in 2001. One thing about my dad, he loved life.

The greatest gift of living in the Canal Zone was the people. Most of us did not have blood relatives living there, so we became family to each other. And even though it has been almost 30 years since my parents left the Canal Zone, we continue to be in contact with "our childhood families." What amazes people in my life now is that I still am in contact with grade school friends. Our homes are always open to each other. Jay and Llori Gibson make it possible for us to go home to Gamboa which is a gift that is priceless. They even live on the street Llori and I grew up on. They are part of the wonderful people who became my Gamboa family. My childhood and the effect on my life continue to amaze me. That my parents took a risk, packed us up, and took off to a foreign country in 1958 was a true leap of faith that has paid off so many times over.

THE ANNA JACOBS KING FAMILY

AS TOLD BY TARA KING

Anna Jacobs King was born to Native American/Scottish American parents in Holdenville, Oklahoma, on April 2, 1925. Her mother was full blood Creek Indian; her paternal great-great-grandfather came to Canada from France in the 1600's. His fur trader descendant married a Native American. She was raised on what they called the "Home Place" from age one through 13 near Little River, eight miles south of Holdenville. Anna didn't learn English until after she started school. After her mother's death when she was 12, Anna was sent to the Chilocco Indian Agricultural Boarding School, where life was extremely regimented.

Anna received a scholarship to attend Hillcrest Memorial Hospital Nursing School in Tulsa, Oklahoma, in 1943. She graduated from nursing school in 1946 and was a cadet student nurse assigned to the Veteran's Hospital in Albuquerque, New Mexico. After working at several other nursing jobs in Oklahoma, in 1949, she joined the United States Army Nurse Corp as a 1st Lieutenant serving as an operating room nurse. After basic training, she was assigned to Madigan General Hospital in Fort Lewis, Washington. She met her future husband (Captain James Francis King III) at the Madigan General Officer's Club in 1949, just before he shipped out to Korea.

Lt. Anna Jacobs King

In 1950, she received orders to join her unit on the 121st Evacuation Hospital troop train to Camp Stoneman, California, and they boarded the troop ship *Wiegel* — with destination "unknown." The nurses all suspected they were bound for Korea since news of the "conflict" had been received. It was a harrowing 14-day voyage over rough seas from San Francisco to Yokohama, Japan. There she joined the 121st Evacuation Hospital in temporary quarters at Camp McNeely. The nurses were issued fatigues, steel helmets, other items, and their class A uniforms were sent to storage.

Captain King (Anna's future husband) had recently returned from Korea to Japan for an orientation and happened to visit a friend in the hospital at Yokohama. In talking with the chief nurse, he discovered there were three units of nurses newly arrived from the States. The chief nurse called her counterpart at Camp McNeely 155th Station Hospital who confirmed that 2nd Lieutenant Anna King was among her crew. Furthermore, the chief nurse said that she was bringing the nurses down for inoculations in a few minutes. Captain King was waiting there with a big grin on his face when she walked in the room. Anna was so surprised to see him! Anna's unit left via the troop ship *Randell* in August 1950 for Korea. Upon arriving in Inchon, the tide receded and nurses sat for 10 days waiting before they and their gear could be unloaded. They set up in an old school building, with open toilets, dirt floors, and unsanitary conditions and subsisted on c-rations. They took care of wounded South Koreans; some were literally bubbling with gas gangrene. They had no equipment yet, but found some hand saws and Japanese instruments. People received Pentothal with their chins being held up manually. Limbs were amputated with handsaws. Most lived!

They were then moved overland up to Seoul, Korea, where they set up the hospital across the river at Yong Dong Po. Wounded U.S. troops were coming in before they even finished setting up the hospital, so they used an abandoned building as a hospital. The nurses slept in tents at temperatures below freezing.

After remaining there for some time, they received word there was going to be a push in North Korea with many casualties and were flown to the 3rd Field Hospital in Pusan, Korea, after stopping in Taegu and Taejon. They worked with the 8th Field Hospital to treat patients. Next, her unit flew to Tak Song Dong, Korea, and set up the hospital in the Won San area. They were next moved to Ham Hung, off the coast, from September until the evacuation of all U.S. troops from North Korea. It snowed in November, and although the sun was shining every day, the snow never melted! Cold, injured troops were coming back with frostbitten toes and fingers.

Her unit boarded the USS Ainsworth in the harbor and took the wounded aboard to treat them while anchored. They also treated wounded Chinese prisoners. The ship sailed on Christmas Eve, 1951, returning to Pusan to work at the field hospital there. Rumors of another push from North Korea circulated, so they were sent by air again to be near the battlefield. Troops again flooded the hospital before they could even unpack their gear. Care involved treating wounds and sending the soldiers on the 3rd Field Hospital to be stabilized and flown back to Japan. Anna then returned to Japan and worked at the 155th Station Hospital.

Captain King was serving with the 2nd Infantry Division and continued his courtship of Anna while they were overseas. He again located the nurses while they were serving in Korea. Finally, his Irish humor won her over and they went to the 8th Corp Headquarters where General Almon signed their request to be married. They were married in Ascom City, Korea, by a military chaplain in front of a packing crate in October 8, 1951. They were the first American couple married in Korea, and her family heard the news when the news article made headlines around the world. She then followed his career by moving to Washington, D.C., Saipan, China, Japan, and many other locations until after his retirement. Her son, James IV, was born at Walter Reed Memorial Hospital in Washington, D.C., in 1951. Her son, Jeff, was born in Yangmingshan, Taiwan, in 1953. Her daughter, Nancy, was born in Saipan, Northern Marianas Islands, in 1955. They stayed in Saipan for three years, living about 25 yards from the beach in a Quonset hut.

After leaving the Army to start a family, Anna began working again as a nurse in 1957. In 1958, Major King was assigned to Mineral Springs, Texas, where Anna found work in the Nazareth Catholic Hospital Labor Room and Nursery. After five months, they moved again to Ft. Shafter near Oahu, Hawaii. Next assignment was to 500th M.I. Group, Camp Drake, Japan. They lived in Grant Heights, later moving to Momote Village. In 1963, the family moved to Fort Hood near Killeen, Texas, where they lived in Copperas Cove.

> *They were the first American couple married in Korea, and her family heard the news when the news article made headlines around the world.*

Lt. Col. James King retired in 1964, and they settled again in Oklahoma. Anna was employed at a variety of nursing jobs within the state and had another daughter, Tara, in 1964. Sadly, her husband was killed while serving as a supervisor of probation officers in 1969. Seeking to move as far from Oklahoma as possible, in 1972 she chose as her next residence Guam, Northern Mariana Islands, where she worked for the Government of Guam at Guam Memorial Hospital.

While in Guam, her love of golf led her to participate in the South Pacific Games — their version of the Olympics. Anna was very proud to win the individual silver medal in golf.

After super Typhoon June hit Guam in 1975, the entire island went without running water for six weeks and without electricity for three months. At the time, June was the strongest typhoon on record, with maximum sustained winds of 185 mph. While still recovering from the typhoon, an earthquake measuring 6.2 on the Richter scale hit the island on November 1, 1975, producing damage on Guam of $1 million. In May 1976, super Typhoon Pamela crossed Guam with sustained winds of 140 mph, causing $500 million in damages. Again, water and electricity were out for months.

Tiring of environmental disasters, and hearing from

friends about the beautiful weather and great golfing in Panama, in 1976 she decided to apply for a job at Gorgas Hospital, at the time still being run by the Canal Zone Government. Relocating from Guam to Panama in January 1977, Anna lived in Gamboa at first, then moved to Diablo and finally to Balboa in 1979.

While in Panama, she made many lifelong friends and enjoyed each of the many golf courses to the fullest. Many Canal Zone people would remember my mother by her nickname, "Jake," and she is also remembered as one of the best woman golfers in Panama. She remained at Gorgas Hospital (later known as Gorgas Army Hospital), working in all the wards except pediatric until her retirement with 21½ years civil service in 1992. She then moved to Sherman, Texas, and lived there until 1997, when she moved back to her place of birth in Holdenville, Oklahoma. In 1999, she moved to Oklahoma City, which remains her present home. She stays active visiting her children all over the world and golfing every chance she gets!

PHOTO COURTESY OF PANAMA CANAL MUSEUM

Ancon Hospital (now Gorgas Hospital) Sections A and B, Ancon, Canal Zone.

THE HERMANUS KLEEFKENS FAMILY

AS TOLD BY HIS GRANDCHILDREN

Virginia Kleefkens Rankin - Grand daughter

Our grandfather, Hermanus (Herman) Kleefkens, was born in Rotterdam, Holland, in November 1887. He left home as a teenager and went to sea as young boys did back then. Herman became a seaman machinist who traveled the Atlantic Ocean with his ship and made several trips to the United States with his vessel, when he decided to stay in America. He jumped ship on one of his ports of call in New Jersey. He found himself a job as a machinist. His wife, Maria, and their first daughter, Johanna (Jo), remained in Holland until he was able to bring them to the States. He became a naturalized citizen in New Jersey in 1915, and by that time, his family had grown by three more children: Wilhelmina (Minnie), Maria (Marie), and Hermanus (Louie), all born in the Untied States.

In 1916, Herman found himself a job with the Panama Canal and moved his family to the Canal Zone. His first job was as a machinist with the Marine Division assigned to the tugboats.

Herman's wife Maria became ill when the children were still young. Her mother and sister, Leida, came to Panama to help out with taking care of the children. After Maria died, her mother returned home to Holland, and Leida stayed on to help raise the kids. Herman and Leida eventually married in 1924. The entire time the Kleefkens family resided in the Canal Zone, they lived on the Atlantic side. All four of the children attended Cristobal schools. The story goes that young Louie had a pet duck that would follow him to school each day and wait for him to follow him home at the end of each school day.

Three of the kids played musical instruments — Johanna on the piano, Minnie played the violin, and Louie played the trumpet. The family lived in the Fort DeLesseps area, near the Hotel Washington. As a teen, Johanna was hired to play piano for the silent screen movies that were playing in the local theaters. On Sundays, local bands held concerts outside the theater for the public. Johanna played the piano at the concerts and afterwards headed inside the theater to play for the movie that was showing that evening.

Herman Kleefkens rose to the position of Senior Chief Engineer and served on the tugboats *Tavernilla* and the *Alhajuela*. He was a licensed steam and diesel engineer and also held a license for all waters as assistant engineer for the Panama Canal 15-yard dipper dredges.

During World War II, the United States knew the German U-Boats didn't have enough fuel capacity to make it from the coast of France to the Caribbean to do any extensive patrolling. It was suspected they were meeting large fueling subs called milchecows and refueling in the coves up and down the Central and South American coasts. The U.S. military contacted both Herman, who spoke Dutch and some German, and his tugboat captain, who spoke German, to try and help locate such refuelings that were taking place. Herman answered the call. The military set up

Herman and his second wife, Leida, and his four children, L-R: Marie, Louie, Johanna, and Minnie.

Herman and his tugboat captain with a yacht loaded with electronic gear and had them sailing up and down the coast trying to find which coves the Germans were using. They never caught any and found out after the war that the U-Boats were refueling in the open seas. For their service, Herman and the tugboat captain were awarded the Atlantic War Zone Merchant Marine Ribbon Bar. Because his West Indian crew member on these excursions was not similarly recognized for his time and effort in this mission, simply because he was not a U.S. citizen, Herman refused to ever wear or display his ribbon, thinking that this was not right.

Herman was also involved in the rescue of a torpedoed vessel that had left the Canal and was hit by an explosion near a mine field located a few miles outside the Cristobal breakwater. At the time, it wasn't known if the vessel was torpedoed or hit a mine, or if a U-Boat was waiting to attack any rescue vessel. Herman, the captain, and the oiler of the tugboat, *Tavernilla*, volunteered to take the tug out to tow in the damaged ship. The crew had previously been rescued by the U.S. Navy. For taking on this dangerous mission, the crew of the *Tavernilla* was recognized by the Navy and by the Canal Zone Governor.

Herman had a small shipyard in the area known as Folks River in Colon. He ran a turtle shell business there, as well as other "wheelings and dealings." We, his grandkids, were never told just what all went on there, but we had heard stories about some of Herman's wartime spy escapades involving boats upfitted in that shipyard. Many of his post war dealings were related to his war time service coasting the Central and South American coasts.

Herman retired in November 1949 and moved to Tampa, Florida. His children stayed in the Canal Zone, married and raised their own families there.

> *Herman had a small shipyard in the area known as Folks River in Colon. He ran a turtle shell business there, as well as other 'wheelings and dealings.' We, his grandkids, were never told just what all went on there, but we had heard stories about some of Herman's wartime spy escapades involving boats upfitted in that shipyard.*

Louie married Virginia Lee Sanders, whose father, Bruce Gordon Sanders, was a Roosevelt Medal holder for his years during the construction days. Marie married Harold Fraser, who came from another Roosevelt medal Zonian family. Minnie married Danny Rudge, who also worked for the Canal for a while, and Johanna married Walter Freudigmann, who served in the U.S. Navy and was stationed in the Canal until he retired. The grandchildren of Herman Kleefkens all grew up in the Canal Zone and attended schools there. Some of these grandchildren married other long-time Zonian families and remained there for years until the U.S. jurisdiction of the Panama Canal ended.

Alexander M. Fraser
Roosevelt Medal No. 3367 with Two Bars

Bruce G. Sanders
Roosevelt Medal No. 4180 with Two Bars

See also the Bruce Gordon and Grace Aloise (Meister) Sanders; the William Gonzalez and Bernice A. (Sanders) Hill; the William Andrew Barnard and Alvin Monroe Rankin; the Alvin M. Rankin and Paul D. Thompson; the Gerald DeLeo Bliss, Sr.; and the Reinhard A. Boggs and Max Reinhard Boggs family histories.

The Krziza Family — Esther, Ethel, Leo

AS TOLD BY LEO KRZIZA

Looking for a Warm Place of Adventure

We three Krziza kids owe our development and maturity to the Panama Canal experience. We were orphaned when our mother was killed by a drunken driver, and our father went to the "Big City" for a better paying job. I was 15 years old and without a penny in my pocket. We kids were trained well — we were Christians — so we made it.

Esther, the oldest and a teacher, was the first to go to the Canal Zone in 1939 via a Form 57 — the Federal Employment Application. Seeing a good deal, she sent one to Ethel, a nurse; and, bingo, another Krziza was on her way to the Canal Zone. In mid-June, I was on a United Fruit banana boat going to Panama. So now we had three happy Krzizas on the Panama Canal payroll.

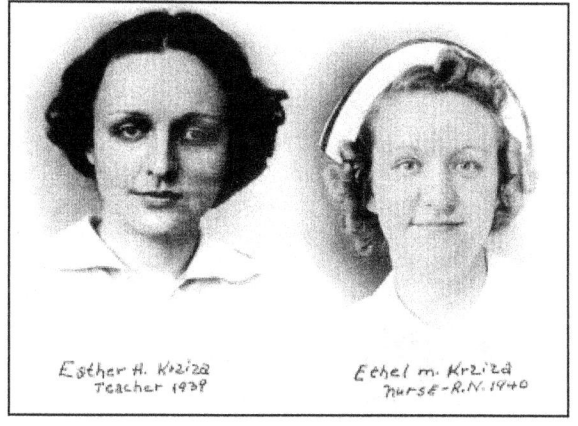

Esther and Ethel Krziza

We were all in Balboa on December 7, 1941. What next? We waited. Ethel, the nurse, could not leave her job at a U.S. Government hospital — the job code prohibited this. Esther, the teacher, single, was urged to stay with the school children. The Panama Canal needed skilled teachers. I was kept on by Buildings Division for various U.S. "gold roll" jobs. I had three plus years of college, so I had good training. There was a shortage of skilled workers, but everybody had something to do for the World War II effort at the Panama Canal.

Esther was moved to the U.S. Navy censor office, 15th Naval District, Fort Amador, to make sure Canal Zone activities were not being leaked to the enemy. After a few months it was observed that the sailors did not write home anyway so Uncle Sam moved her to the U.S. Army Air Force with the same pay and a few free flights. This move gave Esther dual service – Navy and Air Force (Army), two war department discharges — quite an honor for a World War II woman.

Ethel worked at Gorgas Hospital caring for the wounded — there were a few G.I.'s who got drunk and cracked up their motorcycles, but no wounded — no shots were fired. Her big event was being the bond queen. She and the staff at Gorgas sold more war bonds than any other group in the Canal Zone. It went into the lower millions. It was a great fundraiser — Army, Navy, Panama Canal, Air Force, and shipping agencies. Our Canal Zone salaries were not made for war bond purchases. Ethel located some local construction firms that had big bucks and she talked them into buying war bonds. She was married to Jack H. Hoarn, a Canal pilot.

We had little or no war activity in the Canal Zone. We were too remote and not worth the cost or effort. A B-17 aircraft hit Taboga Island during a rain storm and a small German submarine waiting outside Canal waters put one torpedo through a United Fruit banana boat out of Cristobal heading for the U.S. Down it went – bananas and the U.S. mail. The sub was from a German base at Belize, Honduras. It had been developed by Hitler some 20 years before. All the bananas and some U.S. Army mail and Army payrolls were gone. No bodies were lost.

I finally got into the Army — I was on my way to becoming a fighter pilot! I was stationed at Fort Amador; and after completing basic training, I was scheduled for some special training. What and why I did not know; you do not ask questions. I was an excellent shot with an M1 rifle and knew the jungles because of my hunting experiences. And I spoke Spanish. The training was rather extensive, but I kept my mouth shut. A few days later the drill sergeant asked me to take my clothing and M1 rifle up to the transportation office. I thought: Oh! Oh! Something's going to happen. It did. I was sent to the supply depot, repair shop and a finance office on the post of Corozal. The driver helped me carry my stuff to the 40-year-old French construction building, a new area for me. I was still wondering what was happening. It was an office – typewriters, adding machines, addressograph and safes. Well, the next morning I found out the good news. I was a member of the cadre assembled to retype and replace the 1,000 or so payrolls that were lost on the United Fruit ship that was on the bottom of the Atlantic Ocean near Cristobal. Some office girl had noticed a line in my 201 file — special skills. I had put down that I could type 50 words per minute. Boy, what a find! So, I ended my war career not using a parachute but driving a Royal typewriter. I found out postwar that I would be trained for the Bushmaster Brigade, paratroopers who were to go behind the Japanese lines in the Philippines. The Filipinos spoke Spanish, so with my Spanish background I would be useful. Thank God I learned to speak Spanish and type before I joined the Army and went to war!

My feeling on working with the Panama Canal was that it was a 32-year vacation with pay. God, it was fun! I would not quit until all the benefits were counted. I was 55 and had 32 years of service. I got an annual salary, my paycheck was above average, I had no debts, no credit cards — you name it, I had it. Where can you find this? Work a year and get two months paid vacation —

Leo Krziza and black marlin, Panama Canal office, December 1961.

Ruth Baumbach Krziza with 370-pound black marlin, 1964.

free trips to the United States.

Here are some highlights of my Canal tours of duty:

I earned two college degrees by going to night school. I learned Spanish because it was a handy thing to do and earned a Bachelor of Arts degree from Florida State. Before going to Europe, I studied French. I did very well and went to the University of Panama and received a B.A. in French. It was fun and easy to learn to speak French from Spanish. This gave me a graduate degree in French from a Spanish university. My brain would shift rapidly from Spanish to French and then into English if I happened to be in the Canal Zone. I was an American because I had a few bucks and I used American money. I was considered a foreigner, probably working as an FBI agent or "G" man for the Government. My blond hair threw them off. It would confuse people as to who I was. Spanish, French, or English – my speech was clear with no hesitation.

Working for the Panama Canal was not interesting. Same, same each day, 365 days a year. The sea level canal studies to expand the size and depth of the Panama Canal via the Darien area and other routes were considered by the Corps of Engineers and the U.S. Congress. Panama Canal loaned the Corps an office staff. The backbone consisted of three loyal old timers — Joan Ridge, Alba Hutchings, and me. We knew if anyone could do it, we could do it better. We all spoke Spanish. Joan was the secretary for Colonel Sutton, Alba was the administrative officer, and I was the banker. We were starters and Colonel Sutton's supporters and made sure things went smoothly. We had gangs of workers up and down the area for the surveys. My finance duties ran north to Jacksonville, Florida, and south to Bogota, Colombia. Yes, I covered that area — about 6 to 8,000 miles — but not on the same day. Most of the transactions were in U.S. currency, but some were not.

The petty cash fund was one of my big operations; it amounted to $35,000. My Panama Canal supervisor told me that it should "not be petty."

The Balboa Yacht Club and fishing boats helped me dispose of my money. Black marlin was my game. In our 15 years of fishing my wife Ruth and I caught over 100 marlin and sailfish. We always had lots of vacation time. Average size for black marlin was 300 pounds; a nice fish was 400 pounds; a good sized one 500 pounds and a real fish would be 600 pounds or more. Ruth's largest was 510 pounds; mine was 590 pounds.

Personal Observation of the Panama Canal and Its People

As told by Irene McCracken, Leo's friend and caregiver

Having visited Panama several times, I want to share my views of the Canal Zone and the many people who worked there. For all the years that I have known Leo Krziza, I never tire of hearing about important events and the fine accomplishments of many in the Canal Zone.

What a wonderful place it had to have been during Canal Zone days — my experiences in meeting many people who worked for the Panama Canal are like no other — so many gracious people whom I met on cruises. A trip to visit the Panama Canal is well worth the journey. Everyone said that building this grand waterway couldn't be done, even the French. But a century later the Panama Canal still proves the power of hard work in successfully completing a tremendous undertaking.

Irene McCracken and Leo Krziza.

The San Blas Islands — a place hard to explain — you have to see it to believe it. There isn't a place like it that I have ever seen, anywhere. The islands were discovered by Christopher Columbus on Saint Blaise's birthday in around 1525. Saint Blaise was the Roman Catholic patron saint who blessed their throats, a tradition for over a thousand years.

The native Kuna Indians of the San Blas Islands are sensitive about having their photograph taken — not wanting their "traditions stolen," and so I was very cautious. The Kuna women look to their own lives and the world around them for mola design ideas. They can translate just about anything they see in the world into a mola. The word mola can mean cloth, clothing or blouse. These brightly colored reverse appliqué panels, hand made by stitching layers of cloth together, form interesting patterns. The authentic molas are made only by the Kuna Indians of the San Blas Islands of Panama, and their sale is one of the most important sources of income for the Kuna households in San Blas.

As a result of all the experiences I had and knowledge I gained as a visitor to the Panama Canal and in meeting many special people who spent years there, I have many memories as if I lived and worked there myself.

The Walter R. Lindsay Family

As told by Judith R. Lindsay

My parents were married on Thanksgiving Day in 1930 in Pullman, Washington, where both were students at Washington State University. Ten days later, my father left for the Canal Zone where he began work for the Panama Canal on January 1, 1931. He had completed his college work; but my mother had not, and it was ten months before she joined him. My mother had never been more than 11 miles from home, but she traveled alone to the Canal Zone. When her ship reached port, she looked down on the pier hoping for a glimpse of my father, but she could not find him. The sun had bleached his hair from brown to blond.

My father was lucky to find a job in those days. He had been born as a first generation American to parents who had emigrated from Scotland. His Uncle Jimmy opened the first nursery on Maui, and it was he who recommended my father to Mr. Higgins, who was the director of Summit Gardens. Mr. Higgins needed someone who knew tropical plants.

My parents' first home was one built up on stilts during French Canal construction days. My father and Roy Sharp enclosed the basement area. Our Summit home had come on rollers and had been placed half up a hill near the gardens. Dad built cabinets from old wooden boxes and Mom worked her magic touch, turning the place into a tropical haven with tall potted plants and wicker furniture.

Because our house was up on stilts, we could look out over the grounds around us. The hibiscus and poinsettia grew up high by the windows, and we could look into bird's nests. Once I watched in horror as a boa constrictor raided a nest. I ran screaming for my father who made short work of the snake. I wanted to save the eggs (too late), and my father did not want a snake so close to the house.

On one side of the house filling an entire hill side was a grove of palm trees. Botanists from all over the world came to study them. On the other side of the house was a bohio for entertaining. Beside it was a large, noisy mango tree. The mangos were small and tasted like turpentine so only the flocks of parakeets ate them. Nearby was a tall tree with pungent leaves. Dad loved to crush the leaves and challenge people to identify the smell. It was familiar, but most could not name the smell.... It was bay rum used in shaving lotion. Behind the house was my father's orchid house, and behind that was a large grapefruit tree that he turned into a citrus tree by grafting on oranges, lemons and limes.

Walter R. Lindsay

At the top of the hill behind our home was a long, low building. In the front part my father stored chicken feed for the chickens he raised during World War II. In the back was an Army camouflage unit my mother and I were not supposed to know about. My Dad told us there were snakes around the place, so Mom and I stayed away. We did not learn about the camouflage until after the war.

My father was too young for World War I, too old for World War II; but he worked with the Army camouflage unit, and he worked with military pilots by giving them jungle survival training in case they were ever shot down over a jungle area. In later years, he often wondered if he had helped to save any pilots' lives.

After Mr. Higgins died, my father became the director of the Summit Gardens. The Gardens were established in 1923 with the purpose of improving and introducing new plants in Panama. I remember Dr. Julius Matz, who worked with sugar cane. Dr. Floyd McClure came to study bamboo. I know work was also done with corn and dry land rice. My father often said that the best customers for new seeds were the Chinese farmers with the roadside vegetable stands.

My father loved to put on his hip boots and pith helmet, strap on his machete and head out into the jungle. He would return with bags full of new plants for the garden. He also loved to go camping with friends. Lynn Cook was a favorite because Lynn knew the jungle, and he could spin a good yarn. Somehow their cayucos always overturned, they would lose all their supplies, and they had to live off the land. I often wondered if the cayuco "accident" was preplanned. My dad would return unbathed, unshaven, grinning from ear to ear, and bringing exciting tidbits of food like smoked, jerked monkey.

After World War II when our country was recovering from the war, Summit Gardens became more of a show place, and a small zoo was established. My father became the Canal's agronomist. Agronomist? He said it meant he was involved with anything to do with the land in the Canal area.

Walter with a beloved orchid.

Those of us who grew up in the Canal Zone know that friends became family and remained so for all our life. During World War II, we did not leave the Canal Zone for seven years. Dad and I took vacations in El Volcan; mother was chief clerk for U.S. Army engineers, and she received no vacations. Enduring friends were good Frog Alley friends like the Corns, Leisys, Kleasners, Stockhams, Barlows, and Barnards.

The last years of my parents' lives were spent in the Pacific Northwest of the United States. Both died while living at Panorama, a beautiful retirement community where I now live. Dad spent the last year of his long life (1906-1997) in a nursing home. I bought him many orchid plants and sometimes we would go outside and repot the orchids. On the last day of his life, my father became mildly agitated and kept telling the staff that he had business with the Army. I wonder if he was thinking about the pilots. He was a tropical botanist to the end.

PHOTO COURTESY OF NELLREE BERGER

Summit Gardens, established in 1923 with the purpose of improving and introducing new plants in Panama — a real tropical paradise of plants, trees and flowers. Near the Canal at Gaillard Cut.

The Ludwig Theodore Maenner Family

AS TOLD BY EDITH LOUISE HUFF WILLOUGHBY

Railroad Expertise Contributes to Building the Canal

In preparation for his departure from New York, 43 year old Ludwig T. Maenner walked into the office of the Panama Railroad Company, No. 24 State Street, New York City, with the required oath of office notarized, a satisfactory medical certificate and a completed personal questionnaire. With all requirements filled, he would be sailing for the Isthmus of Panama on July 3, 1906, on the Panama Railroad steamer *Allianca* from Pier 57. The Isthmian Canal Commission in the Canal Zone was looking for American citizens, physically sound and preferably under 45, for engineers and draftsmen for duty with the United States Government.

Ludwig Maenner had come to America through Ellis Island in 1883 after graduating from Technical High School in Munich, Bavaria. He settled in St. Paul, Minnesota, where he met a German girl, married, and had three children. He was a German with American citizenship by the time of his employment and wrote, understood, and spoke English and German.

As with newly hired employees at the time, he was required to go to the isthmus alone. Mr. Maenner was a family man with wife, Frieda Frech Maenner, 40; daughters, Selma, 17, and Mabel, 12; and son, Theodore Henry, 15. The family, from Chicago, Illinois, followed him at a later date.

L-R: Mabel Maenner (Schwarz) in rocking chair, Ludwig, wife Freida, Theodore Henry, Selma Maenner (Huff) in doorway, circa 1907.

The history of his stateside railroad employment as a draftsman explains his reason for going to the isthmus. From July 1890 until March 1903, he was employed by the Great Northern Railroad Company at St. Paul, Minnesota, as a draftsman and, later, chief draftsman in charge of map work, yard design, estimates for authorities for expenditures and all other drawing-room work, under the jurisdiction of the chief engineer, John F. Stevens.

Mr. Maenner's and Mr. Stevens's paths would split when Ludwig moved to Chicago, from March 1903 to June 1906, to work as chief draftsman in the office of the chief engineer of the Chicago, Rock Island and Pacific Railroad Company. Meanwhile, Mr. Stevens remained in Minnesota until he was named chief engineer of the Panama Canal in 1905 and arrived on the Isthmus of Panama to find the place in much need of repair and sanitation improvements before Canal construction could even begin.

Stevens needed to fill the position of draftsman with someone qualified and meeting the eligibility of the classified civil service position of draftsman. Remembering

Mr. Maenner's education and his 16 years of experience as a draftsman in the United States, Mr. Stevens did not hesitate to personally invite Ludwig to work for him on the Panama Canal project. A letter from Stevens, on the letterhead of the chief engineer of the Commission, soon arrived at Ludwig's place of employment in Chicago, Illinois. In the letter Mr. Stevens requested Mr. Maenner to join him on the Isthmus to get the job done. Ludwig was provisionally employed, without examination, as a chief draftsman in the engineering department of the Isthmian Canal Commission at the rate of $225 per month. He was employed "for the performance of such duties as may be determined by the Chief Engineer of the Commission," John F. Stevens.

In June 1906, Mr. Maenner accepted the position of chief draftsman with the ICC under the chief engineer with headquarters at Culebra, Canal Zone. Upon accepting the position he soon found himself in New York City awaiting his departure for Panama. The trip, which was free, would take seven days. His wages would begin at the time of his departure, and bachelor quarters would be available for about $8.50 a month upon arrival. John F. Stevens's letter was to be presented, upon his arrival in the harbor of Colon, to the authorized representative of the ICC who would board the steamer *Allianca* at the port and advise him as to his duty assignment. Ludwig sailed at 3 p.m. from Pier 57 at the foot of West 27th Street, New York City, on Sunday, July 3, 1906, and arrived on the isthmus on July 10, 1906.

Most employees were overcome by the amount of preparation, construction, railroad and sanitary work required to build the Panama Canal. But Stevens and his men began preparing the environment as best they could for the employees. Buildings were cleaned up, and appropriate railroad tracks were laid to replace French rails which did not coincide with the American machinery. Much work was necessary to go along with Stevens's plans for moving the earth to construct a lock canal.

As chief draftsman, Ludwig Maenner installed a system for filing maps and records and prepared forms for cost-keeping and excavation statements. He also prepared statistical statements, in graphic form, of various expenditures and supervised the preparation of topographical maps of the Canal Zone and land maps in connection with lawsuits and boundary disputes.

Soon after his arrival, Ludwig applied for married quarters for his wife and three children. Next, he received permission for his family to sail for the isthmus on the steamer Colon, leaving New York on September 1, 1906. Upon their arrival, the family was quartered temporarily in house No. 235 in Culebra, which had been vacated by Mr. Sullivan.

Mr. Maenner's gratuitous and tactful letter to the supervisor, Branch of Labor and Quarters, Culebra, following the very trying disembarkation of his wife and children from the steamer upon its arrival in Colon, could be found in his personnel file. He wished to express his "very sincere thanks of myself and my family, for your good work in getting us fitted up at Culebra. We had been through so many disheartening experiences in Colon, in getting our baggage, etc. put through, that it was a pleasure to find that you had done such good work for us here." Upon receipt of Mr. Maenner's letter, Mr. Stewart sent an interesting personal letter with a copy of Mr. Maenner's letter to the manager of Labor and Quarters. "Attached find a communication which I think deserves a place in the files in your office for curios. I could not resist the temptation of sending this to you, as it has appealed to me, and I know it will appeal to you the same way, like a spring of pure water in a desert."

A few months after the arrival of his family in November 1906 and settling into their pre-furnished housing, President Theodore Roosevelt visited Panama to in-

Daughter Selma (left, wiping her cheek), awaiting arrival of President Roosevelt's train, Culebra Railroad Station.

spect the Canal's progress. This was the first trip outside the United States by a sitting president and everyone was hoping to get a glance of him, his wife and his son. Even Ludwig's older daughter, Selma Maenner endured the heat of the tropics to await the arrival of President Roosevelt's train at the Culebra station.

Ludwig Maenner rendered his resignation on August 2, 1909, in order to accept a position in the States. His service would be ending with the ICC effective August 17, and he would embark aboard the *SS Ancon* to the United States with his wife and three children on that same day.

He was employed as chief draftsman for the Missouri Pacific Railroad Company in St. Louis when his Roosevelt Medal arrived in early September 1909. For his two consecutive years of ICC service from July 10, 1906, to July 10, 1908, Mr. Ludwig T. Maenner was awarded Roosevelt Medal No. 1602.

Mr. Maenner continued working with the Missouri Pacific Railroad Company and advanced with an appointment in 1912 to chief draftsman in the office of the chief engineer. In 1917 he was promoted to the position of chief office engineer, in which capacity he served conscientiously and faithfully until September 30, 1933, when at the age of 70, he retired. He died on April 6, 1937.

In the meantime, the second generation to follow in Ludwig's footsteps at the Panama Canal was already in progress with the marriage of his older daughter Selma to Mercer Huff, who had met her and courted her for several years on the Isthmus. After a marriage in the United States, they would return, raise a family and eventually retire from service.

Ludwig Theodore Maenner
Roosevelt Medal No. 1602

See also the Mercer Blanchard Huff; the Donald Thompson Baker; and the Maenner Blanchard Huff (two) family histories.

The town of Culebra.

PHOTO COURTESY OF PANAMA CANAL MUSEUM

Tony and Anna Mann

AS TOLD BY TONY MANN

Tony and Anna Mann had no ancestors among those heroic Americans who came down to the tropical jungles and marshes of Panama. Those first Zonites took on many dangerous jobs and faced serious risks of disease. We sometimes remembered this as we lived our easygoing happy lives in the Canal Zone.

We arrived in the Canal Zone in March 1952, fresh from college with my degree in civil engineering. I was hired, along with a group of several others, to plan and design new housing communities for Canal employees. I had worked for two years on the grading and earthwork for that project. Then in March 1954 I was suddenly sent out to Contractors Hill on the west bank of Gaillard Cut.

Tony Mann

The hill, about 400 feet high and forming a precipice on the very edge of the Canal, was cracking across the top. Blockage of the Canal was certainly a threat. Canal employees were working around the clock and I, with several others, was examining the cores of rock being drilled from the top to find out at what depth the hill was shifting. This information would be vital to the stabilization of the hill. Meanwhile, Canal forces and local contractors were starting to blast and cut the hill back. Soon, however, a large U.S. contractor was engaged to finish the job more rapidly.

I continued to work in Gaillard Cut until retirement, but I was also involved in a wide range of other civil engineering activities, especially after I became Chief of the Civil Engineering Branch in 1968. However, my greatest interest and concern continued to be the viability of the Canal through Gaillard Cut where, with the possible exception of the locks, the Canal was most vulnerable to a disaster which could close it, seriously affecting world shipping.

The widening of Gaillard Cut was done between 1957 and 1970, except for Culebra Reach, which was widened earlier by the Dredging Division. The underwater excavation was done by the Dredging Division and the above water mostly under four major contracts with U.S. firms. Several previous studies on Canal widening, dating back to World War II, provided adequate guidance for the design of most slopes, along with in-depth assistance from the Corps

Anna Mann

of Engineers and several private consultants. One area, the upper slopes on Las Cascadas Reach, extended into hillside areas beyond all previous studies, and did require computerized design procedures, which we carried out with assistance from the Massachusetts Institute of Technology.

The Canal through Gaillard Cut has a history of landslides dating back to the French excavations in the 1880's. The geology of the area is extremely complex, including a wide variety of rocks from hard volcanics to very weak sedimentary rocks which are the obvious cause of the landslides. To make matters worse, the weak rocks, when relieved of pressure by excavation, are prone over time to gradually weaken further by absorbing water and swelling. Thus a slope stable for many years can suddenly become a landslide. For these reasons, the weakest rocks tend to be in the worst possible place, right at the base of the slopes and the edge of the Canal. The inescapable result was landslides.

Three of the landslide problems were very serious and deserve mention. On Hodges Hill, at the north end of Culebra Reach, cracking and shifting of the slope became an imminent threat to the Canal. Total, prolonged closure was possible. It was controlled largely by perforated drainpipes drilled into the base, but required permanent, expensive monitoring. On Empire Hill, sliding and slumping developed during the cut widening and continued for several years, actually encroaching into the Canal several times and requiring dredging. On the east bank of Culebra Reach in 1974, a block of rock about 1,000 feet long partially blocked the Canal; tugboats were required to help the ships get by. Our efforts to drain the slope had been inadequate. It took the Dredging Division to restore the channel.

Important to many of our tasks in the Civil Engineering Branch was the use of Army helicopters. Flying with both sides open, usually with four or more branch members with cameras clicking, we covered Gaillard Cut weekly, and from time to time many other projects and problems — the railroad, the saddle dams on the fringes of Gatun Lake and the breakwaters.

Soon after arriving in the Canal Zone and becoming known for my SCUBA diving, I was being asked to inspect various underwater facilities, and was joined by our geologist Bob Stewart and occasionally a few others. For safety, we were given training by Burt Powell who headed the U.S. Navy divers on the dry dock in Cristobal. This led to over 20 years of frequent diving assignments throughout the Canal, the ports and Gatun Lake.

My most extraordinary assignment in all of my 27 years of Canal service was during my first locks overhaul at Gatun. With the dewatering a large number of very large tarpon were captured in several lateral culverts. The assignment was to shoot the fish with spear guns and haul them out. We gave almost all the fish to Panamanians working on the overhaul.

When we arrived in the Canal Zone, Anna and I had been married for seven years, having met in an Army hospital where she was working and I was healing. When I graduated with my engineering degree, we jumped at the offer by the Panama Canal, lured especially by the two months vacation per year. Change and adventure were what we wanted and what we got.

We found the Canal Zone a wonderful place to live and an even better place to play. We were fascinated with the jungles and beaches and offshore islands, with Gatun Lake and all its islands, the Canal itself and especially the locks, all of this within a few miles of where we lived.

We were soon familiar with Panama City and the countryside up to the beautiful beaches at Santa Clara. After I had accumulated a few weeks of leave we went to El Volcan and traveled first by launch up the Tuira River to Yavisa and then by cayuco up the Bayano River, to mingle with Choco and Kuna Indians. Among our strange experiences were Anna dancing with a Choco Indian — it was he who asked — and the two of us awkwardly rejecting an offer by the Kuna to take an adorable little girl home with us.

A few months after arriving in the Canal Zone, Anna and I took up skin diving and I was soon spear fishing too. After a few months my main interest turned to underwater photography. Within a few years I had built an underwater housing for taking 16 millimeter movies. Soon I was showing my undersea films around the Canal Zone and Panama and a little later in the U.S. Underwater movies were something new at the time and my audiences were impressed. In 1963 I won a prize for my films at the International Underwater Film Festival in Santa Monica, California. Later the same year, on SCN televi-

> *The Canal through Gaillard Cut has a history of landslides dating back to the French excavations in the 1880's. The geology of the area is extremely complex, including a wide variety of rocks from hard volcanics to very weak sedimentary rocks which are the obvious cause of the landslides.*

sion in Fort Clayton, I put on eight half-hour evening shows using the films. Incidentally, Anna, I am quite sure, was the first woman to become a skin diver and SCUBA diver in Panama.

In 1958 our daughter Deborah, or Debbie, was born. Anna had her swimming as a baby and we soon had a playpen for her on our launch, *Explorer*. By age six she had been skin diving in both oceans and was joining us in hunting for bottles in Gatun Lake. At nine she learned SCUBA diving and did a lot of it while on many cruises in the Perlas and especially the San Blas Islands. She certainly was a child of the sea.

In 1970 I sold much of my best undersea film to Time-Life Films for their TV movie, *Should Oceans Meet*. We used the money, with some to spare, for a wonderful two-month tour of Europe.

Debbie Mann

During these same years we also had lots of more relaxing fun. Anna became an avid shell collector. When the water in Madden Lake got low, we strolled the still existing Las Cruces Trail, built by the Spaniards to get their gold across the isthmus. At Fort San Lorenzo in shallow water we fanned the bottom for beads and coins and other things such as the gold ring I found. Debbie went there, too. But our favorite diversion was bottle collecting, sometimes from ship wrecks, but mostly in Gatun Lake where we cruised the shores of the islands. We found most of our bottles by diving on the submerged villages from the French era. Our remembrances of those many enjoyable days on Gatun Lake are kept especially keen by the hundreds of old bottles here and there in our home, along with many other remembrances of Panama and the Canal Zone.

PHOTO BY DON GOODE

The USS Paraiso, brought out of retirement to assist in the clean-up operation of the 1986 Gold Hill landslide in the Panama Canal.

The Francis T. Mayo Family

AS TOLD BY FRANCIS L. MAYO, LT COL US ARMY (RET)

Francis (Frank) T. Mayo was born September 13, 1897, in Somerville, Massachusetts, to Francis L. and Anne Shea Mayo. He attended parochial schools in Somerville until he had an altercation with his teacher, Brother Fabian, at which time he left home at the age of 14 and shipped out in the Merchant Marine as a cabin boy. Thus began his love of the sea. After he returned home, his mother forced him into Fisher Business School, where he learned to type and take shorthand. Being from a musical family, he also learned to play both the piano and organ.

When World War I broke out, Frank enlisted in the Navy on May 11, 1917. Due to his clerical capability, he was inducted as a second class yeoman. He served on the *USS Montana* and went to France with the first U.S. convoy. He was promoted to chief yeoman in his second year of service and transferred to the *USS Oklahoma*, where he served as the captain's writer. It was then that he ended up at Coco Solo submarine base and tasted the water of the Chagres River. After his discharge in 1921 he returned to Somerville and met and married Ruth K. Cronin, a telephone operator and one of fourteen children.

By 1923 the allure of the Canal Zone took its effect. He went to Panama without a job. While waiting for a civil service appointment, he supported himself by playing the piano at Mamie Lee Kelly's (Kelly's Ritz). He went to work for the Electrical Division in Balboa in a clerical status. An avid sports fan, he was shortstop for the Electrical Division in the Twilight League when they won the championship. He also exercised his musical ability as the organist at the Ancon Chapel.

Frank brought Ruth down to the Zone, and they lived in the Flats. His love of the sea and all connected with it led him to a transfer to customs and the shipping commissioner's office on the Atlantic side, where the family lived initially in a four-family house in New Cristobal, then in a cottage directly across the bay from Coco Solo. A daughter Ruth was born in 1926 and daughter Patricia (Pat) in 1928. Six years later Ruth returned to the States during her pregnancy, and their son Francis (Frank) was born in Medford, Massachusetts. Ruth returned to the Zone with her new six-week old son. Both daughters attended St. Mary's Academy, while Frank was enrolled in Cristobal grade school. Many fond memories linger from the days in New Cristobal — sitting on the steps in the evening, watching the Navy patrol bombers, PBYs, doing "touch and go's" in the bay; playing hide and seek among the tall palms; flying kites; visiting Billgray's Garden; and waving at the carametas as they drove by with sailors on shore leave.

This all came to an end on December 7, 1941, when

Ruth and Frank, August 1925.

New Cristobal cottage, 1942.

the military police came by to pick up military personnel. The following day Frank went over to Coco Solo and tried to enlist in the Navy; since he was a chief when he was discharged in 1921, he felt he should be reinstated as such. The Navy did not agree with his thinking and felt that he would be more valuable for the war effort serving in his current job. He had to accept their decision, but it broke his heart as his first love was for the Navy and the sea. In 1943 he was transferred to Balboa as special deputy shipping commissioner. The war years were horrendous, with a constant flow of both naval and Merchant Marine vessels transiting the Canal. On many occasions Frank's old Navy shipmates would visit while their ships were in port. The Mayo household was a home-away-from-home for several in the shore patrol unit stationed at "The Limits." Most of these personnel were police officers in civilian life, and some of them knew Frank and Ruth from their childhood days in Somerville. During that period the family lived in quarters next to St. Mary's Church and later moved to a cottage on Plank Street. Pat and Frank graduated from Balboa High School.

On Mother's Day in 1950 Frank, a boarding officer, was thrown and killed while climbing a Jacob's ladder to board a ship. He had 31 years federal service — four with the Navy and 27 with the Panama Canal. He is buried in Corozal Cemetery, where he always said, "There are as many heroes buried there as in Arlington." He instilled in his children a lasting love of country and patriotic spirit. Ruth left the Zone in 1951 to settle with the Cronin family in Somerville.

> *He is buried in Corozal Cemetery, where he always said, 'There are as many heroes buried there as in Arlington.'*

Looking West, aerial view of Colon and New Cristobal, January 1919.

PHOTO COURTESY OF PANAMA CANAL AUTHORITY

The William David McArthur Family

As told by David McArthur

William David McArthur, Sr., a 21-year-old soldier from Cass Lake, Minnesota, went to Panama in 1921. He was a member of the 4th Coast Artillery Band stationed in Fort Amador. He played the tuba, and in addition to playing in the Army band, became a member of a dance band in Panama City. During this time he met Eusebia Betsabe Samudio, a 23-year old Panamanian lady that would become his wife. Even though he didn't speak Spanish and she didn't speak English, they courted and sign language prevailed; they were married in Balboa, Canal Zone, in 1924. After he was discharged from the Army, he was employed as a signalman with the Panama Canal Company.

William was a member of the Chippewa Indian tribe from White Earth, Minnesota. His great-grandmother was a full blooded American Indian. His birth certificate indicates that he was one-eighth Chippewa Indian. Eusebia Betsabe Samudio McArthur was of Colombian descent; together they had six children — William David McArthur, Jr., was born at Gorgas Hospital; Ruth Mary, Edgar Robert and Edna Alice were born at Santo Tomas Hospital in Panama City; and Charles Albert and George Adam at Gorgas Hospital.

William loved Panama. He learned to speak Spanish fluently and got along perfectly with all classes of Panamanians. He never returned to the United States because he could not afford to do so; a signalman was a low paying job. There were no computers then, so ships approaching the Canal for transit were given instructions on how to proceed by means of a manual signaling system. Located at strategic points throughout the Canal, a signalman would relay messages to a ship using a combination of flags, large canvas balls and cones that were hoisted up on poles.

William, Sr., panned for gold in the 30's in the northern provinces of Panama and occasionally would find as much as $350.00 worth in a three week period. Gold at that time was valued at about $37.00 an ounce; unfortunately, this did not provide a steady cash flow. He loved children and would go out of his way to provide recreation and entertainment for them. He was the founder and leader of a Catholic Boy Scout troop, which he marched across the Isthmus of Panama via the railroad tracks. He also organized a baseball team for boys 14 years and younger. The uniforms were made from flour sacks, which he purchased with his own money. He would ask professional baseball players for their used equipment, nailing broken bats, taping old gloves and balls so the kids would have the equipment to play with. The team played in a league in Panama City and fared well. The children from the neighborhood would always be waiting after school for him to return from work so they could play a marble game under his house or some other type of sporting event.

William David McArthur, Sr., died of cancer on October 25, 1946, four days after his 46th birthday, in Gorgas Hospital with his wife and all his children at his bedside. His death was a severe blow to Eusebia, who now had to raise her six children with no work experience whatsoever. The situation caused her to suffer from severe depression and stress from which she never recovered. She was hospitalized for the last nine years of her life in Santo Tomas Hospital where she died in 1980 at the age of 81. Her daughter Edna visited her at least once a day, and on many occasions two or three times a day, either because of an emergency or to make sure the nurses hired to care for her were doing their job properly. The expenses

> *Located at strategic points throughout the Canal, a signalman would relay messages to a ship using a combination of flags, large canvas balls and cones that were hoisted up on poles.*

for the hospital and the around the clock care were all shared by her children. She was a loving, caring and devoted mother and was the driving force behind the good upbringing of her children. She was an excellent money manager and a genius at stretching a penny. She would not allow any of her children to speak English, forcing them to learn to speak Spanish fluently. She insisted on teaching her children morals, courtesy, good manners and respect for adults. Religious training and moral values were held in high esteem. She was a disciplinarian and did not spare the rod and spoil the child. Spanish proverbs were a normal part of her vocabulary, which she would routinely recite to fit the situation. Some of her favorites were: Pay your debts and you will always know what you have; tell me who you hang out with and I will tell you who you are; give and you shall receive; be good and there is nothing to fear; do something wrong and you better be aware.

After the death of William, Sr., his eldest son, William David McArthur, Jr., became head of the household at the ripe age of 21. He had just completed his apprenticeship as a ship fitter and wasn't earning very much money. Ruth Mary, who had recently graduated from Balboa High School and was working as a secretary, shared the household expenses with William, Jr. The family moved from a 4-bedroom cottage to two 1-bedroom apartments in a 12-family apartment complex in Diablo. Housing was managed by the Panama Canal Company and based on seniority. As the oldest member of the family married, the next in line would take over as the head of household. The married members would contribute whatever they could afford to help the others out. The youthful years were tough, but very enjoyable as they hung together as a family taking care of one another.

PHOTO COURTESY OF PANAMA CANAL REVIEW

Cucaracha Signal Station, 352 feet above sea level at Contractor's Hill, served for almost half a century as an aid to navigation of ships through the Canal.

The James McCartney and James Lewis Phillips Family

AS TOLD BY BEVERLY PHILLIPS GROSS

This is the story of how three children of James and Katherine McCartney — James, Zedock and Kate (Phillips) — helped shape the Canal Zone. The McCartney kin became part of the Pacific side Canal history from 1908 to 1972.

In 1907 James (Jim) McCartney (Roosevelt Medal No. 3792) was helping his father and brothers farm in Alabama. After serving in the U.S. Army, Jim believed there was more to life than working in the fields. On February 22, 1908, Jim became employed as a policeman in the Canal Zone. On May 14, 1910, he became a collector on the Panama Canal Railroad. His career and life were cut short when he contracted tuberculosis in 1911 and returned to Alabama to pass away.

In April 1911 Zedock (Zed) McCartney joined Jim in Panama. He worked as a cashier in the Hotel Metropole, Panama City. From August 1918 to July 1947, he worked in the Motor Transportation Division, Panama Canal Company. Uncle Zed rose to become the Governor's chauffeur. One highlight of his career was to chauffeur President Franklin D. Roosevelt during his visit to the Canal Zone.

Zed McCartney married Ethel from New York and had two children, Catherine and Daniel. In May 1916, while playing house, Catherine ate as play food the black tar poison used on the house stilts to kill crawling in-

President Roosevelt, back seat; Governor's chauffeur Zed McCartney, Madden Dam.

sects. Before her parents realized what happened, she had swallowed enough to take her life. This small part of the McCartney family will forever be a part of Panama soil.

James Lewis Phillips, Kate's son, my father, became the third McCartney kin employed by the Panama Canal Company. After graduating from high school in Alabama, Lewis volunteered for the U.S. Army, but wasn't accepted because of a heart murmur. Realizing he did not want to do farming or saw-milling like his father, Lon, Lewis looked elsewhere for work. In November 1939 Lewis received a letter from his Uncle Zed saying, "Get to the Zone within two weeks to be available for truck driver positions." Lewis seized the opportunity, and joined his Uncle Zed in the Zone, where he took a truck driver position.

In August 1940 Lewis joined the Canal's fire department. Diablo, Pedro Miguel and Balboa were the fire stations he served. After being transferred to Pedro Miguel Fire Station, feeling settled, he returned to Ohatchee, Alabama, to marry Vera Calhoun in October 1941. Vera's father liked Lewis, but was not too happy about them living overseas since World War II was in full swing.

A twelve-family apartment in Frog Alley, Pedro Miguel, was Lewis and Vera's home. Vera worked for the Clubhouse until I (Beverly) arrived on March 4, 1944.

While Mom was delivering, Lewis was sent home, as Gorgas Hospital preferred not to have expecting fathers hanging around the hospital.

My brother, George Kenneth Phillips, was born on October 7, 1947. Our family now lived in a four-family house on the same street as the Pedro Miguel Union Church, where the family was active. Our family later moved to a cottage across from the swimming pool and movie theater, much to our delight.

Between my fourth and fifth grade, all the gold employees were told the piping was bad in Pedro Miguel and everyone had to move. We moved to a duplex in Diablo. Later, Pedro Miguel was rebuilt for the silver employees.

When I was fifteen years old (1959), the Panama Canal Company had a fire department reduction in force so they could hire local people to replace the American employees. My father, Lewis Phillips, with 17 years service was cut because he was not a veteran. He then was hired as a lock guard, first on the Atlantic side, then Pacific side. We continued to live in Diablo since my mother was a substitute teacher. Because of this action, I realized the U.S. Government was in the slow process of turning the Canal over to Panama. I had an insight that few of my friends experienced. I knew I needed to prepare myself for a life in the United States. In 1962 after graduation from Balboa High School, I headed to Montevallo, Alabama. I received my degree in Elementary Education in 1966 and married Henry Gross.

My brother Ken played football at Balboa High School, graduating in 1965. After attending Canal Zone Junior College, he attended the Marion Institute in Alabama. George Kenneth became "George" when he joined

Lewis, Vera, Ken and Beverly Phillips.

Zed McCartney and Lewis Phillips.

the U.S. Army as a helicopter pilot. He served in the Vietnam War twice. George was also stationed in Korea; Fort Clayton, Canal Zone; and the United States. After Warrant Officer George Phillips's retirement, his family settled in Ohatchee, Alabama, living in his father's childhood home with our grandmother, Kate McCartney Phillips (died 1989), until he died in 1993.

While Ken and I settled into our stateside lives, our parents were finishing up their work careers in the Canal Zone. Lewis transferred back to the fire department in 1967. At retirement time in 1972, Lewis was the Fire Inspector Supervisor in Balboa, and Vera was a substitute teacher. They retired to Selma, Alabama. After they left the Zone, I heard several people comment how they missed hearing Lewis's laugh. He really loved life. Enjoying traveling, visiting friends and family, church work, watching many television stations (with commercials) and air conditioning, they felt satisfied that they had had a full life. Lewis said, "I was born in the horse and buggy days and lived to see a man walk on the moon. Who could ask for more?" In 1991, Lewis died eleven months after the death of his mother Kate Phillips. Vera died in 1992.

Thus ends the story of how three McCartney children served the Panama Canal Company. Even though Kate never stepped foot in the beautiful country of Panama that serves so many countries in the transit of ships from one ocean to another, she influenced and loved those who lived and worked there.

James McCartney
Roosevelt Medal No. 3792

The Maurice Lee McCullough Family

As Told by Judi McCullough

The story of the Maurice Lee and Hazel Margaret Heim McCullough family's migration to Panama began with my grandparents, George W. and Carolyn Paulson Heim. They first arrived on the Isthmus of Panama on September, 27, 1908. They traveled from Cincinnati, Ohio, aboard the *SS Allianca*. At that time they had two girls — Anniel and Dorothy. My aunt Louise was born in October 1910. Granddad worked as a machinist for the Mechanical Division, Gorgona shops. He began his work on October 1, 1908, and left March 3, 1912. Except for a brief return to Panama of three months, the family remained in the United States for about twelve years, until September 1923. The Heims were blessed with Mary, Charles, Lois, Hazel, Jean, and Frances. My mother, Hazel Margaret, was born prematurely in 1918, due to the flu epidemic. My grandmother was very sick with the flu, so Mom was separated from the Heim family and raised by an aunt and uncle, the Schimmels, with whom she lived until she married.

Grandma Carrie, tired of the difficulties of a large family and Granddaddy moving around in his jobs with the Chesapeake and Ohio Railroad and the Baltimore and Ohio Railroad, filled out an application for reemployment with the Panama Canal, signed Granddaddy's name to it and mailed it to Washington, D.C. In time, Granddad was accepted for reemployment. Grandma casually told him, "Oh, by the way, we are going back to Panama." From September 28, 1923, to March 31, 1946, Granddaddy worked as a machinist and lived in Cristobal, Canal Zone.

Sometime during this time, my Uncle Bill was born. Now there were ten in the family. Aunt Dorothy died when she was 19 and is buried in Corozal Cemetery.

Life, as life will do, moved along nicely for the Heim family. In 1938 my mother married Maurice Lee (Mac) McCullough. Dad was a student at Kenyon College in Gambier, Ohio. Mom was a student nurse in a hospital in Newark, Ohio. They had married secretly because married women were not allowed in nursing school. Soon Mom had to leave nursing school because I was to be born in January 1939. Life for the young McCullough family became financially difficult. Granddaddy Heim invited Mom and Dad to go to Panama. He promised to get Daddy a job. In August 1939 Maurice (Mac), Hazel (Snookie) and Judi McCullough arrived in Cristobal, Canal Zone.

> *In 1938 my mother married Maurice Lee (Mac) McCullough ... They had married secretly because married women were not allowed in nursing school.*

Eventually my father got a job with the Panama Canal. He worked for the Motor Transportation Division, where he was employed for 30 years, starting as a mechanic and working his way up to superintendent.

The best thing about Mom and Dad moving to the Canal Zone was that Mom was able to have a joyful relationship with her mother, father, and siblings. She enjoyed her family and the Canal Zone so much she did not want to leave. She was the last of the Heim children to leave Panama. Mom worked for the Schools Division for about 18 years. She was a recreational specialist, working at several schools, and "Snookie" became well known in the schools of the Canal Zone.

My brother Tom and sister Joan were born in Colon

Hospital and my brother Don in Margarita Hospital. Our family left the Canal Zone in August 1952 and returned in March 1957. During our time in Ohio my sister Susan was born.

We spent most of our years on the Atlantic side. Our first home was in Old Cristobal next to the clubhouse. My grandparents lived nearby. It was a fun place to visit. Our next home was on Eighth Street in New Cristobal. Hot afternoons were spent playing in wash tubs filled with water, other times at the beach at the end of the street. World War II was in progress, so there were blackouts and times to go into the air raid shelters. Our next home was in Coco Solito. Dad had joined the Navy. My cousins Wendy and George Cotton lived next door. We enjoyed fun filled days roller skating and riding bikes. When the war ended, Navy ships returned to the East Coast via the Canal. We went down to Gatun Locks to welcome the sailors back home. They would throw their hats to us. Some kids were able to get a lot of hats — they were bigger than I was! Dad had to go to Chicago to get out of the Navy; he went first. The Navy put Mom, Tom, and me on an Army transport ship to New Orleans. From there we took a train to Ohio. In August 1946 we returned to Panama and lived in Margarita. I attended second through sixth grade at the old Margarita Elementary School.

My fondest memories are those years living in Margarita. Our days were filled with play, play, and more play. We went to the gym for after school and summer activities, rode the chivas to Gatun for all day swimming, and sometimes we walked on the bulldozer trail to Fort Gulick to swim in the pool. We performed reenactments of Roy Rogers and Tarzan films in the jungle across the street.

I believe that living in Panama gave many of us a sense of being in a very special place. Panama is a country with pre-Columbian history, great for archaeology exploration — a country with a diversity of people and cultures.

As an American, I can be proud that my grandparents played a small part in bringing to the world a waterway that changed history. It was hard to leave our way of life in the Canal Zone and turn it back to the people of Panama, but it gives every one of us a chance to see the strength of the Panamanians. They are doing a great job running the Canal.

George W. Heim
Roosevelt Medal No. 4428

PHOTO COURTESY OF PANAMA CANAL REVIEW

Motor Transportation Division: Geared to keep Canal operations on the road. April 1964.

Robert R. McMillan, Chairman of the Panama Canal

AS TOLD BY ROBERT R. MCMILLAN

First non-Defense Department Chairman in Canal History

The furthest thing from my mind when confirmed by the U.S. Senate to serve on the Board of Directors of the Panama Canal Commission was to become Chairman of the Board. But as time moved forward, there was a growing sentiment on the Board that the right message for Panama, after Noriega, was to have a Chairman who was not connected to the U.S. military.

While for the most part it was mostly private talk amongst board members, there was one period in 1991 when I held some discussions with congressmen about the desirability of electing one of the non-military board members as chairman. Before any real action would be taken, Michael P. Stone, the Secretary of the Army, was named to the Board and became chairman.

Phoebe and Bob McMillan at Gatun Locks.

The behind-the-scenes chatter continued through the balance of George H.W. Bush's tenure as president. Then, with the election of Bill Clinton, other issues confronted those of us on the Board. Should we resign so the new president would select his own board? Who would become the new chairman? Mike Stone had stepped aside as secretary and chairman on the day of Clinton's inaugural.

We did not have to wait long. Actually, before the inauguration of Bill Clinton, the new administration asked all Board appointees to remain on the Board until replacements had been selected. That would take some time because the process, including FBI checks, could take several months or longer, even if new designees had been selected.

The reason for asking us to remain on the board was simple. If we had all left, the board would have been controlled by the remaining Panamanian board members. Our next question was who would become chairman. That became clear when John W. Shannon, a Bush appointee in the Department of the Army, assumed the position of Acting Secretary of the Army. He also immediately took on the chairmanship of the Canal.

Next came an event which shook the Canal's board. John Shannon was forced from his new post because of an incident causing his arrest in August 1993. Added to the confusion was the fact that there were no other civilians in place in the Department of the Army, and the Defense Department was totally preoccupied with the serious challenges in Somalia.

During the period after Shannon left his post, there was a great deal of board chatter. Could this be the chance to set an example for Panama and elect a civilian as chairman? There seemed to be a growing consensus that the time was right. As the next few months unfolded, I received a great deal of support from both U.S. and Panamanian board members to be the chairman. The next board meeting would be at the Port of Baltimore in October 1993. Election of a new chairman would be the first item on the agenda.

As October approached, it looked as though I had the votes to be elected. Arriving in Baltimore, we were greeted by four-star general Gordon Sullivan, Chief of Staff of the United States Army. General Sullivan was designated to be on the board in place of John Shannon. Since there were no civilians in place and General Sullivan had been confirmed as a general by the Senate, he was eligible to serve on the board.

From the beginning, he realized that there were not the votes for him to become chairman — unless he invoked his "nuclear" option. A never-used part of the law, creating the Panama Canal Commission, gave to the Department of Defense nominee on the board the right to cast all the votes of U.S. members. Would he use that authority to counter the votes of the board members who were ready to elect me? With nothing to lose, and wanting to carry out the theme of not having the U.S. military run the Canal — something all U.S. and Panamanian board members wanted to see carried out — I decided to stand firm.

Before the vote, I indicated that if General Sullivan were to vote all of our votes to elect him as chairman, I would immediately go to Washington, D.C., and hold a press conference to denounce what had taken place. The vote was taken — General Sullivan ended up voting for me. I was elected chairman unanimously. After the meeting adjourned, General Sullivan could not have been more gracious. He was a gentleman, and he really understood the importance of the message we were sending to Panama.

> *His first question was, 'What qualifies you to be chairman of the Canal?'... I quickly added that there were probably some 10,000 other people in the United States with similar credentials — but not one of them ran against Pat Moynihan for the U.S. Senate!*

Soon after being elected chairman, I had an interesting conversation with a newspaper reporter. His first question was, "What qualifies you to be chairman of the Canal?" Thinking for a few seconds, I responded by citing a few reasons. I told him that I had been to Panama several times before becoming Chairman, understood its history and while at Avon even shipped goods through the Canal to Asia. I quickly added that there were probably some 10,000 other people in the United States with similar credentials — but not one of them ran against Pat Moynihan for the U.S. Senate! With that, both of us had a good laugh and the interview was completed in a most amicable way.

As I reflect back to that period, I am so appreciative of the support I received from the board and am glad I stood my ground. Staff personnel at the Canal were also appreciative of the outcome. Many said to me that the election sent a message for a new era at the Panama Canal — one which they were hopeful would extend even past the transfer to Panama six years later. Serving as chairman of the Canal is something which I shall never forget.

Panama Canal leadership, 1994. Front row (L-R): Deputy Administrator Raymond P. Laverty; Administrator Gilberto Guardia F.; Board Chairman Robert R. McMillan; and Secretary Michael Rhode, Jr. Middle (L-R): Walter J. Shea; Joe R. Reeder; Alfredo Ramirez; Joaquin J. Vallarino; and Cecilia Alegre. Top row (L-R): William E. Carl; Luis Anderson; and John Danilovich.

John A. McNatt

As told by Blanca McNatt Schield

In 1933 when he was 23, John Andrew McNatt sailed south to Panama from Nashville, Tennessee, browbeaten by the economic disaster of the Depression and drawn by the lure of jobs in the Panama Canal enterprise. His parents, John and Winifred McNatt, of Irish Scottish descent, stayed behind with his younger sister Marye.

His first job, however, was not for the Canal organization but as manager of the Cecilia movie theatre in Panama City. Boyishly handsome and extroverted, he met my mother, Sonia Guizado Valdes, and delighted her girlfriends with free passes to the movies. They married in 1935, one year after my grandmother Blanca Sonia Guizado's death.

Blanquita was born in 1936, John in 1939 and Richard in 1943. Until 1951, home was the spacious house of John McNatt's father-in-law, Comandante Juan Antonio Guizado, first chief of the fire department, in Bella Vista, Panama City.

Eventually, John "Jack" McNatt joined the Panama Canal Company, becoming for a few years a signalman on Sosa Hill, followed by working as a contraband inspector at the Pacific side commissaries making sure everyone shopping had a card.

Sometime around 1950 he took a correspondence course in accounting; and when he passed the Panama City's certified public accountant's test, he was promoted to the fledging internal audit department of the comptroller's office. He was already fluent in Spanish, albeit with a soft Tennessee drawl.

In 1951 internal audit was transferred to Cristobal, and for nearly two years we lived across the street from Cristobal High School with the Atlantic Ocean on the other side of our house. Transferred to Balboa in 1953, we lived in lively Balboa at 813½ Empire Street — we shared a driveway with the neighbors. He took mandatory retirement at 62, and "Jack" and Sonia built a home in Panama City's Obarrio district. In 1978 at 67 his death came suddenly, and he is buried in Corozal Cemetery. Sonia lived for many years and died in 1994 at the age of 84.

Grandfather Comandante Juan Antonio Guizado was a heroic figure in the Panama Fire Department. He was born on February 12, 1867, and at age 20 became one of the founding fathers of the Bomberos, as the firemen are called, on November 28, 1887. The Bomberos' marching band frequently participated in Veterans Day and Fourth of July parades in Balboa.

> *His career included three important landmarks. Three months before the inauguration of the Panama Canal, a cataclysmic explosion of a powder magazine occurred. The magazine was part of the munitions factory of the National Police. . . . Six of his men died.*

His career included three important landmarks. Three months before the inauguration of the Panama Canal, a cataclysmic explosion of powder magazine occurred. The magazine was part of the munitions factory of the National Police. This is known historically as "El Polvorin." Six of his men died.

A second landmark was the Colon fire during the early 1940's. Chief Guizado was at a luncheon at the British Embassy when he received a frantic telephone call that "Colon was burning out of control, complicated by acute water pressure problems." Chief Guizado enjoyed an excellent friendship with the Balboa fire chief, and he quickly contacted Canal authorities for support. The Panama Railroad was put at Chief Guizado's disposal. Two fire trucks were loaded unto flatbeds and the firefighters entrained toward Colon. The United States Army's demolition team was deployed from Fort Gulick to the scene. He instructed them to dynamite homes in a desperate attempt to control the fire. With water pressure restored the conflagration was extinguished. With the singular cooperation of the Canal Zone Fire Department, the U.S. Army, and Panama's Bomberos, Colon was saved. The use of dynamite to combat fires made history.

The third landmark came during World War II. The defense of the Panama Canal was critical. Sightings of German U-boats were commonplace. History indicates that 14 ships were sunk by U-boats operating off Colon, and a Japanese miniature submarine ran aground on an island in the Gulf of Panama.

Because of his well-trained fire brigades and his reputation for organizing the fire departments throughout Central America, Canal authorities designated him as Chief of Civil Defense for the entire Republic, including all U.S. military bases, until the end of the war. He was issued proper military identification; and his official car was decorated with bright military emblems, facilitating his frequent meetings with the brass. Chief Guizado was fluent in English and Italian.

He ended a fabulous career in 1950 after 38 years of being Commander, First Chief, serving from 1887 without honorarium. He died in 1951 at the age of 84.

Old Cristobal Fire Station.

PHOTO COURTESY OF PANAMA CANAL MUSEUM

THE MELLANDER FAMILY

AS TOLD BY GUS MELLANDER

Mellander is a rather unusual surname, but three Mellanders wound up in the Canal Zone — Karl J. Mellander (Balboa High School, 1950), Gus A. Mellander (Balboa High School, 1952), and their mother, Adela Maria Navarro.

Adela Maria was born in Panama on November 7, 1903. She was one of the first Panamanians born after Panama declared its independence from Colombia on November 3, 1903. A serious student, she became proficient in French, English and Spanish. She graduated from the Instituto Nacional and worked as a teacher. In 1927, she left for California to continue her education. There she met Harold Mellander, a native of Kalmar, Sweden, who had been drawn to Hollywood in hopes of becoming a movie star. They married; she never finished her university studies, and he never became a major movie star. Karl and Gus were born in Los Angeles.

In 1940, Mrs. Mellander and the boys visited Panama. They planned to stay two months. But events beyond their control dictated a different reality. They would remain on the Isthmus for over 15 years. They lived in Bella Vista and later in Balboa.

Karl and Gus benefited from the excellent Canal Zone schools. Both attended Ancon Elementary School, Balboa High School, and Canal Zone Junior College.

They remember December 7, 1941, very vividly. They were at Farfan Beach on that Sunday. It was a typical day at the beach. All of a sudden all the adults became quite agitated. Kids were called out of the water while mothers quickly packed up their picnics. It was the first time the boys had ever seen grown men cry. Everybody rushed home. Pearl Harbor had been attacked.

There was an eerie sense of uncertainty and fear. But there was also an unexpected pleasant outcome for the children. Two days after the attack, every child on the block where the Mellanders lived received their Christmas presents. Clearly, the adults realized how vulnerable the Panama Canal region was to attack and decided to have Christmas a bit earlier that year. Luckily for the children, they received gifts again when Christmas rolled around. Unlike their parents, they were quite happy that December. Unaware of inherent dangers, growing up in the Canal Zone as it mobilized to a wartime status was exhilarating for children. A warm sense of patriotism and purpose permeated our days.

Looking back, we realize that the Canal Zone was a very privileged part of the world. We enjoyed the benefits of a secure American lifestyle in a tropical paradise. We did not have snow, but we had palm fronds and many a grassy hill to scoot down. Ours were halcyon years.

Dire rumors were weekly occurrences while drab olive colored military vehicles zoomed along the highways. Bomb shelters were built on school grounds and air raid drills were held periodically. It was fun and exciting rushing out of the classrooms into the underground bunker. Sandwiches were served, but only one — no seconds!

Other memories include scary but patriotic war movies, blackouts at night, volunteer civilian patrols, food rationing in the commissaries, collecting aluminum foil and rubber bands, war bond drives, Christmas tree bonfires, movie stars and war heroes such as Bob Hope and the red-headed General Eisenhower.

Looking back, we realize that the Canal Zone was a very privileged part of the world. We enjoyed the benefits of a secure American lifestyle in a tropical paradise. We did not have snow, but we had palm fronds and many a grassy hill to scoot down. Ours were halcyon years.

Karl Mellander distinguished himself at Balboa High School as a serious science student. He excelled in chemistry and physics and enjoyed spending hours on end in the school's science laboratories. After graduating in 1950, he attended Canal Zone Junior College where he continued his interest in all matters scientific. He worked as a doorman at Balboa Theater. Unable to pursue college in the States because of financial constraints, he became an apprentice electrician with the Panama Canal Company. After completing the four-year program, he worked in Balboa while saving for college. Karl eventually enrolled at the University of California at Berkeley and earned a Ph.D. in astronomy. He worked at the university until his retirement and lives in the San Francisco area.

Gus Mellander was not as serious a student in high school as Karl, and he certainly never distinguished himself in the sciences. Instead, he developed an abiding interest in history and was active in extracurricular activities. He was president of several student clubs. He recalls fondly the diligence and competence of the teachers, the grisly Reserve Officer Training Corps films, the airplanes that bumped the high school roof periodically, football games with arch rival Cristobal High School, train rides to Colon, and school dances at the elegant Tivoli Hotel. A few beers at ten cents a glass at the El Rancho invariably capped those evenings. We lived an electrifying dichotomy — a very staid and conservative environment in the Canal Zone juxtaposed with the exotic stores, bright lights and urban delights of a bustling Panama City.

The many military ships traversing the Canal provided thrills and fodder for adolescent dreams. Gus remembers going on board the *USS Missouri* when it docked in Balboa, fresh from hosting the signing of surrender documents that ended the war with Japan. He sold Fuller brushes, personalized door nameplates, and insecticides door-to-door and newspapers every afternoon at the Administration Building. At night he worked at the Diablo and Balboa theaters. He accelerated his studies and finished high school at age 15. But his college plans stateside were torn asunder. In 1952, the year Gus graduated from BHS, one of his cousins, Norberto Navarro, who had been Minister of Public Works in several governments in Panama, decided to run for president. Navarro was the leading contender until the chief of police, José Remón, decided to run. Navarro declined to be his running mate, and ultimately Remón was elected. In those days when you lost, your future was very dim. Unable to secure contracts for his construction firm, Navarro had to leave the country. The family's finances collapsed; there was no money for college. Gus was unable to get employment with the Panama Canal Company or the Canal Zone Government because there was an 18-year-old age requirement. He worked with a construction company in Curundu for seven months digging ditches for 40 cents an hour. The Panama Canal Company made an exception and hired Gus as a machinist apprentice in Mount Hope on the Atlantic side. Colon was a new, exciting and cacophonous city to explore. A reduction-in-force ended those adventures. He then worked at the Payroll Division in Diablo Heights.

> *We lived an electrifying dichotomy — a very staid and conservative environment in the Canal Zone juxtaposed with the exotic stores, bright lights and urban delights of a bustling Panama City.*

Gus attended classes part time at Canal Zone Junior College. In September 1955 he quit his job and enrolled full time. He worked thirty hours a week at night and on weekends with the Postal Division. The next two years at CZJC were enlightening and spirited ones. Gus was very involved in student government and a variety of clubs. He ran successfully for the Balboa civic council. He wrote a weekly column for the *Panama American* and later for the *Star & Herald*. He was one of the first Americans to interview Juan Perón after his overthrow when he passed through Panama.

After graduating from CZJC in 1957, Gus transferred to The George Washington University in Washington, D.C. As a student, he worked on Capitol Hill, at the National Archives, and at American University where he co-authored two books — one on Panama and one on Malaysia and Singapore. Both books were used by government agencies to train personnel assigned to those countries. Gus had previously conducted research in Canal

Zone and Panamanian libraries and government archives on relations between the two countries. In Washington he was able to study records not previously available to the public. Gus earned three degrees at G.W.U. culminating with a Ph.D. in history.

After graduation he wrote *The United States in Panamanian Politics: the Intriguing Formative Years*. It covers key historical events from 1903 to 1906 and was nominated for several awards. It became a best seller at the United Nations. More recently, Gus and his wife Nelly wrote *Charles Edward Magoon: The Panama Years*. Magoon, in Panama from 1904 to 1906, was the only person ever to serve as a U.S. ambassador to the Republic of Panama and the governor of the Canal Zone — he held both jobs at the same time.

Gus taught and served as a university administrator at institutions nationwide. He wrote several local history booklets and lectured widely on Panama. He also served as president of three colleges for twenty years. Recently, he was a graduate school dean at George Mason University. He lectured in many countries including Taiwan and helped write higher education master plans for several Latin American countries as well as Hungary.

At Nelly's — "she who must be obeyed"—insistence, he retired, screaming and kicking, on February 1, 2006. He has yet to adjust. They live in Palm Beach, Florida, and in Puerto Rico.

Canal Zone Junior College, 1932-1964, located next to Balboa High School. Photo dated 1939.

Balboa High School with R.O.T.C. Building (to right), circa 1955.

GILBERT MORLAND

AS TOLD BY DAVID CORRIGAN

A Special Class of Man

My grandfather, Gilbert Morland, was the quintessential English gentleman. He was well known throughout the Canal Zone, especially on the Atlantic side, and was loved by all who knew him. Unfortunately, I did not get the chance to know him well. I was five when he passed on. My brother, Gil Corrigan (named for Grandfather); my mom, Carol (Morland) Mead; and I lived with my grandmother and grandfather in Brazos Heights for a time before his death. My uncles, Robin and Peter Morland, also lived in the house at that time.

Gilbert Morland

Gilbert was born in 1901 and hailed from Birkenhead, Cheshire, England. As a young man he worked for White Star Shipping Lines, famous because it had the *SS Titanic* in its fleet. He came to Panama in 1929 to take a job with the Payne and Wardlaw shipping agency. During the Depression, which hit Panama hard, Gilbert returned to England for about a year. He loved Panama; and by this time he had met Virginia Woodhull, my grandmother, and longed to return to the Isthmus. Gilbert got his opportunity when C.B. Fenton, President of the shipping agency of the same name, contacted him about returning to Panama to work for his agency. Of course, my grandfather accepted the offer and some years later would become President and General Manager of C.B. Fenton, which is still in operation today and is one of the largest shipping agencies in Panama.

My grandmother Virginia and grandfather Gil married in 1934. They first met in 1929 when my grandfather came to live in the guest bedroom of my grandmother's parents' home. Initially, Virginia was quite amused at Gil's decidedly English accent and mannerisms. For instance, he would cut his hamburger with a knife and fork, never quite getting the hang of picking up his food with his hands. When he went to the beach (not his favorite activity) he would wear a cotton long sleeve button down shirt, trousers, and dress shoes. Eventually they married and had six children: Muriel "Sunny" Mizrachi; the late Sally Williams; Mary Coffey; Robin Morland; my mom, Carol Mead; and Peter Morland. Grandma (deceased April 2, 2005) and Grandpa (deceased December 1, 1971) have both passed from this life as has Aunt Sally (November 23, 2005), but the many of us that remain are a fairly close bunch.

Gilbert and Virginia at Galeta Point (Ft. Randolph, Atlantic side), 1940's.

The people who knew Gil Morland knew him as an avid sportsman. He played rugby as a young man, but his passion was golf. He served as President of Brazos Brooks Country Club for several years. He was a golfing visitor to Pinehurst, North Carolina, for 30 years. He eventually bought a house in Pinehurst in 1964 and became a member of Pinehurst Country Club. After purchasing the house in Pinehurst, he and Grandma became part-time residents of that lovely resort town. Grandma spent the last 10 years or so of her life in Pinehurst.

Almost equal to his love of golf was his acquired love of baseball. My great-grandmother, Agnes Moran Woodhull, and great-grandfather, Wellington Warren Woodhull, a Panama Canal pilot, moved back to New York after Captain Woodhull took ill. He died in New York in 1933. Agnes lived until 1961. It was during visits to New York with my grandmother that Gil was first introduced to baseball. Virginia and her brother George would take Gil to Ebbets Field to watch the Brooklyn Dodgers back in the 1930's. Being from England, he had never seen baseball and liked it very much. The more he visited Brooklyn and watched baseball, the more he grew to love it.

After World War II, Panamanian professional baseball was beginning to gain popularity. My grandfather decided to pursue the possibility of bringing U.S. players to Panama and began looking for a sponsor. He found a willing party in his friend, businessman Alberto Motta. Thus, the team Cristobal Motta's was born; they took their last Panama championship in 1950. After that Mr. Motta pulled his sponsorship for business reasons, and Grandfather then solicited Carta Vieja to sponsor the team. The Motta's were renamed the Carta Vieja Yankees and went on to win several hotly contested championships.

Gilbert Morland, awarded highest honor (Knightship) by the King of Norway for his services to Norway.

Perhaps the most famous player my grandfather was able to recruit to Panama was Tom Lasorda. He pitched for the Motta's from 1948 through 1950. According to the stories, he was not very effective. He was hit hard by Panamanian batters. Another friend of my grandfather, Al Kubski, who was manager of the team, convinced him that Lasorda was good for team morale and kept the boys loose. Kubski went on to become a renowned major league scout for the Braves, Orioles and Angels. Another famous ballplayer that my grandfather attempted to lure to Panama was the great Yogi Berra. In those days the team would pick up the expense of bringing the players to Panama and returning them to the United States. In addition to that, they made $250 a month. Well, Yogi thought he was worth $300 a month, contending he was going to be a major leaguer one day. My grandfather declined to make him a "special case," and so Yogi didn't come.

In addition to being the owner of C.B. Fenton, President of Brazos Brooks Country Club, and General Manager of his baseball team, my grandfather was also the Norwegian consul to Panama for 10 years. In recognition of his service, he received Norway's highest honor. My grandfather was decorated a Knight in the first degree in the Order of St. Olav. The honor was bestowed in the name of King Olav V of Norway.

I am honored for the opportunity to write this passage about my grandfather. This writing brings me closer to him. And I am very grateful that the future generations of our family will have a chance to read this story about Gilbert Morland and visualize who he was and learn that he was indeed a special class of man.

See also the Virginia Morland history.

Virginia Morland

AS TOLD BY ROBERT MIZRACHI

Fesser Cox joined New York's Tryon County Militia in 1775 and fought in the American Revolution. Nearly 150 years later, his great-great-grandson would take part in another exploit that helped to shape the world.

"He was a seafaring man; he had a license to sail in any water in the world," Virginia Morland said of her Utica-born father, Wellington Warren Woodhull. "Every time he went back to sea, my mother was pregnant. My mother got fed up with him being away so much, so when the opportunity to work on the Panama Canal came up, my dad applied for a job there in Washington, D.C., and was appointed right away."

Wellington soon left for Panama to become a pilot in training. In 1916, his wife, Agnes Moran Woodhull, originally of Hartford, Connecticut, sailed from New York to Panama with their five children. Virginia was four years old.

"My brother George and my sister Florence, they were all over the ship. My mother was seasick. She never knew where they were. They were down in the engine room. . . . They had a wonderful time. My oldest sister Pamela, she looked after me and Muriel, who was six."

Though Wellington relished the opportunity to become a pilot on the Panama Canal (the 45th hire), Agnes felt differently about the move. "My mother always said she'd rather be a lamp post in New York than a millionaire in the Canal Zone. She was transplanted; she had all her friends in New York."

As it turned out, Agnes had no trouble adapting. "She loved to party; she had a party at the drop of a hat."

Virginia Morland

(Virginia's sister once speculated that her mother was perhaps a vaudevillian performer because of her knowledge of many off-color songs.) "She was very social. My father was very laid back, but he adored her. She enjoyed a glass of beer — nothing else, though."

Poor health forced Wellington to retire in 1932. Agnes and Wellington moved back to Brooklyn. Of their five children, Virginia was the one who went on to spend her life in Panama.

"I was 18 when I graduated from Balboa High School in 1930. I never went to live on the Atlantic side until I got married to Gilbert Morland. I met him in '28 or '29. He was from Cheshire, England. He worked for the White Star Line in England. I think he was a photographer."

When Gilbert Morland arrived in Panama, he began working for the Payne & Wardlaw shipping agency. "Every three years they got four months' vacation to go back to England. That awful Depression came in '29. Gilbert went to Captain Payne, and he said, 'Look, things are bad. I'm going to England to see my family. If you don't want me to come back, I want to know now because I'll have to start looking for something to do while I'm over there.' Captain Payne said, 'Oh, no, no, no, Morland. We want you to come back. You come back after your four months.' So, after he was over there for two months, Payne told him not to come back."

"The Depression hit Panama hard — all over the world. Then Gilbert had to stay there. By that time, we were really quite engaged. He missed Panama. He loved golfing and everything that was there for him in Panama.

. . . I was working then, up at the Administration Building. I was working for the postal accounts. One day Captain Fenton of the C.B. Fenton shipping agency came up there. He was an old man then, even. He asked me did I believe Morland would be interested in coming back. So, of course I said, 'Of course!' Captain Fenton cabled Gil to come back."

Virginia Woodhull and Gilbert Morland were married in 1934 and raised six children: Muriel (Sunny), Sally, Mary, Robin, Carol, and Peter. As Gilbert Morland climbed the company ladder at C.B. Fenton, he distinguished himself apart from the unfair practices that existed in the Canal work environment.

"He was very fair, very upright. The West Indian people all loved him. One of them came to him one time and said, 'Mr. Morland, I have been asked to speak to you because maybe you can help us.' Of course, my husband said, 'I'll do anything I can.' The man said, 'We had two gangs [of dock workers] and we were unloading such-and-such ship and we had to transfer explosive materials from the ship to the pier.' Pan Canal paid extra for those gangs — not the regular rate; they got a whole lot more money unloading explosive materials. The man said, 'They didn't pay us; we got the regular rate. I can prove we unloaded these things on such-and-such a ship.' Gilbert went right away to the man that was responsible and, boy, he not only bawled him out and demanded that he pay this crew what they were entitled to, but he also wrote him a letter and sent a copy to the Executive Secretary of the Panama Canal."

Virginia and Gilbert, with ship captain, on cruise, June 1967.

After Gilbert passed away in 1971, Virginia began devoting more time to philanthropic endeavors for the poor of Colón. Each Thursday she'd help distribute dry goods such as rice, sugar, and canned milk at the old fire station on Roosevelt Avenue. The Cristobal Women's Club, of which she was a member for decades, raised the money to facilitate the weekly offerings. It's been said that she helped support the Women's Club itself in its waning days of existence.

A longtime congregant of St. Margaret's Episcopal Church in Margarita, Virginia eventually became a benefactor there as well. For her contributions to the Atlantic community, she was recognized with the Panama Canal Honorary Public Service Award, presented to her in the rotunda of Balboa's Administration Building in 1979. After moving to North Carolina in 1995, Virginia researched her ancestry and became active with a chapter of the Daughters of the American Revolution. She passed away in 2005.

See also the Gilbert Morland history.

The Milton Lee Nash Family

As told by Andra Lee Nash English

My father, Milton Lee Nash, was born in Berkley, Norfolk County, Virginia, on July 29, 1901, four months after the untimely death of his father, Milton Lee Nash, Sr., at age 37. Milton Lee, Sr., had been employed as a clerk in the Steam Engineering Department of the U.S. Navy Yard in Portsmouth, Virginia. His mother, Ida Lydia Sykes Nash, now had the responsibility of not only her newborn son, Lee, but two older little girls. Family members moved in to support the struggling family.

Margaret Kaeffer Moore, "Peggy," my mother, was also born in Berkley on May 8, 1903. She "enjoyed" three older and two younger brothers. Her parents were Woodbury Langdon Moore, Sr., and Cassandra "Andra" Deane Cory Moore. Peggy's father operated a lumber mill. At the time of his death in 1925 at age 59, he was secretary and manager of Industrial Savings & Loan, Inc., and a member of Sons of Confederate Veterans.

After graduating from Maury High School, Norfolk, in 1921, Peggy attended the two-year program at State Normal School in Harrisonburg, Virginia (now James Madison University). Since Mom wanted to teach on the secondary level, she continued on to Farmville State Teacher's College (now Longwood University), graduating in 1925. She was involved in many activities at college, particularly dramatics. Returning to Norfolk, she became a junior high school teacher in 1925 and was also employed in a part-time secretarial position at the Norfolk Division of the College of William and Mary (now Old Dominion University).

Family necessity forced Lee to begin his working life in 1918 with the Old Dominion Marine Railway Corp., and he became an apprentice machinist the following year, completing training in 1924. He was employed with the company through 1931.

As a teenager Lee participated in Boy Scouts and later played end position with the semi-pro football team "Berkley Braves." He was team captain and "All Tidewater." Lee loved to hunt and canoe with friends throughout the Great Dismal Swamp area and out into the ocean.

Lee Nash, Berkley Braves, 1922.

Peggy and Lee were married on June 15, 1929, at which time Peggy's teaching career ended since married women were not permitted to teach. She continued secretarial work and for one year worked for FDR's Works Progress Administration (WPA) in the payroll section. From 1925 to 1930 she had summer employment at the Martha Washington Hotel in Virginia Beach as a stenographer (She knew shorthand.) and desk clerk. From 1931 to 1934 Lee was employed by several companies including the Norfolk Navy Yard. Work ranged from two-year stints to only months at a time. As the Great Depression wore on, Lee, Peggy, and other family members opened a restaurant near the college called "The College Inn" which closed when a competing venue opened inside the college.

Back in 1930, Lee had applied for a machinist position with the Panama Canal Company, and on October 15, 1934, he was fortuitously hired with orders to report to the Superintendent, Mechanical Division, in Balboa. He sailed from New York on the *SS Ancon* on October 23, 1934. Peggy arrived in the Canal Zone shortly after and was employed briefly at Gorgas Hospital until Lee was transferred to the Locks Division on the Atlantic side, where they happily moved into their first home in Gatun.

Gatun had a fire station, gymnasium (playshed) with a swimming pool, post office, commissary, service center with theatre, bowling alley, cafeteria, lunch counter, barber shop, beauty salon, tailor shop and library. Scouts met at the Trefoil House. Classes at Gatun Elementary School went through sixth grade. There was a Masonic Temple, a Union Church and a Catholic Church, health dispensary, dental office and a train station. On the lake was the "yacht" club, and across the locks was the Tarpon Club. There was also a golf course across the locks which later moved to Brazos Brooks during the war. The Spillway was one of the hazards to launch a golf ball across! It was a busy little town with launches bringing in residents of islands and shores of Gatun Lake to waiting "chiva" buses and the train station for transport into Colon or the "other side." It was small town U.S.A. with a Panamanian accent.

Peggy worked as a clerk for U.S. Army Quartermaster Corps at Ft. Davis until 1940. She resigned that position when she found she was expecting me — Andra Lee Nash, born January 23, 1943, in Colon Hospital. Mom was 39 when the unexpected happened!

Throughout the 1940's Peggy was a substitute teacher, sometimes long term. As there was a 40-year old age ceiling for new full-time teachers, she was never hired as one. She worked from our home for nine years, however, as Atlantic side Society Editor for the *Panama American* newspaper. Typing her column, she placed it in a special envelope with red printing and handed it to the train conductor at Gatun Station every evening for transport to Panama City and the newspaper office. Her journalistic highlight was formally meeting Queen Elizabeth II at the Governor's Reception in 1953 and reporting on the affair for the paper. She practiced in front of her mirror to get the curtsey just right!

Lee Nash (in shorts), overhaul, 1950's.

Lee joined the Masons and Peggy the Eastern Star. She participated in the Inter-American Women's Club and American Pen Women. She helped with Brownies and Girl Scouts. They were active members of Gatun Union Church and Brazos Brooks Country Club.

Lee trained to be a towing locomotive operator and later was promoted to tunnel operator in 1946. In 1957 he advanced to lockmaster until his mandatory retirement in July 1963 at age 62. He received an award for preventing industrial accidents as a supervisor of lockages for a five-year period. Dad often told us how proud he was to work on the Canal with his friends and associates, particularly during the critical World War II years.

> *It was a busy little town with launches bringing in residents of islands and shores of Gatun Lake to waiting "chiva" buses and the train station for transport into Colon or the "other side." It was small town U.S.A. with a Panamanian accent.*

Mom was employed in a full-time clerk/typist position in 1954 in the Commissary Division at Mt. Hope. In time she was promoted to accounting clerk and later produced the Sales Promotion Circular. In 1957 she and others were transferred to the Pacific side and so began the shared six-year daily commute across the Isthmus.

Employment in the Canal Zone led to a higher quality of life for my family than would have been open to them in Virginia. We enjoyed voyages back to the U.S. on the Panama Line. Extended vacation time permitted cross-country drives and visits to national parks as well as urban cultural attractions. We explored Panama, as well, especially Playa Santa Clara. We lived in three homes in Gatun; one overlooked the 3rd locks excavation waterfall, and the last had a view of Gatun Lake and Colon Harbor. As "Andy," I enjoyed piano and dance lessons, scouting, Rainbow Girls, the pool and gym, cayucoing on the lake, horseback riding, playing in the "bush" and observing tropical wildlife with the freedom kids enjoyed in such a unique place. Our lives were enriched by long-time housekeeper and cook, Eulalee Reid. "Lady," my dog, was a constant companion. She usually followed me to school and met me for the bike ride home.

Gatun Elementary School years were idyllic. I was so happy that the school with the bell I used to ring is still in use for Panama Canal Authority training when I visited in 2007. As a Cristobal High School "tiger" I have wonderful memories of football games, sock hops, teen club, Girls State, and friendships. I graduated in 1960.

When Dad retired in July 1963, we moved to Diablo Heights. He died there in January 1964. Mom stayed until her retirement in May 1965 and then returned to Norfolk, where she died December 26, 1992, at age 89. They rest together with family members in Norfolk's Cedar Grove Cemetery.

After the University of Tennessee and receiving my M.Ed. at the University of Virginia, I married 1st Lt. Edward B. English in 1965. We met when he was stationed at Ft. Clayton. Our married life began in Canal Zone vacation quarters and a Gamboa apartment as I was employed as a speech pathologist on the Pacific side and he at Ft. Davis. After later residing at Ft. Gulick, we drove back to the United States in September 1966. We spent 30 great years roaming the world with the Army. I taught as a speech therapist and later a computer system administrator and school librarian. We settled in Williamsburg, Virginia, where I proudly display my license plate "GATUN CZ." It is wonderful Ed and I can share our love and memories of the Canal Zone and Panama.

At this writing, Peggy and Lee have three grandsons: Lt. Col. Edward Lee English, U.S. Army, married to Alicia Morris; David Nash English; and Matthew Cory English married to Jessica Williams. Edward Nash English and Elizabeth Lee English are their two great-grandchildren (children of Edward Lee and Alicia). Our sons and daughters-in-law, grandchildren, and, hopefully, great-grandchildren will be proud to read this story of the grand adventure of Lee, Peggy, and Andy in the Panama Canal Zone.

Lee, Andra Lee "Andy," and Peggy Nash, 1949.

Gatun Elementary School.

PHOTO COURTESY OF ANDRA LEE NASH ENGLISH

The Elmer Franklin Orr and Simon Butler Jones Families

AS TOLD BY JUANITA JONES GIRAND

"If You Want a Good Man, You Will Hire Me"

My grandfather, Elmer Franklin Orr, was born on February 16, 1885, in Blackwater, Missouri, near where his father, John Thomas Orr, of Kentucky, and his wife, Ella Roberts of Missouri, had a farm. From 1903 to 1906, Elmer worked as a telegrapher for the Rock Island System railroad. He lived in a boarding house in St. Louis, where he met Olive May Campbell who was working there. She was from Bloomington, Illinois, where her parents, H. Bedford Campbell and Mary Lucretia Toomey, lived. Olive May was the love of my grandfather's life. It is said, when he kissed her in the butler's pantry of the boarding house, she replied, "Why, Mr. Orr!"

In 1906, Elmer wrote a letter seeking employment with the Panama Railroad, and his words were to the point, "If you want a good man, you will hire me." He was hired in New York by the Railroad on August 27, 1906. Before he married Olive on March 4, 1908, in Bloomington, Illinois, Elmer wrote again requesting quarters for a married couple in the Canal Zone. In 1908, they arrived at their first home, a two-story, four-family, wooden building in Gorgona, Canal Zone, a construction day town. One of my grandmother's oldest memories of this town was the coffins stacked at the railroad station.

On June 24, 1909, my mother, Juanita Nile Orr, was born in Gorgona. She was named Juanita after the title of a song by the same name which my grandfather enjoyed playing on the mandolin. On March 11, 1911, her brother, Earl Campbell Orr, was born; and her brother, Elmer Bedford Orr, was born on January 31, 1912. Both were born in the old Colon Hospital. My grandmother once observed, "It was like having triplets!"

Olive and Elmer Orr

Over the years, the family made friends with so many wonderful folks. My grandmother was a member of the C.Z. Federation of Women's Clubs in Cristobal. She can be seen in a photo taken in 1908 at Empire in her beautiful long white dress and large black hat. My grandfather's cousin, Joe Orr, also worked and had a family in Panama.

From time to time, the Orrs returned to the U.S. to visit relatives. At the very beginning of one such trip en route to Colon, Elmer suddenly realized he had left his gold in the drip pan of the ice box in Gorgona. With the help of some friends, he made a swift recovery of the money before the ship left Colon!

In 1913, the Orrs moved to Balboa just prior to the evacuation of Gorgona, which occurred on July 1. Their new home on Morgan Avenue was close to what was then Balboa Elementary and High School, and Elmer's father came from the U.S. to live with them.

My grandfather enjoyed tennis, golf, and swimming

and encouraged his children in these sports. My mother took piano lessons, and her brother Elmer was in the Balboa Boys Band.

In 1924, my mother Juanita, or Nita, as she was called, met a new student in her high school class. His name was Russel Joseph Jones, nicknamed Rusty. Much later he became my father.

He was born June 2, 1909, in Scranton, Pennsylvania, to Simon Butler Jones and his wife, Bessie Elizabeth Grier. They had married on December 11, 1907, in Belvidere, New Jersey. Simon's parents were Simon Butler Jones I and Ella Shimer, and Bessie's parents were John Grier and Mary Davis, both of whom had been born in Wales.

Bessie Grier Jones and Simon Butler Jones.

In 1912, Simon Butler Jones was hired by the Panama Canal Company as a "wireman." He and Bessie traveled to Panama with their two little boys, Rusty, and Hayden Butler, born in 1911. The family lived briefly in Panama City near 4th of July Avenue before moving to Gatun, where they resided for 12 years. Over the years the family grew with the addition of Mary Ella in 1914; Betty Grier in 1917; Simon Butler, Jr., in 1922; and the youngest, Mildred Arlean in 1926.

When Rusty was about eight years old, he and Hayden decided to start an egg farm. They earned enough money to buy a few chickens which they took out in a boat to a small island in the lake and tethered the birds to some stakes so they couldn't run away. A few days later, the boys rowed out, only to discover that the chickens had been eaten by the caimans! As a child, my dad loved to bring home locusts, his pockets, sopping wet on arrival. He was also known to have hatched a number of crocodile eggs in the family's bathtub.

As he grew, Rusty became skilled at playing basketball; he was an amateur boxer in Gatun. And his dad was a proud father who coached and encouraged all the children in athletics.

Once, in elementary school, the principal sent for Rusty and asked, "Where is your brother Hayden?" Rusty replied, "I think he's in school." Whereupon the principal took Rusty to the window and, looking up the hill, asked, "Who is that boy on the swing?" Rusty had to agree it was Hayden, skipping school and swinging to his heart's content!

Early in the summer of 1924, my dad's parents told Rusty the family would be moving to Balboa, but he didn't want to believe it. So off he went to Boy Scout Camp in Boquete where he had a splendid time. (Much later he told me about the sounds of the wildlife there, particularly the howling monkeys.) However, when he came home, he discovered home had moved to 811 ½ Empire Street, Balboa.

In high school, Rusty was on the basketball team and also played on a local team, the Colts. After graduation, he painted for the Panama Canal Company to earn money for college. Nita Orr, meanwhile, started off for college in Chicago, but desperately homesick, returned to Panama and was employed by the Bureau of Clubs and Playgrounds. In 1928, Rusty went to Brooklyn Polytechnic Institute on a basketball scholarship. He was the high scorer for two years on that team and was selected to be team captain for the 1930-1931 season. However, when funds from his scholarship ran out because of the Depression, he returned home in 1930 and resumed working for the Panama Canal Company as an accountant.

Juanita Nile Orr and Russel J. Jones were married at Balboa Union Church on November 17, 1932. Their first home in the Flats was surrounded with many good friends. They joined the Union Club in Panama. Nita had an exquisite pollera made for Carnival. And Rusty played baseball in the Twilight League. His dad, Simon Butler Jones, worked as an electrician for the Panama Canal Company until his death following a heart attack on December 20, 1934, at Gorgas Hospital.

My grandfather, Elmer Franklin Orr, was retired following a heart attack on June 30, 1935. He, my grandmother, and my great-grandfather moved to San Diego, California, where Elmer died on March 12, 1951. He was a member of the Order of Railroad Telegraphers for 25 continuous years. He was the recipient of the Roosevelt Medal and three bars. And over the years of his employment with the Panama Canal Company, he had risen from a position of operator, then dispatcher, then Acting Master of Transportation, to Master of Transportation and Superintendent of the Panama Rail Road Company.

My mother played the piano in a recital at the National Theater in Panama City in April or May of 1937, and I was born at Gorgas Hospital on December 8, 1937. When I was in the fifth grade, my parents built a small cottage in El Valle near several friends' homes. We spent many weekends in that wonderful place. My friends and I hiked, swam and rode horseback. There were potluck dinners at night, and all the kids played ring-a-levio and told ghost stories.

In 1951, my grandmother Orr came to live with us until her death on July 27, 1966. And my grandmother Jones lived with her daughters in the Canal Zone until her death on May 13, 1970.

My dad Russel J. Jones worked for the Panama Canal Company for more than 38 years until his retirement in 1966 as Assistant Chief Accountant. And my mother, Juanita Orr Jones, also worked for the Panama Canal Company as an administrative assistant for 20 years. They retired to Palo Alto, California, where they lived near me. Rusty died on January 2, 1990, and Nita died on November 1, 1991.

All of the above named family members are gone now. They, like many others, were committed to the dream, the realization, and the operation of the Panama Canal. The Canal remains a remarkable engineering wonderment. And those who went to Panama, were born in Panama, grew up there, and worked there, will never be forgotten.

Juanita Orr and Russel Jones.

Elmer Franklin Orr
Roosevelt Medal No. 1807 with Three Bars

Joseph Orr
Roosevelt Medal No. 5501

Panama Railroad Station at Balboa, July 1927.

PHOTO COURTESY OF PANAMA CANAL MUSEUM

The Charles H. and Margaret R. Peterson Family

As told by Thomas C. Peterson

Charles Peterson was born to Louis C. and Gretchen (Maebeck) Peterson in Chicago, Illinois. His given name was Charles, and later in life he added the middle initial "H." Charles's father was employed by the Illinois Central Railroad; when Charles was six months old, his father was transferred to the railroad yards in Decatur, Alabama.

Charles was married to Cecil (Petey) Peterson before marrying Margaret. When Cecil contracted tuberculosis, Charles resigned and took her to the States to recover. She died in Los Angeles, California, in 1925. Charles then returned to the Canal Zone.

Margaret Lucile Reed was born on a farm in Vernon County, Wisconsin, near Viroqua. She attended public schools in that area and went on to graduate in 1916 from La Crosse Normal College in La Crosse, Wisconsin, renamed University of Wisconsin-La Crosse. Upon graduation, she began teaching in Minneapolis, Minnesota. While there she saw an advertisement for teachers in the Canal Zone; she applied and was accepted. At first she was undecided whether to accept the offer since it seemed so far away from her home and family. She talked to her parents, and her father said to go as it was the chance of a lifetime. She and another teacher from the same school were recruited and remained friends for life.

Charles was employed as a blacksmith at Gatun Locks during the construction of the Panama Canal and was awarded the Roosevelt Medal with one bar for his services. Charles was an outstanding golfer and won the Panama/Canal Zone Isthmian Golf Championship in 1919, 1920 and 1921 — the only person ever to win the championship three times. He became a member of the Professional Golfers Association, was the club professional at the Panama Golf Club and the Summit Hills Golf and Country Club at different times, and built the majority of the golf courses in Panama and the Canal Zone.

Margaret taught her favorite class, third grade, at Ancon Elementary School. She was promoted to supervisor of curriculum and education which required visiting other elementary schools in the Zone. The unofficial title given to this position was supervisor, track teacher, or just track teacher (without supervisor), which referred to the fact that she traveled on the Panama Railroad to each townsite to visit the schools.

> *Charles was an outstanding golfer and won the Panama/Canal Zone Isthmian Golf Championship in 1919, 1920 and 1921 — the only person ever to win the championship three times.*

Charles and Margaret were married by Bishop Morris, a close personal friend, on April 9, 1927, at St. Luke's Episcopal Cathedral in Ancon. After the marriage, Margaret had to resign because the policy was not to employ married teachers. Charles and Margaret had one child, Thomas Charles Peterson, born January 14, 1931, in Gorgas Hospital in Ancon.

Charles was employed as a blacksmith by the Locks Division at Pedro Miguel, and for a while he was also a golf instructor in the Division of Schools and Playgrounds. At the same time he was the golf pro and manager of the Panama Golf Club. He served as mayor of Pedro Miguel for several terms. One of the many changes he made to better living conditions was to introduce and require that all government service centers and restaurants in the Canal Zone serve covered drinking straws. This was done mainly for sanitary reasons.

Charles passed away on January 11, 1947, while employed at the Pedro Miguel Locks. After his death, Margaret was employed by the Maintenance Division in Balboa and later transferred to Cristobal. She retired in 1957 and moved to the States. Prior to her death on July 22, 1993, Margaret was recognized as the oldest living alumnus at the University of Wisconsin-La Crosse.

Charles H. Peterson
Roosevelt Medal No. 6234 with One Bar

See also the Thomas C. and Barbara D. Peterson family history.

PHOTO COURTESY OF *PANAMA CANAL REVIEW*

Front entrance to Panama Golf Club (1968). A far cry from the bohio it was in 1918.

The Thomas C. and Barbara D. Peterson Family

As Told by Barbara Peterson

Tom was born January 14, 1931, to Charles H. and Margaret R. Peterson, in Gorgas Hospital, Ancon, Canal Zone. He graduated from Balboa High School in 1949, received an A.A. degree from Canal Zone Junior College, an A.B. degree from Birmingham-Southern College, and worked toward an M.P.A. degree at New York University. Tom earned the Eagle rank in the Boy Scouts of America in 1946. At the time, he was the second Zonian ever to reach this rank. The requirement for Eagle Scout is to be a Life Scout and earn 21 merit badges; Tom went on to earn 21 more for a total of 42. He served in the U.S. Army during the Korean War period, stationed at Fort Jay, Governors Island, N.Y.

Barbara Louise Denier was born July 23, 1934, to Thomas William and Gladys (Fowler) Denier in Jamaica, New York. Tom and Barbara met in New York in 1954 while Tom was in the Army. They were married August 27, 1955, on Governors Island. Tom's tour in the Army ended shortly thereafter, and they moved to Birmingham, Alabama, where Tom attended college and Barbara worked in a local department store.

In 1956, they went to the Canal Zone so Barbara could see where Tom grew up. He was offered a job in the Personnel Bureau. At the time of his retirement, he was Chief, U.S. Recruitment Section and Special Placement Officer. He was responsible for recruiting from the States for all positions that could not be filled locally. On retirement, Tom was presented the "Master Key to the Panama Canal" by the Panama Canal Commission Administrator, the highest honorary award that the Governor or Administrator could confer.

Barbara was also employed, first in the General Audit Division, Office of the Comptroller, and later in the Engineering and Construction Bureau. While with the Bureau, she worked in the Sea Level Canal Support Division and served as Secretary of the Canal Zone Board of Registration for Architects and Professional Engineers. At the time of retirement, Barbara was Administrative Assistant to the Chief, Construction Division/Contracting Officer.

Tom and Barbara had three daughters, all born in Gorgas Hospital: Carol Louise (Heintz) (Krueger), Diane Margaret (Pearson), and Elaine Ann (Little).

Tom was very active in the Masonic fraternity and attained the top office there as well as in affiliated organizations. He received the Joseph Warren Medal and the Henry Price Medal from the M.W. Grand Lodge of Massachusetts. He was appointed as R.W. District Grand Master of the R.W. District Grand Lodge, A.F.&A.M., at the Panama Canal, 1976-1979. During this period he also served as Deputy of the Supreme Council, 33rd, A.&A.S.R., Southern Jurisdiction, USA, and Potentate of Abou Saad Temple, A.A.O.N.M.S. At the time of his appointment, Tom was the only U.S. citizen to be made an honorary member of the Grand Lodge, A.F.&A.M. of Panama, and the Supreme Council, 33rd, A.A.S.R. of Panama, both extremely high honors. Tom and Barbara also served as Worthy Patron and Worthy Matron, respectively, of Orchid Chapter No. 1, Order of Eastern Star. They both worked with the Order of Rainbow for Girls, Balboa Assembly No. 1, and received the Grand Cross of Color from that Order for their services.

Tom and Barbara retired to Sarasota, Florida, in 1981. After retirement, both became active with the Panama Canal Museum. Tom also served as the President of the Panama Canal Society, 1998-99.

They have nine grandchildren ranging in age from 25 to 14: Christine, Jason and Eric Heintz; Kristle, Michelle and Tim Pearson; and Tom, Sidney and Chimere Little. Tom passed away on October 20, 2006.

See also the Charles H. and Margaret R. Peterson family history.

The Russell B. Potter and Ross Cunningham Families

AS TOLD BY DICK CUNNINGHAM

Russell B. Potter was born in Belvedere, New Jersey, on September 28, 1884. He attended public school there, graduated with honors at the age of 16, and was valedictorian of his class. After spending the next year working as a grocery boy for the sum of $3 per week, his father sent him to Rider-Moore Business College in Trenton. There he learned secretarial and stenography skills and took a job with the Pennsylvania Railroad.

After taking the civil service exam, he went to work for the Isthmian Canal Commission of the Panama Canal. He arrived in Panama and started working on July 5, 1905. Malaria was rampant, and he was given mosquito netting, a tin pitcher, and a tin pail. When the new arrival reached his quarters, he found there were no walls, and all he got was a cot. The pay was good at $100 per month, so he stayed.

During the construction years, he lived in Empire, Canal Zone, and worked for the Central Division. He was promoted to the position of private secretary to Colonel Galliard. One of the jobs of the Central Division was the excavation of the "cut through the mountains." Called the Culebra Cut, it was renamed the Galliard Cut after the opening of the Canal.

Miss Linda VanEmster was visiting her sister Bertha in the Canal Zone. Bertha was married to Fred Whaler, and they also lived in Empire. Russell met Linda, and after a courtship he wrote a letter to her mother asking for Linda's hand in marriage. Linda returned to her home in Bay City, Michigan, and Russell traveled there for the wedding, which took place on August 24, 1911. The couple returned to Panama and lived in Empire. The newlyweds took a honeymoon trip on the Chagres River, traveling by cayuco and visiting many of the villages and people along the river. In those days, this was a grand adventure as the river was lined by jungle most of the way since the lake for the Canal had not been completed.

Linda and Russell Potter, 1947.

In 1913 Russell was promoted to "office head," and the next year on January 24 their first child, Janet, was born at Ancon Hospital. The Panama Canal was officially opened six months later on August 15, 1914.

Russell and his family went to Washington, D.C., in January 1915. He was awarded the Roosevelt Medal with three bars for his service during the Canal construction era.

He became a secretary for the Department of Government Research. Some of his assignments included travel with the U.S. Congressional Committee delegated to establish national parks. These trips were made by train and wagon and covered the 48 continental U.S. (some still territories) in order to find appropriate areas for national parks.

During this time, World War I began, and Russell tried to sign up for the Officer Training Corps in the U.S. Army. He was rejected due to "flat feet."

In March 1917 he returned to work in the Canal

Zone. Linda stayed with his parents in New Jersey, where she delivered their second child, Richard Russell Potter, on June 4, 1917, at the McKinley Hospital in Trenton. Linda returned to the Canal Zone with the two children in September 1917, and the family lived in Balboa.

Russell became one of the leading members of the U.S. Governor's staff and worked in the Administration Building until his retirement in 1947. Anyone having a question about things in the Canal Zone could "ask R.B."

While in the Canal Zone, Russell participated in many activities. He won several silver cups in golf, his favorite sport. Fred Whaler, his brother-in-law; Chris Garlington; and Russell owned a boat which they used for sport fishing on Panama Bay. They caught many types of fish, including marlin, sailfish, tuna, and corvina.

While living in Washington, D.C., Russell was on a baseball team made up of men from his workplace. He played baseball and basketball on a team from the Central Division while in the Canal Zone. He also hunted in the jungles of Panama with his friends.

Linda belonged to the Balboa Union Church Women's Club. She was always busy baking pies, cakes, and other delicious food treats for social gatherings, especially for church dinners. She taught herself the art of oil painting and completed many scenes of Panama. One of her works was recently selected by the Panama Canal Museum to be in the 2008 calendar.

Russell's father, Barzillia Potter, born on October 5, 1857, worked for the Pennsylvania Railroad for 46 years, over 40 of which was as a locomotive engineer. He died on October 16, 1936, in Belvedere, New Jersey.

> *Her parents, Russell and Linda Potter, had to walk on a swinging cable footbridge across the Canal bed (no water yet as the Canal was not finished) to board the railroad train to get to the hospital. On their return they repeated the perilous journey, this time carrying baby Janet in a valise with one parent at each end holding the handle grip with one hand and the swinging railing with the other.*

There is quite a story associated with daughter Janet's arrival in the Canal Zone. She was born on January 25, 1914, six months before the Canal opened for ship transit. Her parents, Russell and Linda Potter, had to walk on a swinging cable footbridge across the Canal bed (no water yet as the Canal was not finished) to board the railroad train to get to the hospital. On their return they repeated the perilous journey, this time carrying baby Janet in a valise with one parent at each end holding the handle grip with one hand and the swinging railing with the other.

Janet was enrolled in a private school when she was four years old. After a year of kindergarten she went to second grade, but always regretted the age disparity with her classmates. She resolved that if she ever had children of her own, they would never be permitted to skip a grade. She was 14 years old when she met Ross Cunningham. Because he was six years her senior, her father would not allow them to date. Ross spent many a balmy evening, sitting on the Potters' front steps visiting with Janet while her parents remained close at hand.

Ross and Janet Cunningham family, 1949.

Ross graduated from high school in Sedalia, Missouri, and along with his younger brother, Joshua (JA), went to the Canal Zone during the depression in 1928. Their older brother, Frank, was working in the Canal Zone and had written home that there were available jobs in the Zone.

After working several jobs, Ross became a civil service accountant in the Canal Zone. His job was to travel by pilot boat and board the big ships waiting to transit the Canal. Carrying a black satchel, it was his job to collect the Canal tolls and verify the ship's documentation. He wore a pistol on his belt, but thankfully never had to use it. In later years, the tolls were collected by wire transfer.

After Janet's graduation from Balboa High School in 1931, her parents sent her away to Fredericksburg Teacher's College, now Mary Washington College, in Virginia. After two years, she returned to the Canal Zone to marry Ross. The wedding took place in the Balboa Union Church on February 5, 1934.

The couple's three sons, Richard, Edward, and Tom, and their daughter, Linda Sue, were born in the Panama Hospital.

Ross retired in 1963, and the couple moved to St. Petersburg, Florida, to live with Janet's parents. They became active in the Northeast Presbyterian Church. As charter members, Ross was a faithful choir member, and Janet contributed her talents to the women's activities.

After many happy years in retirement, Russell died at home in St. Petersburg on December 2, 1978, and Linda died on June 28, 1985. They had traveled extensively during their retirement years, making many trips to the West coast to visit relatives and friends.

The Cunningham family purchased a home in North Carolina. While they loved sunny Florida, they also loved their summer home at Connestee Falls, about seven miles south of Brevard, North Carolina. Being there for six months enabled them to visit daughter Linda and her family in Winston-Salem.

Ross passed away on November 27, 1983, at the summer home. A memorial service was held at Northeast Presbyterian Church in St. Petersburg, and another was

L-R: Ed, Tom, Linda Sue, Mom (Janet), Dick, 2004.

conducted at the lake house in North Carolina.

Janet still brightens our days with her cheerful outlook and enthusiasm for living. Her family helped her celebrate her 90th birthday in 2004 by having a family reunion. The many activities included several dinners, a picnic at Ft. Desoto Park in St. Petersburg, Florida; recognition at Northeast Presbyterian Church; a visit to the Panama Canal Museum; and a reception at the home of Dick and Lynn Cunningham. All of her children and their mates, six of her ten grandchildren, and five of her great-grandchildren were in attendance, along with many friends. Janet requested that the family limit the celebration to one day when she turns 100 as she was extremely tired.

Sons Ed and Dick live in St. Petersburg with their families. Tom resides in Castle Rock, Colorado, near Denver with his family; and Linda Sue lives in Winston-Salem, North Carolina, with her family.

Russell B. Potter
Roosevelt Medal No. 649 with Three Bars

The Patrick Joseph and Jane "Jennie" Quinn Family

AS TOLD BY BRUCE QUINN, MARC QUINN AND ELAINE LOMBARD NEWLAND

Soldiers of Erin

Her name was: Jane Quinn. His name was: Patrick Joseph Quinn. He called her "Jennie." Everyone called him Pats. He was born in 1875 in Newark, New Jersey, the first U.S.-born son of Irish immigrants from County Clare. She was born in Manchester, England, in 1879. They met in an area populated by Irish immigrants in northern New Jersey, near Newark. They were married in 1900 — the beginning of a new century, a new life. Little did they know that their future would take them far from the Jersey shores to where both their families had recently emigrated from Ireland following the American Civil War.

Jennie's brother, Joseph Aloysius Corrigan, like the biblical Joseph, led them into a promised land: the Isthmus of Panama. It was Joseph Corrigan who went to work for the Isthmian Canal Commission (ICC) shortly after the United States took up construction of the Panama Canal following Panama's separation from Colombia in 1903 and recognition of the new country by the United States. Joseph got in at the ground floor of the newly formed ICC. He became a supervisor in the Atlantic Sanitation Department. He may well have encouraged his parents (John and Anne Corrigan), his brothers and sisters (Peter Francis, John Paul, Owen, Jane, and May), and Jane's spouse, Patrick Quinn, to start a new and exciting life where they would be well paid and where a better future for their families was a significant probability. May would also marry in the Canal Zone to Edward Welch.

As noted, Pats heeded his brother-in-law's advice. He set sail, without Jennie and four children, from New York City aboard the steamer *SS Colon* arriving in Panama in May 1907. On June 3, 1907, he was employed as an ironworker at 56 cents an hour, "United States currency value." That was more than he was making as a leather grainer at 40 cents an hour with the Rex Imperial Company in Newark.

The future looked good for the young Patrick Quinn family and all the Corrigans.

Upon accepting the job, Pats signed an oath in which he solemnly swore to support and defend the Constitution of the United States. These builders were heroic soldiers, to whom they were later likened by President Theodore Roosevelt during his historic visit to Panama. The President, upon his departure from the Canal Zone, said that recognition due to them would be accomplished by a medal to be struck in their honor. Patrick Quinn became

Patrick and Jane Quinn, 50th wedding anniversary, 1950. Then retired from the Panama Canal and living in Toms River, New Jersey.

one of the 7,400+ American recipients of the Roosevelt Medal, awarded to U.S. citizens who worked more than two consecutive years on the Canal construction from 1904 to 1914. He held medal number 2811. Three Corrigan brothers also were recipients of the Roosevelt Medal.

By 1908, Jennie and four children (Marcus, Alice, Theresa, and Genevieve) were finally joined with Pats in the Canal Zone. In addition, Joseph Corrigan had been instrumental in convincing three other brothers to come to Panama and they were on the ICC payroll. Anne Quinn, Pats' sister, married Peter Corrigan, Jennie's brother.

With the arrival of his family, Pats transferred from the Colon Dredging Division to the La Boca Division in 1908.

Jennie would recall that life changed dramatically for her, arriving at La Boca where coal for her stove had to be retrieved from a railroad car parked alongside the townsite where she and her brood would make their way over to the train yard to obtain their quota of coal. They had no electricity, and perishables were kept in an "ice box." NOTE: Even decades later into the 20th century, a true "Zonian" would always refer to the refrigerator as an "ice box."

Although twice riffed for short periods, Pats continued working with the ICC until near the end of the construction era. Between 1907 and 1913 he served with the Mechanical Division as an ironworker, car repairer, and inspector assigned to Gorgona, Culebra, and Gatun. As completion of the Canal was achieved, his position was abolished and the family returned to New Jersey. By now there were seven Quinn children with the addition of Marjory, James (born in the Canal Zone), and Rita.

During World War I, Pats was a shipfitter at the Brooklyn Navy Yard. He was reemployed by the Panama Canal in 1919 as a shipfitter at 70 cents an hour. He had lived through a bout of "malaria fever." Post war life in the Canal Zone was bustling. The former ICC was now The Panama Canal. The organization had cleared up the devastating "Cucaracha Slide" blocking Culebra Cut. Between 1914 and 1919, children eight and nine, Anne and Regina, were born in New Jersey. The 10th child to survive, Claire, was born after the family returned to the Canal Zone.

Pats (seated left) with his son, six-year-old Marcus, on his lap and brother-in-law Joseph Corrigan (far right, brother-in-law of Patrick) aboard an ICC dredge moored at the future Pacific entrance of the Canal at La Boca (with master of the dredge – unknown).

By 1925 Pats was promoted to shipfitter/boilermaker earning $1.08 an hour. The Quinn Family lived now in Pedro Miguel and would move to Morgan Avenue in Balboa, their residence until Pats' retirement in 1937. The family thought of itself as two families with Marcus, Alice, Theresa, Genevieve, and Marjory the senior members and James, Rita, Anne, Regina, and Claire as the junior family. The Quinn family was known for its beautiful women. The two male members were images of handsome Irish stock, both sportsmen. Jennie taught her family to love the arts. Once she told the story of going to the wooden Balboa Theater, where canvas awnings were lowered to darken the interior of the theater, for the performance of a renowned tenor. She said that there were only six people in attendance to hear Enrico Caruso sing! She also saw Pavlova dance at Pier 18 in Balboa. Being a staunch Irish Catholic, Jennie also gave her children their moral integrity. She loved doing crossword puzzles and reading. She even took French classes. Pats loved music, dancing, and playing pinochle. Their large household was characterized by an atmosphere of *joie de vivre*.

So in 1937, Pats retired after 24 years, nine months, and two days of service. He and Jennie with their youngest progeny, Claire, made their retirement home finally in Toms River, New Jersey.

In a loving gesture to their parents, Pats and Jennie Quinn's 10 children (names italicized) posed for a classic

Our family is the real testimony of the full life of Pats and Jennie Quinn.

photograph (above) in May 1936 taken on Ancon Boulevard with spouses and children sitting in wicker chairs on the lawn in front of a stately French Canal quarters like a cadre of Erin's soldiers passing in review for their parents. Standing from left to right: *Theresa "Tess," James, Claire, Marjory* with spouse James "Buster" Burgoon, *Regina, Marcus,* and *Alice*. Seated from left to right: *Anne* with Marc Quinn, son of *Marcus* and Berta; Berta with son Bruce; Alberta Lebrun (married to *James*) with son James "Jimmy"; *Rita* (married to Robert "Cocky" Crume (not pictured) with daughter Jane; *Genevieve* seated with Elaine Lombard. Not pictured: Eugene Lombard, spouse of Alice, and their son Richard "Dick" Lombard, presumably taking the photograph. Grandchildren not yet born at time of photograph: daughter Patricia to Marcus and Berta; adopted son Gregory to James and Marjory; daughter Lari to James and Alberta; daughter Jo Anne to Robert and Rita; sons Charles and Alan to Walter and Anne; son Tristan and daughters Kathleen Mai, Patricia and Diana to Tristan and Regina.

Patrick Joseph Quinn
Roosevelt Medal No. 2811 with Two Bars

Joseph A. Corrigan
Roosevelt Medal No. 260 with Three Bars

John P. Corrigan
Roosevelt Medal No. 2283 with Two Bars

Peter F. Corrigan
Roosevelt Medal No. 2348 with Two Bars

See also the John Paul Corrigan, Sr., and the Peter Tiernan "Pete" Corrigan, Sr., family histories.

Irish immigrants were the backbone of the construction of many great undertakings in the 19th century and into the 20th century as they escaped the Great Famines. Photo is of great-grandfather John Quinn's family, circa 1880. Standing, L-R: Margaret Quinn (daughter), John Quinn (father – note medals on chest), Brigitte Quinn (nee Conroy, wife). Seated, L-R: Luke (John's brother) with baby in lap, John, Jr. (son); Patrick "Pat" Joseph (son), age 5; and Anne Quinn (daughter).

The E. L. Raines Family

As told by Fred Raines

From Pine Apple to Frijoles on the Railroad

Several times a day, the long, lonesome whistle of a distant train could be heard drifting across the red dirt farmland of Conecuh County, Alabama. It seemed a siren call to a young farm boy as he leaned on his hoe in the cotton field near a rural intersection called Pine Apple, his birthplace. So it was in 1922 that Elma Leroy (E.L.) Raines at age 15 put aside his hoe, walked almost 20 miles to the settlement of Tunnel Springs and apprenticed himself to the telegraph operator at the railroad station. By the mid-1930's, the Great Depression had devastated the economy; and E.L. struggled to care for his young wife, Ernestine Gaston, a former Conecuh County school teacher, and me, their only child, Freddy. His telegrapher skills allowed him to hire on for seasonal work with the railroads; and one desperate day, his telegraph tapped out a message that the U.S. Government's Panama Railroad Company was hiring. He eagerly applied, qualified, signed on and sailed "across the waters" to the Panama Canal Zone to "try out" the job with the world's shortest (fifty miles) intercontinental railroad. In early 1938, Ernestine and Freddy boarded a ship to their new home.

Settling into life in Balboa was dramatic and anxious as World War II loomed and finally broke out. The Balboa railroad station was the Pacific side anchor of the transportation and communication system and the center of much military activity and ceremony. Ernestine took a job with the Army at nearby Albrook Field. It was exciting for a youngster who knew little of the realities and consequences of war but loved the soldiers, planes, searchlights, blackouts and air raid drills. Sometimes, though, it was fearsome.

In 1941, E.L. was detailed to a railroad station in one of the most remote places in the Canal Zone. We moved to Frijoles ("beans" in Spanish), which was located on a peninsula of Gatun Lake deep in the tropical rainforest, isolated with no connection to the outside world except for the railroad. The Frijoles railroad stop served as land access

Frijoles Railroad Station, circa 1941.

and water transfer point for people, supplies and equipment for the Smithsonian Institution's Tropical Research Facility on Barro Colorado Island. Frijoles consisted of a tiny railroad station, our house on stilts at the edge of the jungle on the shore of what was then the world's largest man-made lake, one other house, and a small dock which served the Smithsonian's needs and provided access to the rail line for the Gatun Lake natives and their "commerce."

For a seven-year-old boy, it was a place out of a storybook. Daybreak was heralded by a deafening cacophony of monkey and bird noise. Evenings brought alligator barks and the deep, not-too-distant growls and movements of other denizens of the jungle. Dazzling butterflies and birds danced through the forest clearings, beams of sunlight illuminating their brilliant colors against the cool dimness of the rainforest. Exotic orchids clung to branches in the higher levels of the forest. Moving from the hot, bright sun into the dense dark jungle would always cause a shiver.

Going to school became an adventure, too! Every morning I would board the early southbound train for the half-hour ride to Balboa railroad station. There, I was met by the driver of a taxi — magnificent touring cars with white canvas covered seats and convertible tops — and would be driven the few blocks to Balboa grade school. At the end of the day, the trip was reversed, and I returned by train to the rainforest. There I would swim in the lake off the Smithsonian dock; paddle around in my small cayuco boat to explore the lake and jungle; catch fish; practice with rifle, pistol and knife; and tend to my growing collections of orchids and butterflies as well as an assortment of wildlife and insects that I had captured and cared for.

By 1944, we had moved to Gatun on the Atlantic side, and the scene still reflected the ongoing war. Barrage balloons floated near the Gatun Locks. Machine guns, anti-aircraft guns and searchlights were scattered throughout the town and on the high ground overlooking the Canal and locks. The extraordinary security measures, shortages and accoutrements of wartime had by then become routine; among them: dented cans of food salvaged from damaged or sunken supply ships, powdered eggs and milk, air raid drills and censored mail.

Our family lived inside the Gatun Railroad Station, occupying the entire second floor of that large concrete and wood building. Everything to do and every place in Gatun was no more than a short walk away — school, pool, clubhouse, commissary, post office, gym, baseball field, and all the rest. The entire second floor apartment of the station was open and screened on three sides, which took advantage of the prevailing winds and offered spectacular panoramic views of Gatun Lake and the Panama Canal.

The wildlife at Gatun railroad station could not compare with that of Frijoles, but I remember watching the migration of thousands of large yellow, black and green butterflies as they swept across the lake onto the shore. I remember standing on the railroad station's open platform and easily hand-picking butterflies in flight from the crowded air before they began to rise and disperse overland.

Fred and E.L. Raines, with pet parakeets, Gatun Railroad Station, circa 1944.

The palm trees around the station were roosting areas for little green parakeets. Every evening as the sun set, they would flock in by the hundreds and noisily settle in the palm foliage for the night. The palm could be scaled, allowing me to actually reach into the mass of sleeping birds and collect a few as pets. Despite their sharp beaks and a sometimes ornery disposition, they were people-friendly, easily-trained, and provided great entertainment until rereleased into the wild.

The railroad itself was a handy source of entertainment and education. Some days would find me sitting in the cab with the engineer as he allowed me to operate the controls and whistle. Other days I would hop into the baggage car of a train and ride across the isthmus and back, sitting in the wide open doorway within arm's reach of the flora, fauna, activities, and people along the way. I practiced Morse code and telegraph techniques by sending and receiving messages to and from other stations on the line, and I was often utilized as the paid telegram delivery boy.

In those days, train crews communicated via visual signals and train whistle code, while more complicated messages were attached to six-foot bamboo poles to be "caught" by the train as it thundered by. My dad (or I, preteen boy) would stand at the very edge of the tracks and hold the pole so the engineer could lean out and grab the message; a second copy of the order was grabbed in the same way by the conductor in the last car or caboose. Seldom have I experienced more thrilling moments than standing firm and holding the delivery pole as the huge, hot, steam-belching, engine thundered past only inches away.

After VE day in 1945, our view of the Canal offered a massive parade of warships and transports passing through the locks from the Atlantic Theater to the Pacific. Our close proximity and easy access allowed us to stand at the locks' edge, sometimes close enough to touch hands and exchange gifts and souvenirs with those on the passing ships.

As darkness fell each day, the southbound lighthouse on Gatun Locks cast a glow into our railroad station home. Throughout the night, we could hear the faint, and somehow comforting, sounds of the locks, mechanical mules, and muffled voices of the operators and crews as ships passed through. Periodically, these comfortable night sounds would be punctuated by the rumble of a passing freight train. And so it was at the Gatun railroad station home of the Raines family.

In 1947, E.L. would be diagnosed with tuberculosis and retired on disability from U.S. Government service and the Panama Railroad. The family left our Canal Zone home fully expecting an early return, but this was not to be. For years after his successful treatment and recovery, E.L. sought, to no avail, to be reinstated and return with the family to the Canal Zone. Back home in Conecuh County, he tried farming; but he was called once again by the whistle of the train and took a job with the Rock Island Railroad. I was, by then, away from home with my own family. E.L. and Ernestine were living in Cedar Rapids, Iowa, when his life ended all too soon. He was returned home and laid to rest in the warm, red earth of Conecuh County, finishing his journey near where it began — and where Ernestine still lives — far from Frijoles, but not far from Pine Apple, where the lonesome whistle of a train can still can be heard drifting over the fields.

Fred Raines, flanked by friends Freddy and Sonny Templin, Gatun Railroad Station, circa 1944.

The Erwin and Dorothy Ramsey Family

The Indelible Memories of Erwin and Dorothy's Children, 1941-1963

Erwin F. Ramsey and Dorothy M. Tingley, married on June 26, 1930, spent the first ten years of their marriage working on some of the massive construction projects of President Roosevelt's New Deal Era. Employed by the contractors Wunderlich and Oakes, Erwin operated heavy equipment at Montana's Fort Peck Dam and Colorado's Wolf Creek Pass and Vallecito Dam projects.

In February 1941 Wunderlich was awarded a contract for excavation work for the 3rd locks project at Gatun. As Wunderlich's lead heavy equipment operator, Erwin moved the company's heavy equipment to San Diego for shipment to the Canal Zone. Because housing for the contractor families was not yet completed, Dorothy and their three children, Jim, Nancy and Dan, could not accompany him at that time. Sailing on the Japanese freighter *Maraka Maru*, the Wunderlich crew encountered very rough seas with most men being seasick for the entire voyage. Balboa was a welcome sight after days of seasickness and eating rice and green turtle soup.

In May 1941, when family housing became available, Dorothy and the children left her parents' farm in Minnesota and drove to New Orleans to board the United Fruit ship *SS Taloa* for their passage to Panama. The *Taloa* stopped in Havana, Cuba, and Puerto Limon, Costa Rica, before arriving in Cristobal on May 18. Remembering the trip, the children recall, "There were always large trays of bananas available, and we could eat all we wanted."

Contractor housing was located in Margarita on the hill where Holy Family Catholic Church currently stands. Families were assigned to a one-bedroom frame house, built on stilts with a combined living/dining room and a kitchen complete with an electric stove and refrigerator. Refrigerators were a real luxury as most PCC families only had iceboxes. The houses were relatively small — especially for a family of five — but proved to be comfortable with high ceilings, ceiling fans and hardwood floors. The walls in the living/dining room were framed to a height of about three feet with the rest of the wall being screened for continual ventilation; wide eaves made windows unnecessary.

> *Families were assigned to a one-bedroom frame house, built on stilts with a combined living/dining room and a kitchen complete with an electric stove and refrigerator. Refrigerators were a real luxury as most PCC families only had iceboxes. The houses were relatively small — especially for a family of five — but proved to be comfortable with high ceilings, ceiling fans and hardwood floors.*

The 3rd locks project was a mammoth excavation. Panama Canal Company engineers scoffed at the contractors' chance for completing the job on time, telling them, "You'll never complete this job in five years; just wait until the rainy season comes." Working two twelve-hour shifts, seven days a week, rarely shutting down their machines, the excavation progressed at a relentless pace. Driven by pride and a lucrative financial bonus for early completion, with the tenacity that exemplified American workers of that era, the 3rd locks excavation was finished in less than three years. Erwin's bonus was 10 percent of his total earning during the project — a tremendous amount of money for that time.

With their job complete most of the men looked to return to the U.S. However, the German U-boat threat to allied shipping and transportation priorities made return passage to the States very difficult to obtain. Erwin was offered and accepted a position with the PCC Health Department managing and operating the Atlantic side landfill at Mt. Hope. With the job came a move to a PCC "four family courtyard" on Second Street, just a few blocks from their contractor's cottage.

The serenity of the Canal Zone changed dramatically with the attack on Pearl Harbor. Fears of vulnerability were relieved three days after the attack as Army Air Corps fighters began arriving. As the planes swooped low over Margarita to land at France Field, women and children in the streets waved white hankies and towels at the pilots who were clearly visible in their cockpits. The kids were having a ball; the women were laughing and crying at the same time. Soon bomb shelters were built in the housing areas and a blackout was enforced from 6 p.m. to 6 a.m. Most evenings Erwin was joined by other men to hear the latest "war news" on his car radio. He would always park the Desoto under the house to avoid any light that might draw the attention of wardens patrolling the neighborhoods looking for light sources.

Early in the war Erwin was sent to Nicaragua to assist in the construction of an Army Air Corps airfield at Puerto Cabazas. At this remote location Erwin amazed the local children after lunch by taking out his false teeth to brush them. He returned from this job looking awful, a result of bad food and terrible living conditions.

Of the many interesting war stories perhaps the most poignant is Dorothy's account of the passage of the aircraft carrier *USS Franklin* — a story that after 40 years had passed, she could not tell without her voice breaking and eyes filling with tears. Typically when large warships passed through the locks, they were met by very exuberant crowds. Families were allowed to walk along the lock walls, and the kids would wave to the sailors and throw them bananas. In return the sailors would toss their sailor hats to the kids. The *Franklin's* passage was very different. Severely mauled in March 1945 by a Japanese plane near Japan, she was making her way back to the east coast. Passing through the Gatun Locks, flight deck mangled and charred by fire with servicemen rumored to be still entombed in her lower decks, not a word was spoken as the hundreds of solemn onlookers paid their final respects. The somber crowd did not leave until the ship ghosted northbound out of sight.

Late in the war as the Transisthmian highway was preparing to open for civilian traffic 'Little Mac' MacAlexander, a friend working on the road, asked Dorothy and the children if they would like to take a ride across the isthmus. As they drove the new road, Little Mac opined that "they were probably the first civilians to cross the isthmus on the new highway."

L-R: Dee, Jim holding Steve, Nancy, Erwin, Dorothy and Dan, 1952.

After the war the Ramseys had two more children, Diane in 1945 and Steve in 1951. With the family now at seven, a new house followed with the family moving to an "up and down" on Fourth Street. Both Erwin and Dorothy were very active in the community. Erwin served on the building committee for the construction of the Holy Family Catholic Church and the Knights of Columbus building; Dorothy was very active in the Holy Family Altar & Rosary Society. They also had a great love for the outdoors; Erwin was active in the Gun Club and helped build the Tarpon Club clubhouse, docks, parking lots and marine railway. Dorothy's gardens of orchids, gardenias, bougainvillea and hibiscus surrounded us with tropical splendor. If the Ramsey family had a second home, it surely was their boat, the "Nancy Dee," on the Chagres River. Most Sunday mornings after the seven o'clock mass the family was off for a day of tarpon fishing on the river. After leaving the dock in the lagoon, lines were set and they trolled to the mouth of the river at Fort San Lorenzo. The river and surrounding jungle teemed with parakeets, toucans, and howler monkeys. In 1949 Erwin caught a 151-pound tarpon on the Chagres that received a bronze medal from *Sports Afield* as the fourth largest tarpon caught in their worldwide tarpon contest. If ever a river ran through the lives of a family, the Chagres ran through ours.

While the post war years were generally serene there were some tense moments. During the Korean War when electrical outages at night brought fears of air raids, Dorothy taught her youngest children how to stand on the toilet seat and look out the window toward the Canal; if the Canal lights glowed in the sky, she assured them that all was fine. Once, however, the radio broadcast there were unidentified aircraft approaching the Canal and all employees were to report to work. As he was leaving, Erwin and Dorothy agreed if anything happened she would take the children and go to a place in the jungle where he would find them later. Dorothy then gave each child a pillowcase to stuff with as much canned food as they could carry. Nothing ever came of this alert, but it served as another reminder of our vulnerability and how far we were from the States.

Erwin retired in February 1963 after more than 20 years of service with the PCC. We were allowed to stay in our house until Diane graduated from Cristobal High School, the fourth in our family to do so. In June we boarded the *USS Ancon* for our final trip back to the States. Erwin and Dorothy returned to Minnesota and spent their remaining years farming not far from where they met 50 years before.

PHOTO BY KEVIN JENKINS

Aerial view of Ft. Davis, Canal Zone, looking toward Gatun Lake. Town of Gatun is at far upper right. The Panama Canal's original 3rd locks excavation project (at right), commenced during World War II, but was never completed. The present Canal expansion by Panama will use this old excavation.

The Alvin M. Rankin and Paul D. Thompson Families

As told by Frances Thompson Clinton

Alvin Monroe Rankin was born December 23, 1879, in Knoxville, Tennessee, of Thomas Lee Monroe Rankin and Sarah Bogard, and died August 8, 1952 (age 72), in St. Petersburg, Florida. He left home at a young age to work on the railroad and learn the art of being a blacksmith. He was hired out of New York in September 1906 to work on the construction of the Panama Canal as a blacksmith. For his eight years of service from 1906 to 1914, he was awarded the Roosevelt Medal plus three bars (Roosevelt Medal No. 2043, First Bar No. 1252, Second Bar No. 855, and Third Bar No. 574).

Alvin continued to work for the Panama Canal as a blacksmith and a mechanic, eventually becoming a foreman. He was employed on both the Atlantic and Pacific sides of the Canal during his career and retired December 1942. Alvin did have a couple of breaks in service taking his family to Costa Rica and Honduras, where he and his wife Agnes ran boarding houses.

At age 29 Alvin met Agnes Adelaide Walker (born January 21, 1890, in Culebra, which was still Colombia at that time), and they were married November 16, 1908. Agnes's father, Francis Phillip Walker, came from England through Cuba to Panama as an employee of All American Cable before the French Canal construction. Agnes's mother, who was Rose Wilson, was born in Cuba.

Alvin Monroe Rankin, 1930's.

Agnes Adelaide Walker, 1907.

Agnes and Alvin had four children: Anita (September 6, 1909), Alvin A. (January 11, 1911), Paul (Carlos) (July 4, 1913), and Thomas (April 22, 1915), who all worked for the Panama Canal for at least part of their lives. Anita, Alvin A. and Tom married fellow Zonians and raised their families in the Canal Zone. Two of Alvin's sons continued to work for the Canal through to their retirement.

Anita graduated from Cristobal High School and went to work for the Commissary Division. She married Paul Dana Thompson (born March 13, 1909, in Phoebus, Virginia), the son of Army Warrant Officer James Thompson and Celia O'Malley, on February 6, 1930, in Colon. The Thompson family, which included two other sons, James and Gordon, came to the Canal Zone when the U.S. Army assigned WO Thompson there. All three boys ended up staying and working in the Canal Zone.

When Paul and Anita met, Paul was working for the Foundry. After their marriage in 1933, the newlyweds resigned from the Panama Canal and went to live with Paul's family in Gretna, Florida. They battled the Great Depression and found jobs working in dime stores and on road construction. When Anita found she was pregnant with their first child, Gordon Alvin, they decided to return to Panama to raise their family.

Upon returning to the Panama Canal, Paul proudly joined the Canal Zone Police

Force, and one of his first assignments was as a mounted police at Madden Dam. Anita and Paul had two children: Gordon Alvin (January 2, 1935) and Frances Anne (April 8, 1938). Gordon and Frances were the third generation to be born in the Canal Zone. Life at Madden Dam was an adventure. The story is told that when it was time to give Gordon and Fran a bath, water had to be run out of the hose before filling the tub. This was to flush out the tarantulas that crawled into the hose at night. Fran learned to swim in Madden Lake and, with snakes everywhere around their Madden cottage, became interested in snakes at a very early age. She really enjoyed exploring rivers and jungles with her father and cousins and has many wonderful memories.

Anita returned to work for the commissary and in World War II transferred to Censorship. When the war ended, Anita went to work for the U.S. Navy on the West Bank and then worked for the Panama Canal Company in the Accounting Division. She retired in September 1959. Anita loved to tell the story of how her father took her for a walk in Gatun Locks at the age of four before they were flooded.

Paul retired as a police sergeant in November 1959, and he and Anita moved to Florida. Paul died on December 16, 1980, in Tallahassee, Florida, and Anita died on January 3, 1998, in Austell, Georgia. Son Gordon never married and is deceased. Daughter Frances went to the United States for nurse training after graduating from Balboa High School in 1956. She then joined the Air Force Nurse Corps where she met and married Richard E. Clinton, a career Air Force pilot. They have three children: Richard, Jr.; Victoria; and Pamela. Fran and Richard enjoy going to Panama for vacations.

Paul Dana Thompson, 1959.

Alvin M.'s son, Paul B. Rankin, known to many as Carlos, graduated from Cristobal High School. He won several championships in ballroom dancing and diving. His sister Anita also won awards dancing as his partner. Paul worked briefly for the Canal Zone and then went to work for Pan American Airways in Balboa and in Medellin, Colombia. He attended undergraduate school, graduate school and then entered the seminary in the U.S. and became a priest of the Order of Holy Cross. His ordination was held at Saint Mary's in Balboa, Canal Zone, in 1948. Father Rankin was the first Roman Catholic Priest to be ordained in Panama.

Alvin M.'s son Thomas Rankin grew up in the Canal Zone and married Margaret Barnard. Tom left his position as Chief of the Industrial Division of the Canal in 1951 to work for the American Bureau of Shipping in Michigan. They had three daughters: Janice, Margaret and Colleen.

Alvin Monroe Rankin
Roosevelt Medal No. 2043 with Three Bars

See also the William Andrew Barnard and Alvin Monroe Rankin; the Gerald DeLeo Bliss, Sr.; the Reinhard A. Boggs and Max Reinhard Boggs; and the Hermanus Kleefkens family histories.

The Roy D. Reece Family

BY HIS CHILDREN—FOR THEIR CHILDREN

Registered Professional Engineer, Lic. No. 60E

Roy D. Reece, the youngest child of William Charles Reece and Rula Miller Reece, was born in Terre Haute, Indiana, on January 13, 1907. He was from modest circumstances. His father was a conductor on the Pennsylvania Railroad and also owned a small apple orchard outside of town. His mother was a homemaker and regularly smoked a corncob pipe!

Roy attended public schools in Terre Haute, graduating from Wiley High School in 1924. That fall, he entered Rose Polytechnic Institute, now Rose-Hullman Institute of Technology, a small engineering college on the outskirts of Terre Haute, and commuted there via streetcar. He did well at Rose, graduating in 1928 with a bachelor of science degree in electrical engineering with honors. He also served as commander of the Army Reserve Officers Training Corps battalion and secretary-treasurer of his senior class. Roy was the first in his family to graduate from college.

Roy and Lucille Reece

While at Rose, Roy met Esther Lucille Polk, who was enrolled in the teacher certification program at Indiana State Normal School, now Indiana State University. Lucille, as she was called, was born on June 8, 1907, in the farmhouse on her family's small farm outside of Eureka, in southern Indiana. She was the oldest child of farmer and school bus driver, James Lester Polk and Blanche Myler Polk. The Polk family ancestry coincides with that of our nation's eleventh president, James K. Polk. Lucille attended Rockport public schools, graduating from Rockport High School in 1924.

After graduating from Rose, Roy was employed by Michigan Bell Telephone Company in Pontiac, Michigan. He and Lucille stayed in touch. Lucille received her elementary school teaching certificate in 1925 and took a job in Richland City, Indiana, near her home. In 1928 she moved to another teaching position in Elkhart, Indiana, to be closer to Roy. They were married in Warsaw, Indiana, on June 1, 1929.

On February 18, 1930, Roy and Lucille sailed from New York on the *SS Ancon* to the Panama Canal Zone, where Roy had been hired as an assistant junior engineer, electrical, in the Electrical Division at $208.33 per month. A close friend and former classmate at Rose, Bart Smith, who was in the C.Z. working in the Electrical Division, had told him of this employment opportunity. The future was indeed bright for Canal Zone electricity.

Roy and Lucille were the parents of three children — Royna Claire Reece, born June 2, 1930, in Panama Hospital, Panama City, Panama; James William Reece, born May 18, 1938, also in Panama Hospital; and David Michael Reece, born December 11, 1940, in Gorgas Hospital, Ancon, C.Z. The Reece family lived on the Pacific side in Balboa, Balboa Heights, and Diablo Heights except for about a year in the early 1930's when they lived in Cristobal. The children attended C.Z. schools in Balboa and graduated from Balboa High School. They all effectively left the isthmus for good, except for short visits, to attend various colleges in the U.S. and move on to successful careers and marriages.

One memorable place the Reece family lived was in quarters 579 San Juan Place, Balboa Heights, during the late 1940's. The cottage was close to the jungle that made up a large section of Ancon Hill. Some of the wild life that lived in or ventured into the yard included deer, conejos, armadillos, sloths, iguanas, myriads of birds and butterflies, tarantulas, scorpions and snakes, including the colorful, poisonous and much feared coral. It was like living in a private wildlife refuge.

During Roy's years with the Electrical Division he became Chief of the Power Generation and Transmission System and served as Chief of the Electrical Division from time to time as needed. He often recalled an assignment early in his career during 1934-35 when he worked with the engineering force responsible for the design and construction of Madden Dam and the hydroelectric station. Roy remained with the Electrical Division for 32 years, retiring in 1962 as Assistant Electrical Engineer.

Lucille was a loving and supportive wife, mother and homemaker to Roy and the children. She did volunteer work and was a substitute elementary school teacher, sometimes for extended periods. She passed away at the age of 52 in Gorgas Hospital on November 25, 1959, after a long battle with cancer. Her ashes were placed in the columbarium at the Cathedral of St. Luke, Ancon, Canal Zone.

Roy married Virginia Knowlton Roberts on December 4, 1960, while they were vacationing in New Jersey. Virginia was the long-time secretary to Canal Zone Government Civil Affairs Director Henry Donovan. She was a native of Swampscott, Massachusetts.

The Canal Zone was a wonderful place to live and the Roy D. Reece family thrived in this truly special environment of good jobs, fine schools, abundant recreational opportunities, excellent medical care, and unique wildlife — the list goes on. The incomparable experience of living in an enclave within a foreign country presented almost unlimited opportunities for the self-growth that can occur from exposure to another culture.

In April 1962, Roy and Virginia loaded up their Rambler station wagon and embarked on a pioneering and adventuresome automobile journey over the incomplete Inter-American Highway from Panama to Texas. They claimed to be the first retired couple to be repatriated from the Canal Zone to the United States by automobile. Many happy years in retirement awaited them and they divided their time between St. Petersburg, Florida, and Crystal Lake, New Hampshire. Thus ends the history of the Roy D. Reece family during the American Era of the Panama Canal.

See also the Roy C. Stockham family history.

Madden Dam, completed in 1935, stands 220 feet high and is an integral part of the Canal. The production of hydroelectric power is a secondary function. After the water goes though the power generating turbines, it flows eventually into Gatun Lake for lake level control.

PHOTO COURTESY OF PANAMA CANAL MUSEUM

The Gladys Houx and Joe Richardson Family

AS TOLD BY GLADYS RICHARDSON

Two Lives in Panama

These are stories about the experiences of Gladys Houx and Joe Richardson — two Americans born on opposite sides of the United States who met at the "Crossroads of the World." While Joe has passed on, Gladys is in her 90's and lives in Canton, Ohio, sole survivor of eleven brothers and sisters.

Gladys Richardson (Houx), 1938.

Gladys was born on a farm north of Santa Barbara, California. In 1920, when she was five, her father suddenly decided to buy land in the Chiriquí Province of Panama to raise coffee. In just three years their lives changed dramatically; and Gladys, her father, mother and three of her brothers found themselves sailing toward an unknown country.

Arriving in Panama City, they boarded a coastal boat to Puerto Armuelles. From there they traveled by train and bus until they stopped in Concepción to rest before riding horseback to El Volcan Barú, a beautiful extinct volcano in the northern Chiriquí Province. Here they settled into a hotel before riding on to view their ranch in Barriles, about 20 miles from the Costa Rican border. Over the next three years Gladys's father and brothers built the ranch house and planted coffee trees, while the number of expatriate Americans in the province grew to include Dr. Jensen and his wife and two married couples — Nita and Egbert Jennison and George and Miriam Macoubray.

There were few comforts in this life. When Nita Jennison became pregnant, her transportation options were limited to a bulldozer or a horse. Gladys's father had a smooth gaited Arabian stallion named Lucero that Nita rode for twenty miles to the doctor's home. The baby, a girl named Marilyn, was fine. Lucero sired a black colt. When he was about two years old he was so badly cut by a tangle of barbed wire that Gladys's father felt he would surely die, but he told her that she could have the colt if she could pull him through. She tended him diligently, keeping the wounds clean and free of maggots. He made it and was named Toddy. When he was old enough to be ridden, Gladys put a saddle on him and swung aboard. He was so trustful of her that he merely snorted once, turned his head to look at her from this new perspective, and started walking. He instinctively understood her commands and they became even more inseparable.

Mail was received once a week at a post office in El Ható. On each mail day the American families gathered to gossip and plan outings. Gladys frequently went home with Nita to spend a couple of days with the Jennisons, or Nita and Marilyn would go home with her. Often they would plan a picnic on a grassy area at the base of the volcano. Gladys gathered wild avocados and made guacamole sandwiches.

The homemade bread for those sandwiches was just low-level magic for Gladys's mother. She knew all the tricks of keeping a family fed and healthy in the wild. Often the only medical help for miles, she treated ills throughout the small community. A kitchen garden supplied vegetables, and of course fruits were abundant. They raised pigs, chickens, and turkeys. The sons were good hunters so game was common at the table. She knew how to preserve food without refrigeration; when a hog or cow was slaughtered, the fat under the skin was removed and rendered. Then the meat was stripped from the bone, ground, formed into patties, seasoned, and packed in large containers. The fat was warmed and used to seal the meat.

Housework was performed without electrical appliances, but each year would bring a strange housekeeping staff. At a particular time of the year a sentry army ant would appear. The family would move out while legions would march through the house and outbuildings, all food and other insects disappearing in their wake.

In 1928 Gladys went to Balboa High School in the Canal Zone. When she returned to Barriles, she found the coffee trees infected with a fungal disease. Ironically, the fungus affected only the crop on the western side of El Volcan Barú. Despite all efforts, nothing worked, and the whole colony on the western slope began to abandon coffee farming.

The Houx's nearest neighbors, a Swiss settlement located across the river from the ranch, were friendly but insular. Their religious beliefs dictated against governmental control, and they refused to apply for Panamanian cedulas. One day Gladys's father was warned that the Guardia National (police) was coming his way. Later they discovered the whole settlement slaughtered. One little girl was wounded, the only survivor of this terrible event. She was adopted by a local family.

Under contract, Gladys's brother Allen surveyed that part of the Pan-American Highway being constructed from Concepción to the Costa Rican border. This was a difficult job as the road had to cross very rough terrain. Particularly challenging was the Caisan Break, a miniature Grand Canyon. Based on the reputation earned by completing this survey, he was able to get an engineering job on the Canal. One of his assignments involved an historical Canal problem. Heavy rains often caused the soil on Contractor's Hill to slide into the Canal making it impassible. Allen studied the problem and made a cut in the hill which greatly mitigated the problem.

Gladys and Joe at their 50th wedding anniversary.

Another brother, Kelly, was helping to build a site for the Panama Canal Dredging Division — a new town called Gamboa. A postmaster was needed, and at Kelly's suggestion Gladys was hired. The post office was a busy place, serving also as a bank for the workers' payroll.

At this point, Gladys met Joe Richardson, a Canal Zone police officer stationed at the penitentiary across the Chagres River and about a mile from Gamboa. Gladys found out later that his path to Panama was as interesting as hers. In 1905 he was born in Zebulon, North Carolina, where he lived on a farm. His father was killed when Joe was eleven, and his mother was forced to send her four boys to an orphanage. Joe ran away at 15 and worked on farms until 1922. Growing restless from farm and dairy work and anxious to see more of the world, Joe left North Carolina and began working on ships headed for far off ports. In 1929 he joined a ship bound for Panama. Ironically, it was the same ship that had brought Gladys's family to that country in 1923. He decided to leave the ship during its Canal transit and become a policeman in the Canal Zone. Unfortunately, the Depression was in full force, and there were no openings. He was told, however, that if he could find permanent employment in the C.Z., he could be transferred to the police department when an opening occurred. Joe did so and finally joined the police force in December 1931.

> *The Chagres River presented an obstacle to their romance, despite the existence of the 'black bridge' that carried the famous trans-isthmian train and a single lane of traffic. At the time, the bridge was in questionable condition and only people living on the Gamboa side were allowed to drive across.*

The Chagres River presented an obstacle to their romance, despite the existence of the "black bridge" that carried the famous trans-isthmian train and a single lane of traffic. At the time, the bridge was in questionable condition and only people living on the Gamboa side were allowed to drive across. The first meeting occurred when Joe rode his champion polo pony across the bridge to the post office; Gladys later joked that the horse convinced her to get to know him. After an engagement of two years Joe and Gladys were married in the rectory of Balboa Union Church and settled into an idyllic life in the C.Z., moving into a house on a hill overlooking the penitentiary and the Canal. There was a farm hill next to the housing area where sugar cane and many different kinds of tropical fruit were raised. Every family had a box at the head of the stairs to their house that was regularly filled with wonderful fresh fruit. Mango laden trees grew all around the hill. Beyond was jungle. In this setting Gladys and Joe had three sons — Jim was born in 1937, Paul in 1944, and John in 1949.

Joe trained and led a special gang of prisoners who would tackle any engineering job presented to them, from painting cars to building bridges, docks, and stonewalls. The list of construction projects is impressive, as is the amount of help he gave to his prisoners. He "graduated" many, who entered the penitentiary without skills and left with both a trade and a job. He was also responsible for more than one manhunt through forbidding rainforest.

Finally, after a retirement ceremony and dinner they said their goodbyes to friends and taking a last wistful look at the house where they had spent nearly twenty years, Gladys and Joe left the Canal Zone for Wilmington, North Carolina. Joe never returned to Panama and Gladys visited the country only once, but their lives forever revolved about the time spent in this adopted land.

PHOTO COURTESY OF DEPARTMENT OF THE ARMY IMAGE COLLECTION

Combination railroad and one lane auto bridge crossing the Chagres River at Gamboa. The Chagres River provides the water to operate the Canal and is the only river in the world to flow to both the Atlantic and Pacific Oceans.

The Robbins/VanderWeg Family

AS TOLD BY SUE ROBBINS

In Robbins family lore, one of the great stories of ancestors (only a little less important than Grandma Priscilla and Grandpa John Alden coming over on the *Mayflower*) is about how Grandpa Edmund Anthony (then seventeen years old) accompanied his dad and Theodore Roosevelt on the historic presidential visit to the Panama Canal in November of 1906. Young Edmund's father, Benjamin Harris Anthony, was the owner of one of New England's largest printing companies, E. Anthony & Sons, and publisher of the *New Bedford Evening Standard*; he was also a personal friend of TR. As members of the Massachusetts elite, this event was one of many that the Anthony side of the family recounts of its glory days before the stock market crash of 1929. But this particular trip signaled a connection to Panama that would resonate decades later . . .

Jump forward to the summer of 1978, and Edmund Anthony's grandson, chief mate aboard the District 2/AMO Tanker *Cove Trader*, was mulling over an idea for a career move: The master of the *Cove Trader*, Captain Carl Dingler, a former Panama Canal pilot, made a convincing argument that if a young man wanted to be a truly great shiphandler, the Panama Canal offered the most challenging and exciting environment for learning "how to drive!"

Jeff Robbins (de Anthony) applied for a piloting job at the Canal, was hired, and on October 17, 1978, found himself in Panama. He figured he'd stay two years . . .

Jump back to January of 1976, and Sue VanderWeg was attending court reporting school in Des Moines, Iowa, and lamenting the very long Iowa winter. On a lark, she tore from the school bulletin board a many-times-Xeroxed notice of a job opening in Panama and began a funny letter correspondence with Canal personnel, Bill Cofer and Doug Schmidt. Mr. Cofer and Mr. Schmidt interpreted these letters to be an application for an open shorthand reporter job at the Marine Bureau's Board of Local Inspectors. Ms. VanderWeg, who had never actually seen a ship in her whole land-locked life, thought she had simply found some new pen pals.

In 1976, workers at the Panama Canal were experiencing uncertainty and upheaval as the organization moved towards a lessening of U.S. influence and greater participation by Panama. Consequently, the pickings for new hires from the States were rather thin — or so Sue always thought because somehow that summer she got hired for the job and found herself in Panama. Too young and dumb at twenty years of age to really understand *what* she had fallen into — she was so young that the night she showed up at Gorgas Hospital with a horrific earache, they said they couldn't treat her without her parent or guardian's permission — she eventually developed some proficiency for the job, an understanding of the special Canal vernacular ("They call that hard thing on the end of the locks the soft nose?") and a desire to build a life in Panama.

And so there she was, sitting in her office at the BLI, when the very handsome grandson of Edmund Anthony walked in to take his first pilot exam. At the time, she had been dating poor Christian missionaries in Panama, and her first (mercenary) thought when she saw Jeff Robbins was, "He's a pilot. If I can get him to ask me out, he WILL buy my dinner!" Which he did. And he's been buying her dinner ever since.

A funny cartoon of Canal Pilot Jeff Robbins, drawn by his brother Tom in 1982.

They were married in September of 1979. Jeff continued to move through the steps towards becoming a senior pilot, and Sue moved from the BLI to the U.S. District Court as it was shutting down operations in Ancon. While working with the court, Sue became pregnant and resigned from that job. But in later years during the Canal's transition to Panama phase, she was privileged to take the minutes of the quarterly Panama Canal Board of Directors meetings.

Jeffrey Baker Robbins II (J.) was born in February of 1981 followed in quick succession by Patrick Christian Robbins (Rocky) in March of '82, Linda Katherine Robbins in July of '84 and Jessica Elizabeth Robbins in July of '86. The family sent the kids to a variety of schools in Panama, including St. Mary's, Balboa Elementary, and The International School of Panama. They participated with other Canal Zone families in T-ball, baseball, soccer, cheerleading, Girl Scouts, Boy Scouts (The boys got their eagles in Panama!), and the cayuco races. The extraordinarily close friendships and sense of community they enjoyed — so unique to Panama and the Canal Zone —

Sue and Jeff, Miraflores Locks Control House, circa 1996.

profoundly affected these young lives, as did being reared at "The Crossroads of the World."

Rocky grew up to become a petroleum engineer who dreams of working internationally as a "country manager." Linda is a special education teacher (she married a seaman, like her mother before her), and she would love to work internationally, as well. Jessica speaks four languages, including Mandarin Chinese; so, by design, her work will take her all over the world. And young Jeff Robbins has graduated from Massachusetts Maritime and has put "Panama Canal Pilot" on his short list of career aspirations.

It's in the blood.

Barring an unforeseen calamity, the senior Jeff Robbins will get his picture in the company newspaper in commemoration of thirty years of service — with high hopes to hang on for "a while past that." (Captain Dingler was right; it's a great job for shiphandlers!)

The 1986 Robbins Family Christmas picture (taken by Jack Sanders, back yard of their Croton Street home).

As for seventeen-year-old Edmund Anthony, one can only guess at how he felt as he experienced the phenomenal Panama Canal up close and personal with his dad and Teddy Roosevelt. He probably felt the same way his progeny feel over a hundred years later: awed at the scope of the enterprise and privileged to be a part of it!

THE SAMUEL HARVEY ROWLEY, SR., FAMILY

AS TOLD BY DOROTHY ROWLEY GERHART, JUNE ROWLEY STEVENSON, SAMUEL (SKIP) ROWLEY, JR.

Samuel Harvey Rowley, Sr., our father, was born the youngest of three children in Quogue, Long Island, New York, on June 12, 1906, into a family of sea captains. The family lived on the beach where his father, Ambrose, was a Coast Guard "surfman," rescuing shipwrecked survivors. Dad, like his father before him, had shipped out as a cabin boy during the summer months on "square riggers." He became interested in the sea at an early age. After high school on Bedloe's Island (the Statue of Liberty Island) he attended Ft. Schuyler, which later changed location and became the New York State Maritime Academy. Their ship was the wooden square rigger *Newport*. Upon graduation, he went to sea for several years. While serving as 2nd mate on a ship, its engine failed, causing them to run aground off Cape Maisí, the easternmost point of Cuba. The nearest available dry dock and salvage tug were in the Canal Zone; the tug was dispatched to pull them off the beach and tow them to the Balboa Dry Dock for repairs.

Sara Pearl Gunn, our mother, also the youngest of three children, was born in Birmingham, Alabama, on January 4, 1914. Sara's father, Henry Gunn, had died before she was born. Myrtle Mae Gunn, her mother, needing a job, left Sara with a childless family, and moved her two oldest children, Landen and Ruth, to Panama at the invitation of her two sisters, Fannie Junker and Ruth Bryan, who were married to Panama Canal Company employees. Myrtle went to work as a telephone operator in Balboa in 1914, the year the first ship transited the Canal. In 1919, Myrtle returned to Alabama to bring five year old Sara to rejoin the family in Panama, where she attended elementary and high school in Balboa. She learned to swim early, thanks to Landen, who had challenged her to sink or swim when he threw her off the tower at Balboa pool. Sara, finding that she could swim and was a natural at it, went on to be a member of the renowned Red, White and Blue Troupe Swimming Team. Of the many swimming events that she swam in, the most memorable was the race against the famous "Tarzan" actor, Johnny Weissmuller. She was later heard to say, "I beat the pants off of him."

Mom and Pop Rowley after marriage in November, 1932.

The Gunns, L-R: Sara Pearl (Rowley), Myrtle Mae, Landen H., and Ruth (VanVielt), Balboa, Canal Zone.

Dad's chief engineer, Timmy Mann, had been to the Canal Zone before and had met Sara's sister, Ruth. Tim mentioned that Ruth had a 17 year old sister named Sara, and a double date was arranged. At first, Mom didn't like Dad, as she would say, "He was fresh." She had her brother go along on the second date for protection; but Dad's professional education and his great dancing ability interested Mom, and they soon began dating regularly. They would all go dancing at the Club Miramar in Panama City, and love soon blossomed between Sam and Sara. They would talk while he transited the Canal, and she walked along the locks walls; and there were letters received from around the world while he spent more time at sea. After about a year of playing catch as catch can via the mail, he cabled her from Australia asking her to marry him. She said "yes," and he signed on a ship headed for the west coast and then another ship heading to the east coast via the Canal Zone, where he got off the ship; on September 19, 1932, they were married in the Balboa Union Church.

After their marriage, Dad decided to stay in Panama, but could only get a job with the Colon Abattoir as there were no openings with the Panama Canal Company at the time. Shortly after that he was hired as signalman in Galliard Cut until an opening arose as captain of the tugboat *Favorite*. Their first child, a daughter, Dorothy Joan Rowley, was actually born in New York in December 1933; their second child, June Adrianne Rowley, was born in Gorgas Hospital in June 1936, and their third child, Samuel H. (Skip) Rowley, Jr., was born in Colon hospital in December 1939. Dad's more than 30 years of service was with the Marine Bureau, including one year as a Canal pilot. He retired as the Canal's senior tugboat captain in 1966. There were two four-year work exceptions during those 30 years. One four-year period Dad worked as captain of the MSTS *AKL 17*, a supply ship out of Rodman Naval Station; and a second four-year period, he worked for the Aids to Navigation as chief mate and captain of the Panama Canal *Atlas*, *Taboga*, and *Torro*. He and Mom moved to Clearwater, Florida, where they both lived out the rest of their lives. They were married over 50 years when Dad passed away in 1983. Mom passed away in 1996.

Dorothy, June, and Skip all married and had children. Dorothy married James Gerhart in Balboa Union Church, and their children are Kenneth, Brian and Suzanne. June married Davis Stevenson in the Cocoli Episcopal Church, and their children are Davis and Lori. Lori was also married in the Balboa Union Church. Skip married Beverly Shircliffe in St. Petersburg, Florida, and their two daughters are Renee and Adriane.

The Rowley family, from left: Sam, Sr.; Dorothy; Skip; June; and Sara (circa 1950).

Dorothy and Jim, after living many years in the Canal Zone, and in Berlin and Heidelberg, Germany, settled in Silver Spring, Maryland. Dorothy worked several years with the U.S. Army in the Canal Zone and later transferred to the Panama Canal, where she retired from the Accounting Division. June and Skip continued to live in the Canal Zone until they, too, retired from the Marine Bureau. June retired as the budget analyst in the main office in Balboa, and Skip retired as a senior control house operator at Miraflores Locks. June and Steve settled in Lawrenceville, Georgia, while Skip and Bev settled in Tallahassee, Florida.

This is the story of how the Rowley family came to be in Panama.

The Clifton W. Ryter Family

As told by Virginia Ryter Wennik and Don Ryter

Clifton Whitney Ryter sailed from New York City for the Canal Zone in 1924 with a license as a tugboat captain. He was twenty-two. Cliff had heard of employment opportunities on the Canal from his mother, Margaret Richardson Ryter Boyd, who worked as a nurse's aid in Santa Tomas Hospital in Panama City.

Cliff's mother was of English descent. She returned to the States in the early 30's and died in 1935. Cliff's father, Louis F. Ryter, was a first generation American from Denmark. He worked for the state of New Jersey. The family visited him and his wife, "Doll," in Maplewood, New Jersey. We remember trips to amusement parks and to the local delicatessen for treats not found in the commissary at home.

In September 1926, Cliff returned to the States and married Margaret Mary McBride in New York City. Cliff and Margaret began married life in Pedro Miguel, where they met Minnie and Earl Kent. The Kents became our surrogate extended family in the Canal Zone. Uncle Earl, Aunt Minnie, and their children, Patsy and Lloyd, will always be very special to the Ryters. We spent many Sunday afternoons with them, "under the house" or at the boat club, where we watched Lloyd race speedboats. We were driving home from the Kents' house on December 7, 1941, when we heard that the Japanese had attacked Pearl Harbor.

Margaret's parents were of Irish descent. Her father's name was Thomas W. McBride, and her mother's name was Margaret Wallace McBride. Thomas worked for the City of New York. We visited them in their apartment in Manhattan, where we enjoyed trips to Radio City, the Empire State Building, and the Statue of Liberty.

Cliff and Margaret had three children. Virginia Ann arrived on November 29, 1930. The next child to put in an appearance, on April 28, 1933, was Margaret Louise. Several years later the family welcomed Donald Whitney, born on March 21, 1940. All three children were born at Gorgas Hospital.

The family moved to Gamboa in 1936 when Cliff was assigned to the Dredging Division as a tugboat captain. Except for a two-year stint with the Marine Division in Balboa, he remained in Gamboa for the rest of his career. His Marine Division assignment was as a captain on the *Thatcher* ferry. The ferry carried cars and trucks across the Canal at the present site of the Bridge of the Americas. While the family lived in Balboa, Virginia attended third and fourth grade at a temporary site called The Lodge Hall and Margaret went to kindergarten and first grade at the East Balboa Elementary School.

Cliff and Margaret Ryter with Virginia and Donnie, 1952.

The principal job of the Gamboa tugs was to support the large dipper dredges that constantly keep the channel dredged to the proper depth. Cliff would move loaded scows (barges) from the dredge and deliver emptied scows back. The tugs would, on occasion, be called upon to assist ships that had difficulty navigating the Canal on their own.

Cliff spent most of his Gamboa years as senior captain of the tug *U.S. Culebra*. He did, however, spend several years as captain of the crane boat *Atlas*. The *Atlas* was used to maintain navigation aids, such as buoys, on Gatun Lake and Gaillard Cut. However, for several years it was occasionally employed to carry cruise ship passengers through the Cut.

Gamboa was a beautiful little town, situated at a

point where the Chagres River flows into the Canal. It was an ideal place for a kid to grow up. The elementary school was wonderful with great teachers and perfect accommodations. Organized sports (and disorganized ones as well) were available to all. A community pool, gym, bowling alley and tennis courts were readily available.

A highlight in the life of Gamboa children was their entrance into seventh grade, when they had to take the 17-mile train ride to attend junior and senior high school in Balboa. This was a time for socializing and finishing up homework and for getting into a little mischief. Later, a bus was provided; but this method of transportation was more of a trek than a trip as the roads were not the best, and the speed limit was 45 miles per hour.

Perhaps the most exciting events in the life of the Ryter family were the extended vacations in the States. These trips took place every other year, when possible. A special train, the *Boat Train*, ran from Balboa to Cristobal with stops along the way to pick up folks who were departing that day. Cliff was always afraid we would miss the train and would herd us to the station an hour ahead of time. Occasionally, he would secretly set the clocks ahead an hour to fool the rest of us.

The five-day voyage on one of the Panama Canal Company's three ships ended in New York City. We would spend a month and a half in the States, visiting with relatives and friends around the city as well as enjoying extended stays in rental cottages in New England.

Cliff and Margaret lived in Gamboa until retirement

Cliff and Margaret Ryter with children and grandchildren, 1958.

in 1962. They settled in Leesburg, Florida. Margaret died as a result of an automobile accident in 1968, and Cliff passed away in 1974. One of the children, Margie, stayed on in the Zone as an adult. She married Bob Saarinen and raised a family of four children: Lauri Siltz, Lynn Jahnke, and twins Donald and Albert Saarinen. After a reduction in force due to the transition to Panamanian control, Margi went to work at Fort Jackson in South Carolina. She died in 2003.

Ginny became a schoolteacher and has lived in the Hartford, Connecticut, area since 1955. She has two children, Peter Dow and Wendy Dow Miller.

Donnie earned an engineering degree and, after four years in the Air Force, settled in Central Florida in 1966. His three children are Kim Bryant, Kerry Fein, and Kevin Ryter.

Tug Culebra.

PHOTO BY PANAMA CANAL AUTHORITY

The Earl Salter Family

AS TOLD BY BOBBY SALTER

People came from all over the world to build and operate the Panama Canal. My father, Earl Salter, came in 1944 and stayed 24 years. Living and working in the Canal Zone was a life altering experience for the Salter family. Not only did we greatly expand our horizons, we associated every day with people of many different nationalities. Peoples united are a formidable force for success. The American era of the Panama Canal was one of the most successful international undertakings in modern history. We were fortunate to be able to participate.

Earl Salter was born in 1911 on a small farm in rural South Georgia. He married Aileen Markey in 1933 in the midst of the Great Depression which lasted another 7 years. During those seven years, Earl was a sharecrop farmer barely managing to feed and house his family. The outbreak of World War II created jobs for skilled workers, and Earl found work as an electrician with the U.S. Army at Camp Stewart, Georgia. Earl's brother, an Army Air Corps Warrant Officer, was stationed at Albrook Field in the Canal Zone and encouraged Earl to seek employment with the Panama Canal. He did and was accepted. Earl came to Cristobal by ship from Miami in June 1944. He was employed as a marine wireman with compensation of $1.58 an hour.

Earl's family consisting of wife Aileen, sons Talmadge and Bobby, and daughter Colleen arrived at Mount Hope three months later via a converted Navy PBY seaplane from Miami. We disembarked for processing and were surprised to find that Earl was not there to meet us. Aileen was not pleased! After processing we rode in the back of an open air transportation truck to our new home in Gatun. Not far from Cristobal we passed a group of small black Panamanian children playing alongside the road. Several of the younger boys were wearing little cotton shirts that came down only to their waist and nothing else. Mother Aileen quickly covered Colleen's eyes with her hands to avoid this unseemly sight. Earl was to get an earful that evening when he finally arrived home from work.

Earl and Aileen Salter, Margarita, 1956.

Our Gatun home was in one of the last of the old French six-family units located beside the railroad tracks not far from the train station and within sight of Gatun locks. We had no car, so Earl rode a "chiva" to work at Mount Hope and the rest of us walked to wherever in Gatun we needed to go. Young entrepreneurs that we were, brothers Talmadge and Bobby obtained jobs selling newspapers at the Gatun Clubhouse and a nearby Army Artillery emplacement for a gain of one cent per paper. The Artillery position was located about halfway between our old French quarters and Dorn Thomas's house across from the clubhouse. Thirteen years later Dorn, Talmadge and Bobby would attend Georgia Tech together in Atlanta.

Talmadge and Bobby attended Gatun Grammar School for one year, fifth grade for Talmadge and fourth grade for Bobby. The level of instruction was much superior to that of rural Georgia, causing Talmadge to really bear down and Bobby to attend extra tutoring three afternoons a week for several months. Both of us caught up to the required levels and were able eventually to take advanced high school courses in preparation for college, a true bonus of being educated in the Canal Zone school system.

Bobby's best friend in Gatun was another nine year old, Med Kellum. These two adventurers nailed together some plywood and boards to make a leaky boat that was to return us to the States. We launched into Gatun Lake thinking it was the ocean. No more than 100 yards from shore we had taken on so much water we were sinking. A police patrol boat picked us up, dried us out and parked us in a cell in the Gatun jail. When our parents retrieved us, our little butts paid the price of adventure.

In the spring of 1945 Armed Forces Radio announced that President Franklin Roosevelt had died. A few months later, the war in the Pacific began to wind down and most of the U.S. 5th Fleet returned to Atlantic Ocean ports. They passed through the Panama Canal: carriers, battleships, destroyers, submarines, etc. It was a sight to see: ships in Gatun Lake, ships in every lock, ships in Limon Bay and sailors everywhere. The bars of Colon and Panama City were busy places. Skilled employees in the Canal Zone workforce were stretched to the breaking point, particularly marine technicians. Earl worked six-seven days a week, 15 hours a day, for several months repairing the war damaged ships.

We moved from Gatun to Margarita so Earl would be close to his workplace, the marine electrical shop at Mount Hope shipyard. Earl worked here for 24 years, eventually becoming lead foreman. Our family lived in a four-family unit directly across the street from Margarita Hospital. Aileen obtained a position with the Commissary Division and worked there for 16 years. In the last 10 years of Earl's service in the Canal Zone, he belonged to the Masonic Lodge in Cristobal on Bolivar Street. He was Grand Master of this lodge when Canal Zone sovereignty street riots damaged lodge buildings.

Panama Canal Masonic Consistory, 1962.

Canal Zone schools were a melting pot of nationalities. A review of our high school yearbooks reveals student surnames Wong, Fernandez, Aleguas, Sasso, Hatgi, Platkevih, Constantino, Tagaropulos, Chin, Didier, deBoyrie, Lim, Peltynovich, Penaherrera, Villalas, etc. It was a "United Nations" in miniature. Talmadge, Bobby and Colleen attended Cristobal High School and were very involved in school activities. In a rare coincidence, each of us was voted "Best Looking" in our graduating classes (1952, 1953 and 1958).

In 1950, Talmadge became interested in a young girl whose family had moved to Margarita from Cristobal. Her name was Mercedes Peterson. She was 14 years old and the first born child of Herbert (Pete) Peterson and Adriana (Tita) Galindo. Six years later Mercedes and Talmadge were married in Miraculous Medal Catholic Church in Colon, thereby joining the Danish Peterson and Colombian Galindo families with the American Salter and Markey families.

The Peterson Family — Patriarch of the Peterson family was Walter Peterson, Sr., a Danish citizen who served under Lieutenant Colonel Theodore Roosevelt in the Cuba campaign of the Spanish American War in 1898 and thereby obtained his American citizenship. Skilled as an iron worker, he moved from Jamestown, New York, to Panama in 1905 to work for the Panama Railroad and then on construction of Gatun Locks. He and his wife Minnie Hanson had six children. Walter, Sr., was awarded a Roosevelt Medal with three bars, and he and Minnie were aboard the *SS Ancon* when it made the first seagoing ship transit of the Panama Canal.

Middle child Herbert (Pete) Peterson was born in Cristobal in 1909. He completed an apprenticeship and worked in Canal Zone administration for 30 years. Pete Peterson married Adriana (Tita) Galindo, daughter of Inocencio (El Toro) Galindo, Jr., and his wife Aurora Rios. Pete and Tita had two children, Mercedes born in 1936 and Shirley born in 1938.

The Galindo Family — Inocencio Galindo, Sr., was born in Cartagena, Colombia, about 1845. He was educated at the prestigious College De Sorbonne in Paris, France, where he received his Doctorate of Law. His fluency in both French and Spanish, combined with his capability in government law, drew him and his family to Panama during the French Canal building effort. He provided legal assistance to the French Compagnie Universelle du Canal Interoceanique.

When the United States secured the rights in 1904 to build and operate a Panama Canal, middle son Inocencio (El Toro) Galindo, Jr., had graduated from Cornell University in Ithaca, New York, with a degree in engineering and had been working in Panama for 10 years. El Toro's engineering abilities made him an ideal employee to help the Americans build the Panama Canal. When it was completed in 1914, he returned to his engineering business in Colon and became politically active in the Republic of Panama. El Toro was elected Governor of Colon Province in 1930 and 1936. He married Aurora Rios. Their first child Adriana (Tita) Galindo was born in Colon in 1914. Tita married Zonian Pete Peterson.

Walter Peterson
Roosevelt Medal No. 1598 with Three Bars

Mount Hope Shipyard, 1951.

Dredge Mindi under repair at Mount Hope Shipyard.

The Bruce Gordon and Grace Aloise (Meister) Sanders Family
AS TOLD BY ROBERT HILL

Dux Femina Facti: A Woman Was the Leader in the Deed

The saga of the Bruce Gordon Sanders clan and its Canal Zone adventures begins in Louisville, Kentucky, in 1908 at the U.S. Public Health Service Marine Hospital. He was nominated for the post by an old friend, also from the Marine Hospital, who was then working in the Canal Zone for the U.S. Public Health Service (USPHS) under the auspices of the Isthmian Canal Commission (ICC). Bruce Sanders was hired without an examination by a Major Carter, a surgeon, also in the employ of the USPHS/ICC, as he had been apprised of Bruce's accomplished professional skills as a nurse and pharmacist.

When Bruce ("Pop" as he was always called by his loving family) returned to the States on his first six weeks' vacation in 1910, he "popped the question." On September 14, 1910, he and Grace Aloise Meister were married and she, a mere young sprig of a woman of twenty, was carried off on the great adventure of her life. Grace and her family were then living in a small white cottage across the street from the USPHS Marine Hospital, and it was over a small white picket fence in the front yard that they had courted. This house was still there when visited by family members during the late 1960's.

Bruce Gordon Sanders and Grace Aloise (Meister) Sanders. Early years in the Canal Zone.

Together they sailed from New Orleans, Louisiana, aboard the *Turrialba*, a vessel in service with the United Fruit Company, and on October 6, 1910, they docked at 5th and Front Street in Colon, Republic of Panama. Unbeknownst to the Sanders, the Chinese crew had mutinied during the last two or three days of the voyage, but it had been handled so well by the ship's officers that they were unaware of the happenings until they were told about them by the ship's officers during disembarkation.

As a consequence of an acute shortage of proper housing during early Canal construction days, and as they did not have access to "family quarters," Bruce and Grace lived briefly in a "boarding-house" type of hotel known as the "Station Hotel" in Panama City at what would be the southern terminus of the soon to be constructed Canal. As Grace ("Mom" as she was affectionately called by her loving family), an accomplished and polished raconteur, always recounted the story, it was a place very rapidly and haphazardly constructed but featured a great tin roof and "rooms" which consisted of an assortment of partitions which divided the basic floor space into small compartments with swinging panels of wood serving as doors. As she used to regale her family with her enchanting storytelling, she would say this type of construction precluded functioning locks of any kind and since, with only one or two exceptions, the clientele consisted mostly of rough and tumble Canal construction workers, this made for the first of many subsequent stressful living experiences to follow in their early days of life in the Zone.

As it was that Bruce Sanders was assigned to duty at "'line" medical dispensaries which dotted the construction railroad line, he was often detained for night duties at those dispensaries or, just as often, at the hospital, and thus Grace was often alone in their little "makeshift" room. Despite knowing that it was against the law to carry firearms, that didn't keep Grace, being a southern girl and quite familiar and accomplished with firearms, from keeping a little lady's pistol under her pillow at night while she was sleeping!

On one particular Saturday, as she used to tell the story, "the 'gentleman' who had the room next to ours had lent his room to friends from out on the line for the weekend. With not much else to do, the men next door had been consorting with one 'John Barleycorn' to the point that they were very loudly proclaiming their intent to go out scouting every room in the hotel to see what they could find! All the while I was sitting in my room, scarcely daring to breath, but making sure I had my little pearl-handled revolver in close reach just in case! Well, as providence would have it, they got distracted and eventually moved off to some other nefarious activity for the evening, but you can be sure my little 'equalizer' was under my pillow for the rest of that long and seemingly endless evening!"

Their next housing was in the then recently reclaimed townsite of Caimito Mulato, a place known in Europe long before the settlements of Jamestown and Plymouth Rock, and situated so close to the banks of the Chagres River that during the period of time when they resided there one could fish off the back porch. Across the Chagres and reached by a suspension bridge was the townsite of San Pablo which, local legend had it, was the location of an early plantation run by a collection of Catholic priests. When the French were working on building their version of the Canal, they had placed housing for their laborers at Caimito; and when they eventually gave up and withdrew, the forlorn little buildings of Caimito were quickly and silently reclaimed by the jungle and soon forgotten. It was not to reappear from the jungle until a survey team of American engineers siting the center line of the Canal stumbled upon the hidden buildings in their work. As housing was in very short supply, the Isthmian Canal Commission quickly set about reclaiming them; and just as quickly, the jungle was cleared away and the buildings were "refurbished" and assigned to Commission employees who, despite the building's inadequacies, were very glad to have won the "housing lottery."

> *Across the Chagres and reached by a suspension bridge was the townsite of San Pablo which, local legend had it, was the location of an early plantation run by a collection of Catholic priests.*

Their experiences, trials and tribulations, and many triumphs would fill a book, which her family always encouraged Grace to write, but are simply too many to be captured in such a limited scope as is available here for this recounting. They are fondly treasured and quickly shared by their many family historian/storytellers that survive them today.

In 1949, Bruce Sanders retired from Canal service, and after a short sojourn with their daughter Edith in the "flats" area of Balboa, Grace and he moved to House 0250 on Goethals Boulevard (Ridge Road) in Gamboa; and it can still be seen in some current literature describing the Gamboa Rainforest Resort. It was while living there that the previously established tradition of family gatherings occurring on Sunday at their home continued and expanded to include family reunions and special occasion gatherings being enjoyed at the Summit Gardens Botanical Park on Gaillard Highway (Gamboa Road) with their extended family.

During his well earned retirement years Bruce, a long time member of the Masonic Lodge, took it upon himself to very generously devote much of his free time to serving his lodge in a myriad of ways which always brought him great pleasure and fulfillment. Grace was, as always, at his side shouldering her share of the work. She, always an avid reader and history buff, particularly of earlier Panama and Canal history, spent her time doting on her family, collecting and reading everything she could acquire on her two favorite subjects, and sharing her vast array of construction experiences with any and all that evidenced even a smattering of genuine interest, especially the Isthmian Historical Society.

"Mom" and "Pop" always expressed great pride in their large family who by now had all become Canal employees, some if only briefly, and had embarked on their own distinguished careers with different branches of either the Panama Canal Company or the Canal Zone Government. Those children were Bruce Gordon Sanders, Jr.; Bernice Aloise Sanders (Hill); Maxwell Sheldon Sanders; Virginia Lee Sanders (Kleefkens); Edith Lois Sanders (Diaz); and Philip Ransdall Sanders.

At the time of his death on July 1, 1958, at the age of 71 in Balboa, Canal Zone, Bruce Sanders was the proud holder of Roosevelt Medal No. 4180 with two bars, Nos. 2556 and 1729, respectively. He always maintained that Grace was just as entitled to the honors that went with the medal and bars as he was because without her at his side, he would never have had the courage, given him by her loving support, to have stayed the course!

Grace died on March 10, 1987, at the age of 96 in Tampa, Florida, while living with a daughter and her husband.

Dead they may both be but never forgotten by their families will they ever be as their courage and adventuresome natures are carried forward by their proud progeny and their respective families.

Bruce Gordon Sanders
Roosevelt Medal No. 4180 with Two Bars

See also the William Gonzalez and Bernice A. (Sanders) Hill; the John M. Davis and Michael Kenny; and the Hermanus Kleefkens family histories.

Suspension bridge, San Pablo.

Suspension bridge being dismantled.

The John E. Schmidt Family

AS WRITTEN BY HIS SON JOHN E. SCHMIDT, JR., IN 2007

Our Life – Our Honor & Privilege & Pleasure — John Schmidt Jr, 4 July 09

What made John Schmidt make this entry in his diary? *October 5, 1926, Baltimore, Maryland, "I Plan to Go To Panama."* What was it that almost exactly eight years later to the day had John arriving on the Isthmus to start his career in the U.S. Army Signal Corps working as a cable splicer?

You could say that John had some "fiscal premonition" of things to come on the state of the economy in the United States and made a very fateful decision when he was just 23 to travel to this distant land and start a career that led to his most successful life in the Panama Canal Zone, raising a family and culminating in his retirement from the Panama Canal Company in 1966.

With 16 years working in the electrical trade, John's application to the U.S. Army Signal Corps was accepted, and he arrived in New York with his War Department special orders in hand directing him to report to the Chief Signal Officer in the Canal Zone. He sailed from New York in October 1934 on the Panama Pacific Line ship, *SS Pennsylvania*, and reported for work in the Canal Zone on October 20, 1934.

On July 24, 1935, Catherine (Kitty), Jacqueline, and John, Jr. (Bill), arrived in the Canal Zone; and evidently the Army was not quite able to have "adequate" quarters available as the Schmidt family moved into an abandoned "paint shack" on the post of Corozal right smack in the middle of a quadrangle of Army barracks. Kitty was not discouraged, as she fashioned "walls" into this shack with drapery material and created a reasonable home for the early Schmidt family.

John and Catherine Schmidt, Pasadena, Maryland, June 1984.

After a year of moves into quarters in Balboa and Ancon, John started to work for the Panama Canal Company on August 1, 1936; and two months later the youngest member of the family, Douglas, was born at Gorgas.

In May 1937 we moved to Pedro Miguel into quarters 207-A on "Incubator Row" for a month or two before we moved in the "Alley" (quarters 209X-A) behind in a large two-family with the Van Vliets' sharing. Right next door in 207X-A and -B were the Fred Hatchett and the George Lieby families. It was exciting to have Pedro Miguel lake just 100 feet from our back porch. Pedro Miguel became our favorite and most exciting next ten years.

The years in "Peter Mike" were fascinating as the war brought so many historical moments with the fleet passing through, the barrage balloons on the ball diamond, and the multitude of smoke pots spotted up and down our streets and bringing black soot to the clothes hanging on lines. A good number of times the kids were able to "catch a ride" on carriers and other ships to Gatun and come back by the train later that day.

These were fun years at the Pedro Miguel Boat Club with John playing sax, Fred Hatchett on the banjo, and others making up a lively band most every weekend for dancing and sharing of our "war years" on the side of the Canal. John's hobby was woodworking, and his specialty was fine wood finishing on lamps and other inlaid work like jewelry boxes. He graduated over the years of this "hobby" into full completion of furniture for our homes.

The Schmidt family enjoyed vacations to the States, and our first trip back to Maryland was on May 1, 1938.

The war years prevented another trip on the wonderful liners until 1946, when we were able to book aboard a U.S. Navy vessel.

Life then began again in Cocoli. We lived in 624-A on Nicobar Avenue alongside Lee and Delta Sampsell. Fred and Mary Hatchett and George and Catherine Lowe lived directly across the street — once again very close friends living side by side. Here Douglas collected many animals in his love for the flora and fauna of the surrounding hills and rivers of the West Bank along with hunting almost every weekend with the Hatchett/Lowe hunt club.

Jackie graduated from Balboa High School in 1948 and went to work for the Inter-American Geodetic Survey. John, Jr. (Bill), was very active in his last two years at BHS with increasing leadership positions in the Army Jr. ROTC unit and graduated in 1950 as the battalion commander of the unit. That training came into play in his future in the U.S. Air Force. Douglas worked in summer jobs with the Panama Canal, which prepared him for his full retirement in his later years.

The family moved into Ancon in early 1950 on Guayacan Terrace with the Lowes again living alongside the Schmidt family. Bill went to work for the U.S. Naval Public Works at Rodman Naval Station, joined the U.S. Naval Reserve, and before long decided that his future should lie in a more dedicated military life and enlisted in the U.S. Air Force in April 1951.

Jackie married Arch Dale Bishop, had two children,

L-R: Douglas Schmidt, Jackie Bishop, and John Schmidt, Jr.; Kerrville, Texas, August 2000.

and left the Canal in 1973; Bill married Patricia Blitch, had five children and left in 1954 and again in 1976. Douglas married Sharon Booth, had one child, and continued his employment with the Panama Canal Company.

John and Kitty retired from the Panama Canal in the summer of 1966 and returned to Maryland to live out their years there and in Texas. They both died in 1988.

Jackie and Dale retired to Kerrville, Texas, and live a full life there.

John, Jr., continued his Air Force career until 1981 and retired to Tallahassee, Florida.

Douglas and Sharon continued life in the Canal until 1990, when Douglas retired from the Marine Bureau and moved to Kerrville, Texas.

Pedro Miguel Clubhouse, once located in Gorgona, was torn down during the early 1950's.

PHOTO COURTESY OF PANAMA CANAL REVIEW

The Louis H. "Pop" Schmidt Family

AS TOLD BY SUZANNE SCHMIDT GOLDSTEIN

The Origin of Panama's First Family of Fishing

May of 1999 marked Suzanne Schmidt Goldstein's first trip back to the Panama Canal Zone after 34 years. In a reunion filled with laughter, crying, and shared memories, the Balboa High School alumni returned to participate in the last student graduation ceremony before turnover of the school and the Canal and surrounding area to Panama. After 34 years out of the country, Suzanne's return trip classified as true sensory overload.

Sitting with eyes fixed on teasers skipping over the water, Suzanne reflected on how she arrived at this point. It seemed normal, but how many women at age 52 attempt to catch their first marlin, she wondered? Her childhood in Panama and the words of her father, Theodore Schmidt, and grandfather, Louis "Pop" Schmidt, came streaming back, "The price of a marlin is eternal vigilance."

Louis H. Schmidt (1888-1964), born in Baltimore, Maryland, married Matilda Vorbringer, a German immigrant, and together they parented five children — Carl; John; Louis, Jr.; Theodore; and Marie. Louis worked in various electrical trade positions in Baltimore, including the railroads and shipbuilders in the area. Self-employed from 1920 to 1921, he bought and sold oysters in the Chesapeake Bay area, an experience which heightened his attraction to the water and the sport of fishing.

Louis "Pop" Schmidt, 1930's.

As well-paying work became increasingly difficult to find, Louis applied for a tour with the Panama Canal Company. On April 25, 1923, he began a year-long appointment as a general and marine electrical wireman. He set sail for Panama from Pier 67 in New York City on the *SS Panama*, which was owned by the Panama Railroad Steamship Line. His family followed him on the *SS Panama* just two months later.

Life in Balboa proved difficult for the family due to illnesses and an enlarging family size. Two more daughters, Ruth and Iris, were born in the Canal Zone; and after the death of Louis's mother in Baltimore, his father, John C. Schmidt, also joined the family. In addition to his wife and seven children, Louis then also supported his father. One son, Carl, had health issues, and another, Louis, Jr., had lost an arm and a leg in a railroad accident in Baltimore prior to the move. Matilda suffered numerous illnesses, and Louis himself was once hospitalized with malaria.

Despite a rough start, Louis approached life as a survivor and adventurer. He used his experience in the Chesapeake to continue his boating and fishing endeavors in the waters of Panama. The name "Panama" means "abundance of fish," and it was indeed. Even with financial hardships, Louis acquired his first boat, the *East Wind*, in 1927. Next were the *Otoque*, *Caiman*, and *Caiman II*. By the time he acquired the *Caiman*, Louis and his sons were seasoned fishing guides and experienced tradesmen who worked to convert the boat for fishing. The *Caiman II* was designed and hand-built by the Schmidts in 1941 for one purpose: marlin fishing.

Louis did whatever he could to bring in additional dollars for the family and fishing. With the *Otoque*, he traded caiman (South American crocodile) skins. Hence, *Caiman* became the next boat's name. He also manufactured trolling leaders and sold them in the commissary of the Canal Zone in the 1930's. It is suspected that this rigging included the Schmidts' "Panama belly bait" that is now so famous. His launch called *M.D.* was used to ferry personnel of the U.S. Fleet to and from shore. He did what was allowed to bring in some additional money.

Life for the Schmidts soon became about fishing. Louis taught his sons all about fishing and they improved upon it. While Louis was small in stature, being only 5'4", his sons were big and strong. Even Louis, Jr., with one arm and one leg, was never thought of as disabled and actually caught the largest marlin in the family. The daughters fished, but also rode horses, golfed, and enjoyed all that the country offered. The sons excelled in fishing like their father. It was unusual that three of Louis's sons would have his same passion for fishing and thus became a fishing force to be reckoned with.

Fish were everywhere in the waters of Panama. There were times when the water would turn red with the numbers of red snapper. After conquering most saltwater species early on, they focused on big-game fishing. Marlin was their game, and everything they did was for this endeavor. Son Theodore followed his father and became a Panama Canal electrician living in Balboa, learning everything about diesel engines and marine electricity. Louis, Jr., became the manager of the gas station in La Boca; and John became a marine surveyor who rebuilt the Admiral's yacht, *The Old Man*.

The Schmidts traveled the waters of Panama, discovering the best marlin grounds, and are believed to be the first to discover the marlin potential of Piñas Reef off Piñas Bay, 130 miles south of Balboa. Pop Schmidt called it the "marlin supermarket," while others called it "Schmidt's Reef" or "Caiman Reef." Guide work in the early days took them as far as Ecuador and up to California. They developed, built, and pioneered everything from techniques to tackle, as well as a 40-foot marlin boat (*Caiman II*). And, they caught marlin — lots of marlin, record marlin. And so did those who fished with them.

The Schmidt men were middle-class, blue-collar workers with an expensive hobby. Pooling their resources and skills enabled them to achieve their dreams. The Schmidts were not writers, promoters, or braggarts; but as the outside world came to the Canal Zone and heard the fishing stories, articles were written. Word spread; they became known and sought out by celebrities and the wealthy. Some were lucky enough to be invited on their trips.

Many world records were caught on the *Caiman* and *Caiman II* with the Schmidts, who were eager to teach those willing to learn how to catch a black marlin. The world learned Pop Schmidt's words: "The price of a marlin is eternal vigilance."

Louis's sons, L-R: Theodore (Ted); Louis, Jr.; and John.

After work and school, life for a Schmidt was about fish. There were beers at the Balboa Yacht Club, parties, and awards of the Pacific Sailfish Club (later the Panama Marlin Club). The kids ate shrimp cocktails made with truly "jumbo" shrimp at the Balboa Yacht Club. There were fish fries of fresh black marlin, the only white-flesh, edible marlin.

Son John had one child, also John, who caught marlin. Louis, Jr., had no children but his wife, Marcela, caught her share of marlin. Theodore's wife, Regina, also took up the sport, and he passed on his legacy to their three children: Karen, Suzanne, and Theodore, Jr. While all of his children were taught to fish and did so aboard the *Caiman II*, Theodore died young at age 41, when they were only 16, 13, and 5. Karen, the eldest, did not particularly like fishing. Suzanne loved watching her father clean fish while getting an education on the stomach contents of each one. Using a hand line, she spent her days fishing with the boys from the Balboa Yacht Club pier while her father worked on the *Caiman II*. Theodore, Jr., only five when his father died, returned to Panama as a teenager and caught his marlin on board the *Caiman II* with his Uncle Louis. But Suzanne left Panama before catching her marlin and always yearned to go back and carry on the Schmidt legacy in Panamanian waters.

So, when the 1999 graduation ceremony concluded, Suzanne went looking for the *Caiman II*. Although constructed in 1941, it was well-built and still around; she found it beached in great disrepair in an area of Old Panama. Suzanne knew then that she would return to Panama to catch her marlin.

Piñas Bay, Panama, is an untouched area etched out of the jungle, now home to a fishing resort accessible only by boat or plane. In the old days there was only a village in the jungle, and the bay served as safe harbor for boats during fishing trips and tournaments at Piñas Reef. In Piñas or wherever the *Caiman II* would anchor, the natives would paddle out in their cayucos to visit and trade. They all knew the Schmidts.

When Suzanne arrived in Piñas Bay to catch her marlin, she found that, even after all these years, the history of the *Caiman II*, her family, and their pioneering fishing lived on, handed down by oral tradition. The young boat captain from the village of Jaque in Piñas Bay and other locals called her "Niña de Caiman."

Suzanne caught her black marlin, but catching one was normal . . . at least for a Schmidt it was.

Louis, Sr. (1888-1964); Matilda (1889-1938); and Theodore (1919-1961) are buried in Corozal Cemetery, Panama. John C. (1854-1933) and Louis, Jr. (1916-1973), were cremated and ashes spread in the Bay of Panama. Rumors abound that the *Caiman II* was restored and may still be roaming the waters somewhere in Panama.

Caiman II, Balboa Yacht Club.

The H.C. (Bert) Schroeter Family

as told by Val and Tina Schroeter

Bert was born July 17, 1916, in Chicago, Illinois, and his family moved to Milwaukee, Wisconsin, shortly afterwards. In 1941, at 25 years old, Bert went to work for the Panama Canal Company as an assistant engineering aid at a salary of $168.75 per month. He worked on the third locks project while living with his sister, Frida Lidicker, and her family in the old French quarters in Diablo Heights. The third locks project was canceled after December 7, when Japan attacked Pearl Harbor. Bert then left the Canal Zone in 1943 and joined the Army where he trained as a paratrooper.

After World War II, while working at the Wisconsin Electric Power Company in 1947, Bert took a correspondence course and became an electrical engineer. He met and married fellow employee, Valeria Bertoncini, in 1949. Their first child, Suzan, was born in Milwaukee in 1950. The family then moved to Houston, Texas, in May 1950, where they purchased their first home under the G.I. Bill for a down payment of $50.00.

When Bert decided to return to the Canal Zone in 1954, he was hired as an electrical engineer at a salary of $5,256.25 per year (according to a document signed by the Personnel Director). The excitement of moving to a foreign country and sailing on a ship for the first time made up for the sadness of leaving friends and family behind. On May 18, 1954, they sailed from New York City on the *SS Panama*. At that time, it was a seven-day journey with a stop in the beautiful harbor of Port-au-Prince, Haiti, where many small boats clustered around the ship selling their handcrafted items. They landed at Cristobal, and Val and Suzie were escorted across the isthmus on the train while Bert waited for their car to be unloaded. They spent their first night in the historic Hotel Tivoli. Their first class room had a bare light bulb hanging on a string from the ceiling. The next day, they drove to Gamboa to see their first quarters. It was an old, four-family building on stilts on the Ridge. It had no glass windows — only screens, with old metal furniture and an ice box with daily delivery of ice and milk. The toilet tank was mounted on the wall with a long pull chain. The mattresses and pillows had a horrible musty smell, but soon their own furniture arrived. Val and Suzie watched it being uncrated at Pier 18, and Suzie was very excited to see her toys again.

H.C. (Bert) Schroeter

Living in Gamboa was Val's first taste of the jungle, where she saw Army ant trails while she hung clothes on the government supplied clotheslines in the backyard. On one of her drives to the railroad station where Bert commuted to the Administration Building in Balboa, a huge green iguana ran in front of the car. After only two months in Gamboa, Bert's previous service qualified them to move into a new two-story, 3 bedroom, 1½ baths, concrete duplex at 5651-A Walker Place in Diablo Heights. They were just down the street from where Bert had lived in 1941, where his four-year old niece had painted her name, "Fritzi," which was still visible on the side of a concrete electrical relay station. Bert brought some bamboo and metal furniture from someone leaving Panama, and they found that it was better suited for living in the tropics than their upholstered furniture.

In July 1955, their second daughter, Tina, was born at Gorgas Hospital. The delivery fee was about $50.00 and covered all pre-natal care and a one week stay in the hospital.

Diablo Heights was a very convenient place to live with its own commissary, post office, clubhouse with a movie theater and schools within walking distance. They could see ships moving through the nearby Canal from a second floor window. The girls had fun playing with tulip tree seed pods in the bathtub. They lived there until Bert retired in 1976 as the assistant head of the Electrical Division.

The coming of 60-cycle electricity from 25-cycle power in 1957 brought more dramatic changes with a whole new lifestyle that included air-conditioning, drapes at the windows, and rugs on the floor. It helped to eliminate mildew in their closets and the need for the small heating units in the kitchen cabinets to keep crackers and cookies fresh. It also made lighting along the Canal possible. Bert was assigned the job of helping to design and install the lights that still line the Canal today. On one of his treks along the banks of the Canal, he dashed into a telephone booth to avoid a downpour, encountered a hive of bees, and was stung so badly that his face swelled up to the size of a basketball.

Life in the Canal Zone is filled with memories of the beautiful flowering trees, the amazing rainfall, shopping at the Chinese Gardens and Salsipuedes Street in Panama City. They also remember the ruins of Old Panama with its fascinating history and weekends at Gorgona Beach, pizzas at the Napoli, outdoor parties at Hotel Tivoli and Ft. Amador's Officer's Club and the scraggly but very welcome Christmas trees delivered to our home. Their family also had a few adventures such as surviving Hurricane Camille on board the *SS Cristobal* in August 1969, Val's climbing El Baru in 1975, and taking a Jungle Survival Course through the Girl Scouts. It was an exciting time in their lives and one that will always be fondly remembered.

Bert retired in 1976, and they moved to Austin, Texas, where Bert died in 1982. On October 13, 1985, the City of Austin named a 12-acre neighborhood park in his honor for all his efforts to keep the land as a park. With all of Austin's growth and development, this little bit of quiet green becomes more rare and important in our daily lives.

> *Life in the Canal Zone is filled with memories of the beautiful flowering trees, the amazing rainfall, shopping at the Chinese Gardens and Salsipuedes Street in Panama City. They also remember the ruins of Old Panama with its fascinating history and weekends at Gorgona Beach, pizzas at the Napoli, outdoor parties at Hotel Tivoli and Ft. Amador's Officer's Club and the scraggly but very welcome Christmas trees delivered to our home.*

The Fred and Ruth Sill Family

AS TOLD BY SON FRED SILL

It was 1907, and everybody in the Sill household in the town of Cohoes in upstate New York thought that the only son in the family, young Fred, was out of his mind. Fresh out of Rensselaer Polytechnic Institute with a degree in engineering, Fred decided to risk his future as a civil engineer in New York by heading to Panama to take part in The Great Adventure. He had just turned twenty-two.

When Fred's ship docked in Cristobal, he was asked if he had carved his name on one of the empty coffins in the shipment being unloaded. There were never enough to go around, he was told. If he had not claimed a coffin, his body would be sent back home in a barrel.

Bachelor quarters were waiting for him in Culebra. He would later tell his two children that on the first morning, he woke up to the sound of a bell. Looking out the window, he saw a little red school house. He felt like he was back in New York. Then a young lady came out and shouted "Children! *The bell has rang!*" (Years later, he would tease schoolteacher friends from those early days, Sue Core and Winifred Ewing: "I think it was YOU!")

As an engineer, he joined the Society of the Chagres as soon as he was eligible. By the time the Canal opened in August 1914, he qualified for the Roosevelt Medal (No. 3090) and two bars. Earlier that year, Colonel Goethals had asked him, "What are you planning to do when this is all over, Sill?" "Sir, I plan to return to Cohoes, New York, and continue my work as a civil engineer." "Stick around," said Goethals, "The fun's just beginning!"

Fred Sill (left) in Culebra, 1907.

Fred Sill stuck around and became an admeasurer. In 1912, President William Howard Taft had decreed via proclamation to the Isthmian Canal Commission that the tolls were to be based on tonnage — a ton being 100 cubic feet. That meant that every ship arriving for the first time had to be measured inside and out. Cargo? Passengers? Ballast? U.S. Navy? Hospital? There were different tolls for each and, therefore, an ongoing job for admeasurers.

Within a few years, Fred had become Director of Admeasurement. He broke his service during World War I when he joined the Army and went to Europe as an officer with the U.S. expeditionary forces, serving with British troops. For an act of heroism under fire, he was awarded the Distinguished Service Cross (U.S.) and the

Fred Sill with pals in bunk room, Empire, 1910.

Distinguished Service Medal (U.K.). He always said, "Two for the price of one."

In the late 1920's, Ruth Melgaard arrived from Minnesota, via Wellesley College, to teach French and Latin to the high school students in Balboa. Fred's interest in school teachers was well known. (When I went to Ancon Grade School in the 1940's, I learned he had dated five of six of my grade school teachers!) A confirmed bachelor at the age of 45, Fred proposed to Ruth. She tried to put him off by telling him (it was true) that she was engaged to a doctor in Minneapolis. She said she was not about to write to her fiancé to tell him that the engagement was off; and if Fred was serious about the whole thing, he could tell her fiancé himself. Fred took a ship to New York, a train to Minneapolis, and called on the doctor, saying he was a friend of Ruth Melgaard in Panama. During lunch at the doctor's private club, he told him that he was going to marry Ruth. (He told him over the coffee, he said, rather than over the soup, since he didn't want to risk ruining their lunch.)

Sill family, 1947.

Back in Panama, mission accomplished. They were married and had two children: Mary, born in 1931, and Fred (me!) four years later. As kids growing up, we were bored listening to Dad's stories of the construction days, especially since we had heard them so often before. "A good tale bears telling twice," he always said. But we liked to hear him tell about the parties at the Tivoli Hotel. If party-goers missed the last train back to Empire (there was no road), they would have to find a space to sleep on one of the billiard tables until the first train at daybreak. (An alternative would be to go to the train yard and commandeer a four-wheeled pump car to power yourself back home. Two couples could handle it; but it meant you would have had to pick your female companions for the evening very carefully.)

My sister and I wondered what it would have been like growing up in the United States. What would Dad be doing if he had not gone to Panama? "I would probably be a somewhat-successful civil engineer in upstate New York." That sounded pretty good to us. "But," he added, "how many somewhat-successful civil engineers in upstate New York have met the Queen of England? Twice!" Point taken.

Compulsory retirement due to age came all too quickly for Fred in 1947. Community activities (American Legion, YMCA board, etc.) kept him occupied in retirement, along with screening his original filmed footage of the Canal construction to anybody who was interested. There were also Canal transits at the invitation of local shipping agents and weekly bridge games at the Union Club in Panama City with long-time amigos. He died as he had hoped, at a party in Balboa Heights, surrounded by good friends. At the memorial service held at St. Luke's Cathedral in Ancon, a lady from Rio Abajo came up and told Ruth, "Oh, that Mr. Sill! He'll even have fun among the dead!" Amen!

Fred's wife, Ruth Melgaard Sill, remembered by three generations of Canal Zone students as "Miss Melgaard" or "Mrs. Sill," died in 1994, age 96. Mary Sill (daughter) died 1989, leaving four sons. Fred Sill (son) currently lives in Rio de Janeiro, Brazil.

Fred Sill
Roosevelt Medal No. 3090 with Two Bars

The Harry Grant Smith Family

AS TOLD BY BETTY LOCKWOOD SKOW NEWMAN

History for Future Generations

Harry Grant Smith was born in Westport, Pennsylvania, on May 26, 1867. He married Bessie Taylor Steuart on July 1, 1903, in Lock Haven, Pennsylvania. Their desire to go to Panama and help in the construction of the Canal led them to say they were younger than they were so they would be selected. They had a desire to become progressive pioneers in a place that had many challenges.

Their life on the isthmus started in Gorgona in 1907 when Harry began working for the Mechanical Division as a coach carpenter. His starting wage was 56 cents an hour, and it steadily increased to $1.16 per hour until his death in 1931. Gorgona was a small town where they maintained railroad cars. Their only child, Miraflores Edna Smith, was born at home in December 1909. The family relocated to Ancon in 1912. The town's construction shops had to be completely dismantled before October 10, 1913, when the dikes were dynamited and the rising waters of Lake Gatun engulfed the town. Harry passed away on August 14, 1931, and is buried in Corozal Cemetery. Bessie lived with Miraflores until she passed away in 1952.

Miraflores grew up with her best friends, Matilda and William Van Siclen, Jr., and Dolly Allen Steiner. She quit school before graduation and began working as a cashier in a clubhouse. A former teacher persuaded her to go back to school, and she graduated from Balboa High School in 1931.

Bessie, Miraflores, and Harry: Gorgona, 1911.

On December 30, 1927, Miraflores married Earl Whitman Lockwood in the Balboa Union Church. Their witnesses were Matilda and William Van Siclen. Earl was born in Norwalk, Connecticut, in February 1905. He joined the U.S. Navy at the age of 18, and his duty took him to Panama. He started employment with the Marine Division as a signalman on June 10, 1926, at the rate of $130 a month. From the late 1920's through the 1930's, Earl received the latest hit songs from a music publisher in the States and played them on the piano for a radio station in Panama City. Often times he was accompanied by a singer named Stella. In 1939 he changed professions and became a tow boat master. During this time he remained in the Navy reserves and on January 27, 1930, was commissioned an ensign. He volunteered with the Sea Scouts and worked with other HAM radio operators to put a radio tower on Sosa Hill. His call letters, W7JWE, stayed with him until his death. On October 3, 1940, he was recalled to active duty. Earl passed away in Bothell, Washington, in September 1990.

Earl and Miraflores had three daughters — Betty Lois, Harriet Louise, and Earlane Alice. Betty was born in the Hospital de Panama in July 1929. Harriet, born in October 1931, and Earlane, November 1937, were both born in Gorgas Hospital. For a time the family lived in 760-A, Barnebey Street, in Balboa. Later they moved to a home in the city near Old Panama.

Betty has many fond memories of growing up in the Canal Zone. Life was simple, relaxed, and family oriented. This was a great life for kids growing up, just one big happy family. Tropical weather, flowers, birds, fruit, and vegetables were plentiful. Homes were single wall construction, and all belonged to the United States government. Quarters were assigned according to your job and length of time employed. No one had to fix anything; public works took care of that. The commissary and script books were the means of shopping. The clubhouse was for recreation. This was life in the Canal Zone. Sunday was a quiet day. After church, families would be together, just relaxing. Sometimes they would take a car ride in the afternoon. Old Panama, where the ruins of Morgan's Bridge and Tower of Old Panama were situated in a lovely quiet green park by the ocean, was a favorite of Betty's because she knew that a stop at the "La Lecheria" was also part of this trip. It was a nearby dairy that sold the biggest ice cream cones.

As the war progressed, Earl became more concerned with his family's safety and decided to send them to live with his aunt in Norwalk, Connecticut. Bessie, Miraflores, Betty, Harriet, and Earlane left Panama for the last time and arrived in New York City on November 7, 1941. The family had been in the tropics all their lives, and November was no time to land in Connecticut. Wearing shoes, coats, hats, and mittens wasn't what they were used to. Earl was involved in the D-Day landing at Cherbourg, France.

In 1945, the Navy transferred Earl to Seattle, Washington. Bessie, Miraflores, Betty, Harriet and Earlane drove across the United States in a 1941 Chevy to be reunited with Earl. He retired from the Navy and stayed in the Pacific Northwest.

Betty married Lyman Skow. They have four children — Michael, Alan, Dan and Julie. They grew up in Seattle listening to their mom tell stories about growing up in the tropics. To them it was a far away place that didn't seem real. Michael and his wife Linda have a son Shaun and daughter Dena. Alan and his wife Linda have daughters Kelly and Jamie. Dan and his wife Judy have a daughter Lauren.

In 1982, while living on Camano Island, Washington, Betty met her childhood neighbor from Barnebey Street, Mary Sullivan Young. Captain Jim and Mary Young had recently retired and moved from Panama to Camano Island. Mary invited Betty and her husband to join them when they returned to Panama for the holidays. Julie went with them to see this strange place where Mom grew up. Julie married Jim and Mary's son Thomas (TY). They lived in Cocoli and Gamboa until 1996, when the Panama Canal Treaty transferred their jobs to Miami, Florida. TY lives in Miami, and Julie relocated to Virginia in 2005.

Harriet married William Wolfgang and later relocated to Oregon. They have a daughter Susan Bodle and granddaughter Crystal Norris.

Earlane married Robert DeCamp. They have three daughters — Janet, Theresa, and Nancy. Janet married Chris Lickey. Theresa married Robert Traulsen; they have a son Erik and daughter Sharon. Nancy married Matt Lowell and has a daughter Austin.

Betty's final trip to Panama was at the end of the millennium. She had a dual purpose — to see once more the place that remains in her heart and to return to the French Embassy one of two French flags that flew over their administration building during the time they attempted to build a sea level canal. The French Embassy didn't want the flag so Betty brought it home and donated it to the Panama Canal Museum. Betty was on the last ship to transit the Panama Canal under the American flag. At noon on December 31, 1999, the ship stopped just outside Miraflores Locks for a gun salute. The American flag was lowered and the flag of Panama raised. As a representative of the Daughters of the Ameri-

> *Betty was on the last ship to transit the Panama Canal under the American flag. At noon on December 31, 1999, the ship stopped just outside Miraflores Locks for a gun salute. The American flag was lowered and the flag of Panama raised.*

can Revolution, Betty made this speech to those on the ship:

"I am here as one of four generations who lived in Panama City and the Canal Zone. My grandfather, Harry Grant Smith, and his bride, Bessie Taylor Steuart Smith, came to Panama in 1907 to help with construction of the Panama Canal. His is a part of the history of this marvelous eighth wonder of the world. My grandfather lies in Corozal Cemetery along with many of his co-workers, part of the cemetery where the French also lie. I am proud to hold my grandfather's Roosevelt Medal for service from 1907 to 1913. My mother was named Miraflores — she was born in Gorgona in 1909. I was born in Panama City in 1929. This Canal was worth living, working and dying for. So many young men and women left their homeland to come to this place of disease and hardship to build this amazing wonder. The United States Government took charge of cleaning up the disease that prevailed and made this a new home for so many. It might well prove to be the greatest contribution to the world in the last millennium. As a member of the National Society of the Daughters of the American Revolution, I wish Panama continued success to maintain and run the Panama Canal, now known as Canal de Panama. God Bless your Country and your People."

Harry G. Smith
Roosevelt Medal No. 3497 with Two Bars

PHOTO COURTESY OF PANAMA CANAL MUSEUM

Barnebey Street housing, Balboa, Canal Zone.

The J. Bartley and Mercedes Alegre Smith Family

As Told by Their Children

This account of J. Bartley and Mercedes Alegre Smith is written by their children — Paul, Carmen, Marjorie and Ralph. We simply want to leave a short record of our parents' lives so that as time passes they are included in the history of the great enterprise in which they participated, the American Panama Canal.

Bartley was born in Terre Haute, Indiana, in 1907 to Charles and Edna Lewis Smith. Charles and a brother worked together in their haberdashery and clothes cleaning business in Brazil, Indiana; Edna was a teacher. Dad had two younger sisters, Marjorie and Anna Frances.

At two-year intervals, their young cousins from the Panama Canal Zone took home leave in Indiana, regaling their Hoosier family with wild stories about life in the untamed tropics. Pranks flew back and forth, each side trying to one-up the other over who lived most dangerously. One time Dad filled a red sock with grass, tied a string to it and dragged it slowly through the weeds telling his terrified cousins to steer clear of the giant Indiana chiggers if they knew what was good for them. In spite of the tall tales about Panama, Dad followed in the steps of his father and his uncle Wallace (Duke) Lewis, quartermaster of the Panama Canal Company. Dad attended Rose Polytechnic Institute in Terre Haute and became an electrical engineer intent on working in Panama.

Upon graduating in 1928, Dad immediately applied to the Panama Canal Company but failed the required physical. Despite working summers in a brick factory shoving bricks into the ovens, at six feet in height he weighed a puny 119 pounds. The doctor declared he couldn't stand the rigors of the tropics. Uncle Duke must have intervened, though, for Dad arrived in Panama shortly thereafter and survived the rigors of the tropics for 41 years.

Mercedes and Bartley Smith, circa 1932.

His career with the Panama Canal Company started in the Electrical Division field offices on both sides of the isthmus, and he became part of the Administration Building office staff before we were born. In the early 1950's he became head of the Electrical Division as Chief Electrical Engineer, detouring temporarily to lead the power conversion project which changed the Canal electrical system from 25 to 60 cycles.

Life in Panama must have seemed very cosmopolitan to a kid from the Midwest. Dad spoke of sampling his first martini during the U.S. Prohibition at the Century Club, which was handily just across the street from the Canal Zone in Panama City. While there was probably nothing cosmopolitan about the bachelor quarters and clubhouses, he was a regular customer at the intriguing East Indian shops on Central Avenue which lined Fourth of July Avenue at "The Limits" — the boundary between the Canal Zone and Panama. Over the years he collected several valuable Japanese prints simply by rummaging through piles of prints stacked on floors.

Having studied French and Latin in school, it was

now time to learn Spanish. Inquiring about a teacher, he was directed to Miss Mercedes Alegre, who wrote a daily column for either the *Star & Herald* or the *Panama American* newspapers to introduce Americans to Spanish. She seemed to have other fine credentials as she had attended St. Theresa's College in Winona, Minnesota, and graduated from Panama's Normal School.

Mercedes did indeed teach Dad Spanish. We remember him reading every inch of both daily newspapers which were printed in two languages — one side in English, and turned upside down and over, the other side in Spanish.

Bartley and Mercedes were married in 1934 with their best friends, teacher Sue Core and Dr. John Darcy Odom, as witnesses.

Mercedes was born in Aguadulce, Panama, in 1906. Her father had the contract to build a highway bridge over the Santa Maria River near Aguadulce. He was Celestino Alegre Mateo, born in the village of Camarillas, Aragon, Spain, and served in the Spanish army in Cuba. He moved to Panama after his discharge and married Octaviza Jurado Revilla, who was born in Gualaca, Province of Chiriquí, Panama, to a cattle-ranching family of Spanish descent. Abuelito Alegre was a pharmacist by training.

Mercedes's brothers were Enrique, Rafael and Fernando, all born in San José, Costa Rica, where our grandparents lived for several years.

We Smith kids had the rich good fortune to participate in two cultures, growing up in a bilingual home, attending American schools, and being part of a large Panamanian family.

Mom taught Spanish at Balboa Junior and Balboa High schools and was passionately committed to taking the culture of Panama to her American friends and U.S. culture to her Panamanian friends. She was an accomplished self-taught cook and particularly enjoyed teaching Panamanian cooking. She also enjoyed painting. We cherish her award-winning oil paintings of tropical themes that now hang in our homes.

Pan Canal employees qualified for housing by years of employment rather than size of family. In the very early 1940's, we moved from a bungalow on Bohio Place in Ancon to a duplex close by on Ancon Boulevard. It was a step up but yet a tight squeeze for our large family. Our half of the house was the bottom half — yet we climbed a set of stairs to get to it. Like all houses built by the French, it rested on stilts off the ground. We parked the car under the house and Dad had a wood-working shop under the house as well. He was a skilled artisan and enjoyed making furniture and other home-decorating accessories. Quartermaster-issue furniture was straight and hard and not to his liking.

During World War II, there was well-grounded fear that the Canal was targeted by the Japanese and Germans. Safety precautions were all about us — in the dank and dark air raid shelter behind our house where we and our neighbors scrambled when sirens went off, the searchlights that crisscrossed the sky at night, nighttime blackouts, and neighborhood sheds stocked with gear to address civil emergencies.

But for children, life was pretty normal. Flowers bloomed everywhere, tall almond and mahogany trees shaded our house and lined the sidewalks, and mango trees invited climbing. Our parents' friends were second families and their children were our best friends. We were seldom bored. There were tarantulas to trap in our backyard, doodlebugs to coax out of their holes under the house, and mango trees to fall out of. We slid down slopes on palm fronds, caught fireflies, and occasionally traded blows. We also made periodic visits to Gorgas Hospital's emergency room. Mom and her friends hosted teas, had a sewing club, entertained visitors from the States and tended family.

In the 1940's Mom and Dad built a house in El Valle de Anton. One has to marvel at the grit required to make the trek in those early days of rationed gasoline. Every bit of space in the old Olds was filled with people and provisions and sometimes a parakeet or two. Between the Panamerican Highway and El Valle the road was un-

> *Flowers bloomed everywhere, tall almond and mahogany trees shaded our house and lined the sidewalks, and mango trees invited climbing.*

paved, and in the rainy season Dad often had to stop the car and put chains on the tires to get through the mud. But his biggest challenge was keeping the kids in the car and out of the red goop.

El Valle in the 1940's had no electricity or running water, so we used kerosene lamps and drew water from our neighbor's hand-pump well. Emergencies were a challenge. Dad was bitten in our front yard by a deadly fer-de-lance viper and was driven by neighbors over the winding dirt road to Rio Hato on the coast where the U.S. Army had an airbase. From there he was flown to Gorgas Hospital where he was kept for several weeks. Mom took a similar trip after a serious horseback riding accident.

Mom and Dad lived in Diablo Heights during their last years in the Zone and retired to St. Petersburg, Florida, in 1969 where they joined a colony of close Canal Zone friends. Mercedes passed away in 1973 and Bartley in 1999.

PHOTO COURTESY OF WWW.CZIMAGES.COM

Diablo Clubhouse in the early 1940's. Note the hill to the left which was later removed to make way for new houses.

WILLIAM CHARLES SMITH

AS TOLD BY NELLREE BAKER SMITH BERGER

The William Charles Smith family, residents of Chattanooga, Tennessee, moved to the Panama Canal in 1923. "Bill" (B: May 1891, D: May 1961) received a notice that he had been hired as an electrician. He arrived via ship out of New York and began his career on Gatun Locks in May 1923.

His wife, Clara Means Smith (to whom he was married in February 1916) and their son, Spencer Bibb Smith, who was born in Chattanooga, came down via ship from New Orleans in September. Spencer was 10 months old (B: December 13, 1922). Bill and Clara lived all of their time in the Canal Zone in Gatun. Bill was active in Masons, Eastern Star, Gatun Union Church and other community activities. He worked on rotating shifts as an electrician at Gatun Locks for 30 years. Clara (B: September 1922, D: May 1986) stayed busy being a mother and homemaker. She was also active in Gatun Union Church, Eastern Star, school and community activities. They were in the Canal Zone during World War II and did their part for the war effort. Bill was a World War I veteran.

Spencer attended Gatun Elementary School and graduated from Cristobal High School in 1940. Cristobal High was in the second location on the beach in New Cristobal at that time. After graduation he went to the University of Tennessee in Knoxville, 1940-1941. While spending the summer at home, he became ill and was unable to return to college in September. After he recuperated he took a job until the second semester. But before he could return to the States, World War II started, and he was frozen in his job. He later became part of the apprentice program as a railroad signal maintainer. In 1943 he joined the Army Air Corps, but was in less than a year when he had medical complications and was discharged. His service was all in the Canal Zone.

Spencer and Nellree (B: February 27, 1921) were married June 30, 1946, in Chattanooga, Tennessee. They had known each other all their lives as their mothers had been friends. They went to New York City on their honeymoon. While in New York they attended a radio show called "Johnny Olson's Ladies Be Seated." Nellree was on the show and won the "Singing Housewife's First Prize" for the day.

They arrived in the Canal Zone July 10, 1946. The airport was at Albrook Air Force Base at that time. They traveled by train across the isthmus to Gatun, where they were to live. They first lived on High Street in a 12-family apartment, next to the old third locks channel. They were there two years and then moved to a duplex on Telephone Road for a year before being transferred to Pedro Miguel in 1949.

In Pedro Miguel, they lived on "the Half Circle" off the main highway and then moved to the old Shaw house near the clubhouse before moving to Los Rios, a new town, in 1953. They were one of the first occupants in Los Rios and lived there seven years before Spencer died. He had a heart attack at work and went very fast on April 8, 1960. It was a shock to everyone because he was only 37. Nellree had to move from the family home to Balboa and then moved to bachelor quarters in Diablo. On December 6, 1962, she married Ernest Emerson Berger (B: November 3, 1913), and they moved to a cottage in Diablo on the bluff between Diablo Road and Haynes Street until retirement in 1971.

Ernie's parents (married in 1910) were Claude Emerson (B: 1887, D: 1961) and Zilda Blair Berger (B: 1890, D: circa 1938). Claude was born in New York and Zilda in Massachusetts. Claude was in the Navy and helped build the Coco Solo submarine naval base before World War I. After retiring from the Navy, he returned to the Canal Zone as a civilian employee, circa 1920. He also worked at Gatun Locks as an electrician in the control house. The family lived in New Cristobal and Gatun. Besides Ernie, the other siblings were Nellie Catherine (B: August 13, 1912, D: November 2005) and Claude E. (the younger) (B: January 21, 1915, D: May 2006). Nellie was first married to Charles Walsh; they had two daughters — Shirley (B: October 3, 1930, D: May 2007), who had two daughters and a son, and Marge (B: November 23, 1935). Both graduated from Balboa High School — Shirley in 1949 and Marge in 1953. Marge had a daughter and a son. Claude E. married but had no children. All three of the Berger children graduated from Cristobal High School — Nellie in 1930, Ernie in 1931, and Claude in 1934. The high school was near the Washington Hotel.

Ernie attended the University of North Carolina for two years and later finished at Tri-State College in Angola, Indiana, with an electrical engineering degree. He worked for the instrument department of the Electrical Division after getting his degree and finishing the apprentice program. He had 37 years of service when he retired in 1971.

Nellree was a fourth generation Hamilton Countian from Chattanooga, Tennessee. She attended Chattanooga schools, graduating from Central High School in 1938 and from McKenzie Business College in 1940. She studied voice at Cadek Conservatory of Music for five years with Dr. J. Oscar Miller. She played viola and was a member of the Chattanooga Symphony Orchestra from 1936 to 1946, when she married and moved to the Canal Zone. She was also a member of the Chattanooga Civic Choral Society for eight years. She gave USO musical programs during World War II, twice a week for three years, and sang for many other organizations. She worked at the American Bank and Trust Company as a bookkeeper for almost five years before marriage.

In the Canal Zone she worked in sectional accounting for the Commissary Division at Mount Hope, and then moved to the electrical engineers office in Balboa Heights when Spencer was transferred to the Pacific side. She was in the main office five years, the Balboa field office ten years, and the engineering and construction budget office seven years before retiring.

Nellree is the person in the Canal Zone who started making postcard albums. She made about 1,000 in ten years. While living there she was soloist for many organizations, including United Way fund shows; Memorial Day soloist with the Army; Easter sunrise services on the Atlantic side; soloist for the Fort Davis and Fort Amador Army chapels; Eastern Star chapters; and First Baptist Church, Balboa Heights. She was a clothes designer and seamstress.

Ernie and Nellree retired and moved to Signal Mountain, Tennessee, near Chattanooga in 1971. Nellree has kept up her music and other activities and has added genealogy to them. She has directed a children's choir for 18 years and is a member of the Garden Club, Daughters of the American Revolution, Colonial Dames XVII Century, Signal Mountain Community Guild and Literature Department, and Signal Mountain Baptist Church. She doesn't have much time for housekeeping.

Ernie died on May 16, 1984. Nellree is still in their home which overlooks Chattanooga.

Easter Sunrise Service in Panama, Canal Zone, 1940's.

PHOTO COURTESY OF WWW.CZIMAGES.COM

John Frank Stevens

AS TOLD BY FRANK STEVENS HAWKS

Chief Engineer, 1905–1907

J Franklin Stevens was born April 25, 1853, in West Gardiner, Maine. His parents were John Smith Stevens and Harriet Leslie French Stevens. In his late teens, he decided he did not like the sound of his name so he gave himself a first name and shortened his middle name. Thus by his early 20's he was John Frank Stevens without any formal legal action, and the new name stuck. He grew up in the harsh winter weather of West Gardiner and attended the typical small town one-room school house. He graduated in 1873 from the Farmington, Maine, State Normal School, expecting to become an educator. He found very quickly that he disliked teaching and left to pursue an engineering career. Interestingly, his brother was a life long educator who had a school named for him in Denver, Colorado.

His work took Stevens to Minneapolis, Minnesota, where he met Hattie Townsend O'Brien. They were married on January 6, 1878, in Dallas, Texas. His work required long stays away from home, and Hattie was often left behind first in Texas and then "back east." In Texas he learned surveying and then began working as a junior engineer. By 1886 he was principal assistant engineer building railroads in Michigan and Minnesota. He also worked as a location and construction engineer for the Canadian Pacific in British Columbia, where he acquired many of the skills he would soon take to the Great Northern Railway and, later, to Panama. During this period, Hattie gave birth to their three sons: John, Jr.; Eugene; and Donald.

In 1889 he was hired by James J. Hill to work at The Great Northern Railway. During the next ten years he helped the Great Northern expand its lines into the Pacific Northwest. First he discovered Marias Pass in Montana as the best route across the Rockies without a tunnel. This was accomplished in weather that was, at times, 40 degrees below zero (shades of the winters in Maine). The next year he found a pass through the Cascade Mountains in the state of Washington. This is now called Stevens Pass. In 1895 Hill promoted him to chief engineer and later General Manager of The Great Northern. In 1903, he took a new position with the Chicago, Rock Island and Pacific as First Vice President.

In 1905 when John Wallace resigned the position of Chief Engineer of the Panama Canal construction project, Hill recommended to President Theodore Roosevelt that he hire Stevens as Wallace's replacement.

Stevens was 52 when he arrived in Panama. After familiarizing himself with the situation, he halted the construction work (digging) until he could improve the conditions. His first move was to provide food, refrigeration (ice), and decent living conditions for the construction workers. He also worked very closely with Dr. William

Chief Engineer Stevens at desk.

Gorgas to provide Gorgas with everything he needed to eradicate yellow fever and greatly improve the sanitary conditions on the isthmus. These efforts and their dramatic results, along with his accessibility to all the staff, greatly improved the morale of the men working on the project.

Drawing on his experience as an engineer, he thought of the Canal endeavor as a railroad project and set about providing the means to move men and equipment from one point to another with maximum efficiency. Another function of railroading is to move cargo. In the case of the Canal construction, the freight to be moved was the enormous quantity of dirt to be dug. Much of it was used to create the dam at Gatun Lake.

Although Stevens was not a politician, he was able to convince President Roosevelt and a very skeptical United States Senate that the only way to construct the Canal successfully would be through the use of locks, as opposed to the sea level approach which was favored by Roosevelt, the Senate, and the majority of the Panama Canal Commission. Stevens had favored the sea level approach himself until he saw the effect the rainy season had on the Chagres River during his first year on the job. A sea level approach would have been much more expensive, more dangerous for the ships passing through the Canal, taken much more time to complete, and probably would have been an utter failure.

By January 1907 Stevens had been on the job a total of 18 months. In Washington, changes were afoot. The Canal Commission was being changed and President Roosevelt was considering naming Stevens its Chairman with, perhaps, complete authority. Stevens found out about this possible move and knew at once that he wanted no part of it. On January 31, 1907, Stevens sent the President a letter expressing his displeasure with the proposal and other personal concerns about his continuing his current assignment. He was now almost 54 years old (about a normal life span at that point in history).

Much of his life had been spent away from his wife and family. While the pay was adequate, there was no doubt he could make a lot more in the business sector. In addition, Stevens was a very private individual, used to keeping his own counsel and making his own decisions. He did not enjoy the notoriety that went with the job nor the constant second guessing that is involved in public life. While he never said so, there was little doubt that he was also physically and emotionally drained from the preceding year and a half on the job.

Although Stevens never actually resigned his position, it seems evident this was his desire. In any event, Roosevelt took it as such and on March 31, 1907, Stevens was replaced by General George W. Goethals as Chief Engineer on the Panama Canal. (On March 31, 2007, descendants of both Stevens and Goethals met at The Homestead resort in Virginia to commemorate the 100th anniversary of that event.)

> *He did not enjoy the notoriety that went with the job nor the constant second guessing that is involved in public life. While he never said so, there was little doubt that he was also physically and emotionally drained from the preceding year and a half on the job.*

After his return to the United States, Stevens resumed his career in railroading, first with the New Haven, and then with James Hill to build a line to compete with Edward Harriman's Union Pacific in Oregon. Following that, he went into consulting work until his wife, Harriet, died in 1917 at the age of 62.

Despondent over the death of his beloved wife, he soon accepted an offer from President Woodrow Wilson to go to Russia as Chairman of the Inter-Allied Technical Board, which consisted of railroad experts from eight nations. His job was to straighten out the railroad mess in Russia to enable the movement of supplies for the war effort from the port of Vladivostock to the western front. Stevens remained there, at the President's request, until 1922. He then returned to consulting in the U.S. and served for a year as the President of the American Society of Civil Engineers. Much of his time with the ASCE was devoted to working on Mississippi River flood control.

For his outstanding achievements during a long and distinguished career, Stevens received many honors

from universities, governments and his fellow engineers. Among them:
- The John Fritz Gold Medal of the Engineering Societies
- The Franklin Institute's Gold Medal
- The Hoover Medal
- The Distinguished Service Medal from the United States
- Officer of the Legion of Honor from France
- Second Class Order of the Rising Sun from Japan
- Order of Chia Ho (Golden Grain) from China
- Order of Wen Hu (Striped Tiger) also from China
- The Military Cross, the highest military decoration awarded by Czechoslovakia

Stevens died in Southern Pines, North Carolina, in 1943 at the age of 90. He was survived by his brother, Eugene; three children; two grandchildren; and six great-grandchildren. One granddaughter was killed in an automobile accident in 1936 and a grandson died during World War II. John Frank Stevens and his wife, Harriet, are buried at Mount Hope Cemetery in Boston, Massachusetts.

(Note: This story is a summary of information obtained from The Path Between the Seas by David McCullough; Wikipedia, the free encyclopedia; John Stevens' own personal notes; and family information passed down from our parents and grandparents.)

PHOTO COURTESY OF PANAMA CANAL MUSEUM

Residence of Chief Engineer, Culebra.

Ellis Dayre Stillwell

AS TOLD BY NORMA STILLWELL MARTIN

1888–1961

Ellis Dayre Stillwell's first traceable ancestors came from Surrey, England — an area rife with boiling springs that threw great quantities of water up quietly. Thus the name "Still Wells" came into existence. In the early 1600's, Lieutenant Nicholas Stillwell immigrated to Kings, now Brooklyn, New York. Over the years the "Far West" spirit moved many hardy souls, and eventually Asher Stillwell ended up in Wisconsin. He married Jane Thorp in 1822. David, one of their sons, married Melissa Auger in 1853, and they had three sons — Arthur, Victor and Wendel. As a young man, Victor (Ellis's father), at an old timer's picnic, read a flyer about "Homestead, Pre-Emption and Tree Claim Laws" in the Dakota Territory. Victor and Arthur set out, staked claims and relocated there. Victor eventually sold his claim but not before returning to Wisconsin to marry Eugenia Hoyt. They had three children while living on the claim and after selling it, moved to town. Ellis, the youngest of their children, was born soon after they settled in Alexandria.

Ellis's childhood in that small town on the Midwest frontier was like all youth of that day. Hard winters, hot summers and chores helped maintain the family of six. It was a time of young families starting out in life so there were plenty of contemporaries — among them was Elizabeth, who had been adopted by Samuel and Mattie Phelps. She was a lovely young lady. After finishing high school Ellis went to the university at Vermillion, South Dakota, for two years and then to the University of Wisconsin, graduating in 1910 in engineering.

Ellis accepted a position with General Electric where he became friends with other employees. The Panama Canal had been under construction for several years, and the government was always on the lookout for promising young engineers. Ellis and a few of his friends signed on for two years. What an adventure — new frontiers! Ellis's father had gone to the Dakota Territory — Ellis went to "dig the Big Ditch!"

The construction towns of Culebra, Gorgona and Empire were close to the actual digging sites, and Ellis ended up in Culebra. The bachelor quarters were old French buildings and by 1911 had been screened and the townsites cleared of bush. Each town had a YMCA where the young employees found recreation and a touch of home. The bachelor quarters were sparse, but adequate; and a central mess hall was available for meals. Ellis was hired as a draftsman, promoted to assistant testing engineer, then transferred to the Locks Division in 1914.

Most of the young men on their first tour were bachelors. On weekends, they dressed up in "good clothes," including straw boaters, and visited the sights — Panama City, Taboga Island, Perry Plantation, Fort San Lorenzo, and Portobelo.

> *Most of the young men on their first tour were bachelors. On weekends, they dressed up in 'good clothes,' including straw boaters, and visited the sights — Panama City, Taboga Island, Perry Plantation, Fort San Lorenzo, and Portobelo.*

My father took many photographs and developed them himself. I have an album full of early day photographs. I also have a large book, *Panama and the Canal in Pictures and Prose*, by Willis J. Abbott, published in 1913, stating on the title page "Profusely illustrated by over 600 unique and attractive photographs taken expressly for this book by our special staff." On page 31 is a photograph entitled "San Blas Indian Boys." That photograph is in my father's album. I think the author or publisher put out a call to submit topical photographs, and my father sent this one. Other photographs in his album show visits to Comacho Reservation and Mandingo Stockade. These places must have been near Chorrera. A 1913 photograph shows that the big saltwater swimming pool at the Washington Hotel was then in use.

Ellis's diary, starting January 1, 1913, shows he was on the list for family housing in Gatun. After six long months, he sailed for the United States to marry Elizabeth. They returned to Gatun to start married life in quarters 96-B.

San Blas Indian boys, April 5, 1912. Photo from *Panama and the Canal in Pictures and Prose* by Willis J. Abbott.

Gatun then was not the Gatun of the late 1930's and 40's. All the quarters were the old French wooden buildings raised several feet off the ground with screened porches. Furnished by the quartermaster, they were comfortable dwellings, albeit different from what the young wives were used to at home. Ice was delivered each day for the icebox; the melting ice dripped through a pipe in the floor to a concrete slab under the house and evaporated. The kids begged ice pieces from the accommodating ice men. The commissaries had the usual necessities; and later there was even "Commissary Brand" on things like aspirin, bleach and other standard items.

An Army post, known as Camp Gatun, was located between Gatun and Fort Davis. The married officers and NCOs had housing near the main part of the town of Gatun with the exception of two houses and a row of garages. Several groups of garages were scattered through the town as automobiles became more popular. In the early 1920's, Ellis had an Oldsmobile, an open car with windshield wipers that were hand operated and side windows that were removable.

Before the Canal opened, the Gatun residents swam at the north end of the locks. Pictures show ladies in bathing costumes — short sleeved blouses, knee length skirts, hose, bathing slippers and a cap — and gentlemen in woolen suits with over the shoulder straps and knee length pants.

Many families became lifelong friends. Outings are shown in the many photographs in Ellis's album. One in particular shows ten couples at Las Cruces, the women in long white dresses and large white hats. It must have been a fun day as the women had a banner that read "Suffragettes," and the men had one that read "Suffer Yets."

A daughter Jean was born in 1917 and Norma in 1919. The military presence was lessened in positions of control on the locks when the United States entered World War I. In 1919 Ellis became Superintendent of the Atlantic Locks. The family moved from a cottage across from the clubhouse to the superintendent's quarters, a large two-story house next to the fire station. It stood on a slight rise with a door on each of the four sides — the north looking toward the channel to Limon Bay and the south overlooking Gatun Lake. The railroad tracks separated the main part of the lake from a very small one where the children learned to swim.

In 1932, when the Superintendent of the Pacific Locks, Mr. Holloway, retired, Ellis became superintendent of the Locks Division, and the family moved to Pedro Miguel. H.M. Thomas became Assistant Superintendent, Atlantic Locks, and J.C. Myrick, Assistant Superintendent, Pacific Locks.

After World War I ended, life went on with ships transiting from all over the world. The lock chambers were overhauled regularly — drained, checked and cleaned. Most employees took advantage of paid vacations and returned every other year to the United States on the Panama Railroad ships — the SS Ancon and SS Cristobal — and later the three new small luxury liners — *Ancon*, *Cristobal* and *Panama*.

Ellis was well thought of by his friends and fellow employees. He was quiet and reserved, but with a sense of humor. Because he was quiet, people listened when he spoke. He walked to his office in the control house every morning, came home at noon and returned after lunch. Ellis played golf for relaxation and exercise, was a member of Sibert Masonic Lodge, and attended and participated in the life of the Union Church. He couldn't sing a note but enjoyed good music and played the piano for his own pleasure. On the day Pearl Harbor was attacked, the family had been to church and Ellis got a phone call. I don't know who called him. Life changed on that day. After the war in 1946 Ellis's office was relocated to the Administration Building, and they moved to Balboa, where they lived until Ellis retired in 1948.

Jean married Bruce Crook, who worked in the paymasters office, after she graduated from college in 1939. They lived in Balboa on Carr Street in the Flats and had two children, Bruce and Elizabeth, both born at Gorgas Hospital in Ancon.

Norma graduated in 1941 and worked for the Army at Quarry Heights. She married Lee Martin and had four children — Lee and Dennis born at Gorgas Hospital; Robert born in Yokohama, Japan; and Normalee at Fort Meade, Maryland.

I have always felt that my father and the many talented men and women responsible for the building and success of the Panama Canal were indeed the original "Greatest Generation."

Ellis Dayre Stillwell
Roosevelt Medal No. 6738

PHOTO COURTESY OF NORMA STILLWELL MARTIN
Town of Culebra, April 7, 1912.

PHOTO COURTESY OF NORMA STILLWELL MARTIN
Bachelor quarters, Culebra, April 7, 1912.

The Charles H. Stilson Family

AS TOLD BY PAULINE (SUE) PINCUS SMITH — FOURTH GENERATION STILSON

Stilson Roots Run Deep — Five Generations in Panama (1863–1991)

The Stilson saga begins with Charles H. Stilson, born in Portland, Maine, in 1835. He married Martha Knight in 1855, and they had three sons. After working several years for the Grand Trunk Railway in Portland, Charles left his family in 1863 to seek work with the Panama Railroad. Several years later his family joined him, and three more children were born. Charles worked as a conductor on the railroad for fourteen years until his sudden death in 1877 at Gatun Station.

Among Charles's sons were Frank and Joseph Stilson. Frank worked as a ticket agent and telegraph operator for the Panama Railroad at Gatun. In 1887, he contracted yellow fever and died. His brother, Joseph, became involved in the exportation of bananas and in 1882 married Maria Evers, whose father worked for Wells Fargo supervising gold shipments in Panama. They had seven children.

Joseph was a shrewd businessman who was successful in the lumber and cattle trades. Some regarded him as the pioneer of the Colon import/export business. He also owned a profitable hardware store and considerable real estate. The Stilsons had two family homes. Most of the year was spent in Colon; but during the breezy dry-season months, the family lived in a large home built by the French near the Old Village of Gatun.

Stilson residence demolished to make way for Gatun Locks.

In 1906, Joseph was informed that part of his Gatun land must be surrendered for Canal and railroad construction. This property included his residence and the vast acreage used for his flourishing cattle operations. Before a final settlement was reached, the Panama Railroad Company took possession of the land and removed the fences, which allowed the cattle to wander. Many strayed onto the railroad tracks and were killed, and Joseph Stilson was arrested twice for sanitary violations because his livestock ran loose.

In 1908, the Railroad Company agreed to furnish construction materials to build the Stilsons a new home outside the Canal construction areas. The agreement stipulated that Joseph Stilson would provide the labor and would be reimbursed for his expenses, the cost of the materials and labor not to exceed $5,000. Once construction was completed, however, the railroad's general manager refused to honor the contract. In 1915, this house was destroyed by fire.

From 1906 to 1909, Joseph Stilson and his lawyer were involved in numerous disputes with the Railroad Company and the Isthmian Canal Commission, arguing mainly that the Stilson family had received no compensation for property damage or for land taken over the years. Joseph finally had to accept the payment offered, which was far below the value of his property.

Stilson's Pond, named for Joseph H. Stilson, was formed in 1912 when Gatun Lake was created by the damming of the Chagres River. In 1940, during the third locks excavation, the pond was filled with construction waste. Today, the pond no longer exists, but the term "Stilson's Pond" still refers to that area in Gatun.

Joseph's success in business provided the family with prominence in Colon and a very comfortable life. The family had cooks and numerous maids, and the children were taught at home by private teachers before attending local schools. As teenagers, the boys were sent to Freehold Military Academy in New Jersey and the Pratt Institute. The girls attended the Academy of Mount Saint Vincent in New York.

Joseph Stilson, Jr., eldest son of Maria and Joseph, was born in Gatun in 1889. After graduation he returned home, and in 1909, he was employed by the Canal Railroad organization as a car record clerk at the Colon Station. Five of his working years were during the Canal construction period, so he earned the Roosevelt Medal with one bar. When he retired in 1951, his total continuous service of 42 years (33 with the railroad) exceeded that of any Canal employee at the time. In 1919, Joseph married Anna Dybrro, daughter of a Canal Zone fireman. They had two daughters; the eldest became a nurse at Gorgas Hospital.

Louis Stilson, born in Gatun in 1890, began his Panama Canal career with the Panama Railroad in 1911 as a wireman and later became a steam engineer. This employment earned him the Roosevelt Medal. Louis transferred to the Commissary Division where he was appointed commissary accountant. At retirement, most of his Canal service was with the Department of Finance. In 1917, Louis married Aline Viall, whose father was also a Roosevelt Medal recipient. They had one daughter.

Edith Stilson was born in Colon in 1894. At age sixteen, she was chosen as Colon's first carnival queen. After graduating from Mount Saint Vincent, she met William Frederick who had been working as a machinist for the Isthmian Canal Commission at Culebra since 1911 and was also a Roosevelt Medal recipient. He worked for the Dredging and Marine divisions before retiring from the Engineering Division. Edith and William were married in 1919, and they had three daughters.

The Stilsons watch Canal construction from their new Gatun home.

Louise Stilson was born in Gatun in 1896. Louise and her sister Edith were fluent in English and Spanish; and during both world wars, they worked at the Censor Office where mail was opened and censored — the only job that Joseph permitted his daughters to have! Louise also did charitable work with her parish church. She never married and, in later years, cared for her ailing parents.

William Stilson was born in Gatun in 1898. After his education in the United States, he returned to Panama and worked in the prospering family business. In 1940, the Stilsons suffered a great loss of uninsured property during the devastating Colon fire. Twenty-four city blocks were destroyed, including their hardware store and many of their tenement buildings. A year later, William married Elsie Clark. They had no children. After his father's death in 1942, Bill continued to manage the family's real estate.

Jessie Stilson was born in Colon in 1899. After attending school in New York, she returned to her family's home in DeLesseps. There she met William Hunt, an Army officer stationed at Fort DeLesseps. With encouragement from her father, William left the Army and accepted employment with the Panama Canal, first as a surveyor and later as a civil engineer. He and Jessie were married in 1920 and had five children. In 1940, William was recalled into the Army and was stationed at Fort Sherman. He retired from the military in 1946.

Alice Stilson was born in Colon in 1901. As the youngest, she was the most pampered. Not wanting to remain in New York after her sisters left, she returned home for her senior year at Cristobal High School, graduating in 1920. Alice met her future husband, Arnold Pincus, on the tennis court. He was in the U.S. Navy. After he received his honorable discharge in 1928 and was hired by the Panama Canal, they were married and had two children. Arnold worked for the Electrical Division until his death in 1950.

After the Stilson children married and thirteen grandchildren were born, they continued to gather at the large family home in Colon. During the year, afternoon tea was served daily in the parlor. Holidays were grand celebrations when the cooks prepared and served elegant meals. The grandchildren loved visiting Grandma and Grandpa. Their yard was a mini-zoo with dogs, two fishponds, cages with exotic birds and pigeons, guinea pigs, marmoset monkeys, a deer, a peacock, and a chicken coop. On the porch was a large parrot that often called out the children's names. Hours of fun were spent in a playhouse made out of an old trolley car.

Weekends found family and friends taking a launch to the Stilson Farm on Escobal Island in Gatun Lake. Cows, chickens and horses were scattered about the farm. There were mangoes and grapefruit to pick, a pony to ride, and trails to walk. From the dock, you could dive into the lake for a swim or paddle in a dugout canoe.

Fourth-generation Stilsons were born and raised in the Canal Zone. Most left Panama and settled in the U.S to raise their families. Only Alice's two children, Richard and Pauline (Sue), fourth generation Stilsons, stayed and worked for the Panama Canal. Richard returned to Panama in 1956 with his family and worked as a structural engineer until 1960. Sue returned after college and married Gilbert Smith, son of a retired Panama Canal plumber. Gil was an accountant and retired from the Panama Canal in 1990 with 34 years of service. Sue taught elementary school in the Canal Zone school system, which became the Department of Defense Dependents Schools in 1979. She retired in 1991.

Joseph and Maria Stilson and their seven children, 1912.

The Smiths had five children, born and raised on the Pacific side. They graduated from Balboa High School and held summer jobs. During the 1980's, one daughter returned after college and worked a year for the Comptroller's Office at Fort Clayton, and a son worked for Motorola and later for Hammer in Panama. If the Panama Canal Treaty had not ended the U.S. control of the Canal in 2000, fifth-generation Stilsons would be living and working in the Canal Zone today.

Joseph Stilson, Jr.
Roosevelt Medal No. 4594 with One Bar

Louis Stilson
Roosevelt Medal No. 7400

Harry H. Viall
Roosevelt Medal No. 2001 with Three Bars

William Frederick
Roosevelt Medal No. 6574

THE ROY C. STOCKHAM FAMILY

BY HIS DAUGHTER—FOR HER SON

From Mule Operator to Chief

The youngest of nine children, Roy Stockham was born November 20, 1902, in Washington County, Colorado, the son of George Washington and Mary Forquer Stockham. Most of his childhood was spent on the family farm outside Sedgwick, a small farming community in northeast Colorado. He graduated from Sedgwick High School in a class of two in 1919 and was awarded his diploma by his father, who was the president of the Board of Education.

After graduating from high school Roy spent several years working on the family farm; but in December 1922 he joined the U.S. Navy in order to earn the money to pay for a college education. Roy served aboard the cruiser *USS Cincinnati* as an aviation machinist's mate and narrowly survived a seaplane crash near the Pacific entrance to the Canal during a visit by the ship in 1924.

In 1927 Roy entered the University of Colorado at Boulder and in 1931 graduated with "special honors" and a bachelor of science degree in mechanical engineering. He served as the president of Tau Beta Phi engineering honorary. The Great Depression made jobs hard to find for a young engineer, so Roy returned to his hometown to work as a district manager for Public Service of Colorado as well as serve as the town clerk.

It was here that he met his future wife, Geneva Ellen Davis, the daughter of William Louis and Janet Ferguson Davis. She was born in Coal Creek, Colorado, on February 22, 1909, and grew up in Florence and Canon City, Colorado, with two younger sisters. She had just received her teaching certificate from Colorado State Teachers College when she accepted the offer to teach third grade in Sedgwick.

On May 25, 1935, the day school ended, Roy and Geneva drove to Colorado Springs, where a justice of the peace married them. They needed to wait for school to end because during the Depression, when only "single ladies" were allowed to teach, Geneva would have lost her job had they married earlier. After a short honeymoon in Denver they moved to Fort Wayne, Indiana, where Roy had accepted a position as a student engineer with General Electric. This did not last long as an ad in an engineering magazine for jobs in the Panama Canal Zone caught Roy's eye. In December 1935 he left Fort Wayne headed for New York to catch the SS Cristobal and a job as a towing locomotive (mule) operator on the Pacific locks at a salary of $250 a month.

Roy C. Stockham

Geneva returned to Colorado to stay with her parents until Roy could locate housing. Roy soon secured "vacation quarters" in Pedro Miguel; and on January 4, 1936, Geneva sailed from New Orleans aboard the United Fruit Company ship *SS Sixaola*, headed for the Canal Zone. Eventually, they were assigned to their own permanent quarters, a four-family building on "Incubator Row," so named because of the numerous children that were born while their parents were living there. In keeping with this tradition their only child, a daughter, Janet, was born on November 28, 1938, at Gorgas Hospital, Ancon, Canal Zone.

Janet attended Canal Zone schools in Pedro Miguel and Balboa and graduated from Balboa High School in 1956. She left for college and eventually married her high school sweetheart, Jim Reece.

Being unusually well organized and highly disciplined, Roy began to advance within the Locks Division, becoming Superintendent of the Pacific Locks in 1945 and moving his family to the "big house" in Pedro Miguel. That same year he received the degrees of Scottish Rites, an organization in which he was extremely active, being honored by the Supreme Council with the rank and decoration of Knight Commander of the Court of Honor in 1951 and in 1956 with the Thirty-Third Degree, Inspector General Honorary.

In 1948, Roy was promoted to Chief of the Locks Division, a position he held for 15 years, and the family moved to Morgan Avenue in Balboa and later to Ridge Road in Balboa Heights. A highlight of this part of his life was several trips he made to Japan in the early 1960's in connection with the purchase of new towing locomotives from Mitsubishi Shoji Kaisha, Ltd., in Tokyo. He was also proud to serve as the chairman of the Canal Zone Board of Registration for Architects and Professional Engineers.

During their years in the Canal Zone, Geneva served as a Girl Scout leader, volunteered with the Women Airforce Service Pilots at Quarry Heights during World War II, was active in the College Club, and was a substitute teacher at Balboa Elementary School. She was an accomplished hostess as well as an excellent seamstress, making most of Janet's dresses.

Roy retired in December 1963 and they moved to Denver to make their home; unfortunately, Roy died unexpectedly on July 28, 1964, at age 61. After Roy's death Geneva moved to St. Petersburg, Florida, where many of her Canal Zone friends had settled. She enjoyed being a grandmother for 30 years before she died on October 3, 1994, at age 85.

Roy C. Stockham: "A gentle person, always courteous and friendly, never aggressive and somewhat inclined to keep his own counsel, but obviously a man of extraordinary ability and capacity for work." From a tribute appearing in the *Canal Zone Orient* at the time of his death.

See also the Roy D. Reece family history.

Miraflores and Pedro Miguel Locks, view from Ancon Hill.

PHOTO BY DON GOODE

The Thomas C. Sullivan, Jr., and Joe Ridge Families

As told by Virginia Ridge Dolim

For Granddaughter Tracy Johnson Cota

Thomas C. Sullivan, Jr., was a third generation Irishman living in New York City. His grandfather Timothy was born in Limerick, Ireland, and immigrated to America in the mid 1800's. His father served as a Marine in the U.S. Civil War and later became a New York City policeman. Thomas married Mary Jo Gammell and the couple had five children who all lived in Brooklyn: Thomas C. III, Julia, Edward ("Jake"), Paul and Vincent. After the turn of the century, Thomas found it increasingly difficult to support the family in New York. In 1906 he went to the Canal Zone and found a job in the Dredging Division in Gatun while the family remained in Brooklyn.

For four years, Thomas would go to the Canal Zone in the fall and stay until late spring. He would then resign and return home for the summer. The family would live comfortably while he was home, but come the fall, Thomas would return again to the job in Gatun.

As his eldest son Tom grew older, he, too, went in search of work to help his family. By 1912 there were no jobs in New York City, so he went to the Canal Zone to join his father. Conditions in the Canal Zone had improved dramatically, and soon thereafter Thomas cabled the rest of the family to join them in family quarters in Gatun.

While living in Gatun, their only daughter, Julia, nicknamed "Goldie," traveled across the "continent" to attend high school in Balboa. Her 15-hour train ride carried her from the Atlantic to the Pacific side every day. This "incredible fact" was rumored to have been reported in *Ripley's Believe it or Not*.

Because the climate in Panama was well suited for baseball, a league was organized and four teams were formed: Pacific, Atlantic, Army and Navy. The league was able to pay its players, so it attracted many Americans who wanted to earn money playing baseball, while also hoping to get noticed by big league scouts from the United States. The Sullivans' young son "Jake" moved to bachelor quarters on Balboa Road and became an excellent player for the Pacific team. He was scouted by the Chicago Cubs and even given a ticket to the Windy City to talk to the team. After he learned their plans, however, he came back to Panama: "I went there to play for the Cubs," he said, "not some small farm team."

> The Sullivans' young son 'Jake' moved to bachelor quarters on Balboa Road and became an excellent player for the Pacific team. He was scouted by the Chicago Cubs and even given a ticket to the Windy City to talk to the team. After he learned their plans, however, he came back to Panama: 'I went there to play for the Cubs,' he said, 'not some small farm team.'

Around the same time the Sullivans were settling into Canal Zone life, Joe Ridge came to Panama to work on the Canal. Joe's father, Peter, had emigrated from County Galway, Ireland, and operated a small store in Rankin,

Pennsylvania, near Pittsburgh. Joe was one of six brothers. Their mother had died when they were very young, leaving the father to raise the boys in quarters above the store. Around Christmas 1908, Joe sent for his brother Aloysius P. "Larry" Ridge, a semi-professional baseball player, to join him. Eventually, three more brothers, all baseball players, came to the Canal Zone. Together they formed one of the best known infields in the Canal Zone. Larry was the second baseman while Jack played third and pitched. Steve played short-stop, and Leo, the baby, played first base. Soon the Ridge name became synonymous with Canal Zone baseball, and that legacy carried on through the next generation.

A short time later, the Sullivan family moved from Gatun to Balboa. "Jake" was the baseball player in the family. Paul excelled in basketball, and Vincent focused on tennis. By this time, Larry Ridge along with two of his brothers, Jack and Steve, were working for the Panama Canal and had received the Roosevelt Medal. But it was the baseball connections that eventually brought Larry Ridge and Goldie Sullivan together. They married May 22, 1917, at St. Mary's Church in Balboa. A wedding reception followed at the new Tivoli Guest House.

Larry and Goldie Ridge lived on Plank Street and had four children: Lawrence, Virginia, James and Paul. Eventually, they moved to quarters on the Prado. Like countless other Canal Zone families, they spent many summers in Bayside, Maine, enjoying a complete change of climate. While the family traveled by ship all those summers, it wasn't until August 15, 1939, that they made their way through the Panama Canal aboard a ship, the *SS Ancon*, on its 25th anniversary transit.

The only girl in the family, Virginia, like her uncles and brothers, excelled at baseball and other sports. In high school she was a founding member of the Girls Athletic Association (GAA). After graduating from Canal Zone Junior College, she began working as a secretary for the Schools Division. She worked in the Administration Building often pictured at the top of the stairs from the Prado. She also played short center field on the Working Woman's Baseball League from 1939 to 1941.

Goldie and Larry Ridge, 1917.

In 1939 Virginia met Henry P. "Hank" Dolim, a young lieutenant sent to Albrook Field after completing flight training in San Antonio, Texas. The couple married February 22, 1941, at Sacred Heart Chapel in Ancon. They too had their reception at the Tivoli.

After December 7, 1941, when the U.S. became involved in World War II, Hank and many other pilots were assigned to protect the Panama Canal. It was a natural target as an attack on the Canal could bring worldwide trade to a halt. In 1942 Hank was sent to the Galapagos Islands to help establish a U.S. air base that would further protect the Canal from attack. Virginia, with her new baby, Henry, Jr., moved back home to the Prado where her parents still lived while Hank completed his tour of duty in the Galapagos. She later joined her parents on Morgan Avenue for almost two years while her husband served in Italy as a combat pilot and B-24 Squadron Commander.

Larry and Goldie Ridge's children.

With the end of the war came the end of Virginia Ridge Dolim's stay in the Canal Zone. She moved with her new family to the U.S. When her father Larry retired in 1950, he and Goldie moved to California, where most of the Ridge children made their home. There they joined Hank and Virginia and new daughter Diane, born in St. Petersburg, Florida. After more than 45 years in the Canal Zone, it was the end of an era for the Ridge and Sullivan families.

Aloysius P. "Larry" Ridge
Roosevelt Medal No. 6501

John E. Ridge
Roosevelt Medal No. 5400 with One Bar

Stephen M. Ridge
Roosevelt Medal No. 6924

Henry and Virginia, 2007.

PHOTO COURTESY OF PANAMA CANAL MUSEUM

Balboa Stadium during the ballgame between the Pacific and Atlantic Fleets' baseball teams, Feb. 17, 1921.

WILLIAM D. TAYLOR

AS TOLD BY CATSY TAYLOR SCHAFER

Our Favorite Ditch Digger

William David Taylor was born in Maldon, Massachusetts, on November 26, 1884, son of Joseph and Harriet (Hattie) Roby Taylor. After graduating from high school, he joined the U.S. Postal Service. When he saw an advertisement that postal clerks were needed in the new Canal Zone, he applied and left for Panama. As the ship approached the Cristobal entrance to the Canal, it was learned that President Teddy Roosevelt was on the ship ahead of them going into the Canal.

The date was November 26, 1906, and it was Bill Taylor's 22nd birthday. He felt like the band and bunting were for his birthday instead of for Teddy Roosevelt! They followed Teddy's train across the Isthmus and Teddy had to stop and wave to the ditch diggers along the way, so it took the train three hours to go across the 50-mile line.

Bill's first assignment was in the construction town of Empire. An engineer working on the Canal was Philip Noonan, who lived in Empire with his wife and three children — Alice, Lillian and Jimmy. In order to get to school, they had to walk across the Canal on a wooden bridge and take the train to Cristobal to attend school there. Bill Taylor met young Lillian after she graduated, and he told her that if she "would put up her hair he would take her out." They were married on December 30, 1910.

William D. Taylor

Their first home was in the town of Paraiso, and Bill tells the story that when they returned from their honeymoon, his male friends had removed the labels from all of the canned goods they had bought. What a way for a new bride to start housekeeping.

Their only son, William Norman Taylor, was born in Paraiso on September 10, 1913. Lillian was too shy to go to Gorgas Hospital, and the old German doctor in the clinic was really upset with her. Bill weight 9 1/2 pounds. When the dike was blown to join the waters of the Atlantic and Pacific oceans, Lillian took baby Bill up to see the great event. She and Bill Taylor divorced in 1924, and Bill, Jr., stayed in the Canal Zone with his father.

Bill, Sr., was invited to sail on the *SS Ancon* on the first transit of the Canal in 1914. He received the Roosevelt Medal for service during 1906-1908, Bar No. 1295 for 1908-1910, Bar No. 888 for 1910-1912, and Bar No. 592 for 1912-1914. Philip Noonan, Lillian's father, also received the Roosevelt Medal. While in the Cristobal Post Office, Bill met Charles Lindbergh, who inaugurated the first air mail to the Panama Canal Zone, landing at France Field. Before his retirement, Bill Taylor, Sr., was Postmaster at the Balboa Post Office. He retired in 1946 and made his home in Bella Vista, Panama. His second wife was Adelyn Harte Aubrey, who had also been an employee of the Panama Canal Company.

William Norman Taylor was sent to school in Maryland to the Tome School for Boys. He met his future wife there, Catherine (Catsy) Hopkins. After graduation, Bill returned to the Zone, and after a year with the Grace Company, he went to work in the personnel department for the Panama Canal. He was assigned to the labor office in Cristobal. On his first vacation to New York, his mother invited Catsy to join them at Christmas time.

Catsy and Bill became engaged and returned to the Canal Zone, where they were married on May 12, 1938. Their honeymoon was on Taboga Island at the old Aspinwall Hotel, a former malaria hospital. Tilley Malloy was in charge of the hotel at that time and was a delightful lady. The junior Taylors had three children: Michael Wylie Taylor, born February 20, 1939; Susan Roby Taylor, born June 15, 1940; and Layne Woodall Taylor, born July 29, 1941.

When World War II started, if wives went back to the U.S., they could not return to the Canal Zone unless they had a job. For that reason, Catsy, Pat Leach, Liz Simons, and Isabelle Clemmons started a nursery school for children ages 2½ to 5 so that mothers could work and not have to leave the Canal Zone. The first year, the school was in Pat's basement in New Cristobal, later in the basement of a building belonging to Cristobal High School, and eventually in an abandoned fire station rented by Liz Simons and Catsy Taylor on Cristobal Point by the bay, where it remained through the war years.

Bill Taylor was assigned to the personnel department in Balboa in 1946, and the family moved to Barnebey Street near the elementary school, below the Administration Building, where Bill worked. After a few years in Balboa, they moved to a cottage in Diablo, where the children finished out grade school and part of junior high school. Their final move as a family was to a new house in the "Flats" where they lived until 1958. That was the year that Bill died at age 44. With his untimely death, Catsy, who was a registered nurse at Ft. Clayton hospital, moved to Curundu, which was a town for civilians working for the military. In 1963, she married LTC Vernon Schafer, who was in charge of the Army's post exchange service at Ft. Clayton. They lived in Panama City until the Panama riots forced them to leave, and they decided to return to the States. William Taylor, Sr., left with them and they settled in San Diego, California. Mr. Taylor died in 1968. His ashes were returned to St. Luke's Cathedral Columbarium to remain next to his wife Adelyn.

William D. Taylor
Roosevelt Medal No. 2112 with Three Bars

Philip Noonan
Roosevelt Medal No. 4100 with One Bar

Balboa Post Office, Balboa.

PHOTO COURTESY OF PANAMA CANAL MUSEUM

The Hugh M. Thomas Family

AS TOLD BY HUGH M. THOMAS III

The Heritage of the Canal Runs Deep Within Me

H. M. Thomas (known as "H.M.") was born in Lawrenceville, Virginia, on March 27, 1886.

He arrived in Panama in early 1910, and he began his employment in April of that year as a machinist on the Atlantic side in the Locks Division at the rate of 65 cents per hour. He held several positions during the construction period, but always worked on the Atlantic side. Life in Panama was rough during that period with reports of his earliest housing being in a railroad car located in the Colon/Cristobal area.

He met Grace Bishop, an American girl from New Jersey, who was in Colon visiting her aunt. Grace enjoyed Panama and stayed on, working as a sales girl in the shoe department of the Cristobal Commissary. In 1917, after a seven month courtship, she and H.M. were married. She told stories of dances in Colon during that period. H.M. had a car (she thought it more of a go-cart), and he actually proposed marriage to her while they were driving in his car. They initially lived in Cristobal and eventually moved to Gatun. They had five children: Virginia (married Ralph Harvey); Robert (married Mary Alice Dod); Hugh, Jr. (married Lois Kridle); Grace; and Thelma. Grace and Thelma both moved to California after graduating from Cristobal High School.

Bud and Lois Thomas at suggestion awards ceremony, Administration Building, Balboa.

H.M. worked his entire career at Gatun Locks on the Atlantic side, becoming Superintendent of Gatun Locks (the highest position on the Atlantic side) in 1942. He was very well-liked and respected throughout the community. He was also quite a golfer and was influential in developing the Gatun Golf Course on the west bank of the Canal between the locks and the spillway. Later, he was instrumental in the development of Brazos Brooks Golf Course. H.M. and Grace retired to California in 1947.

Hugh M. Thomas, Jr., (Bud) was born in Colon Hospital on October 27, 1921. A grade-school friend began calling him "Buddy," and the name stuck. Bud recalled moving into The Superintendent's House (#157) as a boy. It was the largest house in Gatun, across from the new Gatun Clubhouse, bowling alley and movie theater. It had a great view of the Atlantic entrance to the locks as well as a view of Gatun Lake. As a child, Bud enjoyed fishing and swimming in the lake. His mother always said that if there was any trouble in town, he was usually blamed for it. She was happy to report once, when a neighbor called to complain that Bud had done something wrong, she got to tell her that he was in Costa

Rica with the Boy Scouts.

Bud began his employment with the Canal in 1938 as a motion picture operator at the Gatun movie theater, making 75 cents a show. He used to tell us that the movie couldn't start until his father arrived: "Rank had its privilege."

In 1940, Bud began an apprenticeship; and, in 1943, he graduated as a journeyman machinist. In November 1943, he and a group of friends terminated their employment at the Canal and joined the military to fight in World War II. Bud and several friends joined the Construction Battalion (Seabees) and were stationed in Guam. They went into Nagasaki, Japan, for the clean-up operation after the United States dropped the atomic bomb. After the war, Bud returned to Panama and, in 1946, resumed his work as a machinist.

Bud ran with a tight group of friends known as the Gas House Gang. Membership into the group was initially restricted to guys who were born in Colon. In general, they drank, partied and hung out together. Reports from those who know are that they were mischievous but not mean. The Gas House Gang guys remained lifelong friends and for years held an annual golf tournament in Dothan, Alabama.

In 1952 Bud, married Lois Kridle, who was a friend of his sisters, Grace and Thelma. Lois was the oldest daughter of Louis and Dora Kridle from Pennsylvania. She had a sister, Helen (Sullivan). In 1940, a Kridle family friend who worked at the Canal encouraged Louis, who was working in Pennsylvania coal mines, to come to Panama to work. Louis got a job in the SIP Canal studies. As an avid fisherman, photographer, and amateur botanist, he fell in love with Panama. After a year, housing quarters finally became available and he called for his family to join him. In June 1941, they arrived on the *SS Ancon* out of New York. His daughter, Lois, was thirteen years old. She remembers their first house, a 12-family unit on High Street in Gatun. What a change from Pennsylvania, with the screen windows, wood walls with cracks to the outside, lots of roaches and not much privacy.

L-R: Bud Thomas, Bill Sullivan, Woody Woodruff, Hugh Norris, Gas House Gang golf tournament, Dothan, Alabama.

Lois raised three children (Hugh III, Debra, and Gregg) and then concentrated on her own career. She had worked for the Navy for about five years prior to her marriage, but she finished her career at Customs Division in the Port Captains Office at Pier 18 in Balboa and at General Services Division in the Administration Building.

Hugh III worked in the Marine Traffic Control Center most of his career. He is an avid boater, fisherman and diver. He loved catching lobsters around the Amador Causeway and spear fishing at the nearby islands. Hugh married Eileen Rose in 1978. Eileen is a descendant of Charles David Rose, holder of Roosevelt Medal No. 1187. She began her career at Gorgas Hospital but got her teaching degree and went on to teach elementary school at Fort Kobbe and Fort Clayton elementary schools. Hugh and Eileen had two boys, Blake and Colin. Both boys grew up with a strong love for Panama. They played sports, enjoyed fishing and diving, rode motorcycles, participated in the Cayuco Race and enjoyed the beaches and other outdoor activities while there.

Hugh M. Thomas
Roosevelt Medal No. 5825 with One Bar

Charles David Rose
Roosevelt Medal No. 1187 with Two Bars

The Murry Walker Family

As told by Lessie Platt Walker

The Walkers' Pilgrimage in the Panama Canal Zone

The Murry Walker "Write of Passage" begins in August 1963, when Murry flew to Panama to be hired locally. He stayed with Ronnie and Bobby Ward until the processing was completed and housing was assigned. Murry began his work with the Panama Canal Locks Division as an electrician.

Lessie, Clayton, and Larry Walker arrived in September 1963 via the *SS Cristobal*. Our first ocean voyage — we were all sick upon arriving in Coco Solo, where we began our pilgrimage at quarters 301-D.

Murry's brother-in-law, William C. Crews, soon followed and established his family, Opal, and her mother, Mrs. Annie D. Walker, in the Coco Solo area. He was employed with the Panama Canal Company's Guard Division for his entire tenure, retiring from the Atlantic side. Willys James Walker arrived in 1965 with his family Priscilla, Kathy, Nancy, and Phillys, and he was employed at the Gamboa Prison.

Our life in the Canal Zone in quarters 301-D, Coco Solo, was rather interesting. We had hardly settled in when the 1964 riots begin in earnest. It was rather scary since this was our first trip outside of the United States. The hallways of the four-family houses were what I would call the gossip center of the building. Today, "gossip center" is referred to as the talk around the water cooler. You heard all the stories about the burning of the buildings, how Americans were dragged through the streets in Panama's interior, how we would be evacuated by ship from the Coco Solo harbor, plus all the daily gossip routinely shared. If you really wanted to know what was going on in the country anywhere, or at anytime, the maids were always up on the latest scuttlebutt of the day.

L-R: Clayton, Murry, Lessie, and Larry, with Lesa sitting. Coco Solo, 1979.

In September 1966 we were blessed with the birth of our daughter, Lesa Carol, a little pink bundle of love and joy to add to our family.

A wonderful year — 1969 — but also a terrifying one as we rode out hurricane Camille on board the *SS Cristobal* on the Mississippi River. We were then flown back to the Canal Zone in time for our air conditioned schools to open. We refer to our copy of "The Journey to the Promise Land or The Parable of the Ark in the Flood, Taken from the Gospel According to Camille," written by Mrs. Elizabeth Rowley in 1969. We also have the official letter from W.P. Leber referencing our trip on the *SS Cristobal*.

The children were active in many sports, to include Little League, baseball, football, soccer which carried over into school activities. Lesa was active in Cristobal

High School's drill team and was co-captain.

As our children grew in age, we enjoyed all the amenities the Canal Zone had to offer — swimming, skiing, boating, snorkeling, and diving in Lake Gatun. After Murry finished work, we would pack a picnic lunch and head for the jungles—yes, the Jungle Operations Training Center (JOTC) areas, where our children were given a workout on the rope bridges over the streams to develop their balance skills. The jungle trails were one of our family's favorite places to spend a Saturday, while ropes for repelling from the top of hills to the bottom were the hit of day. Our life as a whole did not change much from life in the "good old USA." We had our church to be active in, to draw support from, and be involved in all of their many activities to help us remain strong in our faith and walk with the Lord.

With so much water around us, we insisted that all three children earn their lifesaving certificate in case their skills were needed during the many hours they spent with family and friends diving in the Canal waters as well as in other Panama waters. All three children were lifeguards: Clayton at the Coco Solo pool, Larry at the Gatun Pool, and Lesa at the Margarita Pool.

Both Clayton and Larry were active in Little League and played on the Pirates team. Clayton played two years of baseball on the Motta team (1972-1974). Football was another sport he loved, and he played for four years. Clayton and his friend, Stuart Smith, were known as the twins on the football team as they looked so much alike. His love was soccer which he played for four years and was captain during the last year.

Of course, we cannot forget the famous Ocean-to-Ocean Cayuco Races which Larry and Lesa were smitten with the desire to win, win! The endurance, training and grueling process that each had to accomplish just to participate in the race was horrendous. Larry paddled in the boat *Scenic Route*, but loved his boat, the *Victus*, which took fifth place out of 30 boats. Lesa and her crew of girls finished the race in her boat, the *Styx*, and received the award for the best uniforms.

As the children grew and moved into the college world, it then became time for Lessie to return to work. She was employed at the Canal Zone Credit Union in Margarita, where she was manager and was the one to close its doors as the Canal Zone operations began to phase out. Lessie was employed by the School of Americas in Ft. Gulick and the Little Theatre in Fort Davis, where she worked with the renown Andy Lim, the greatest ambassador the Americans had in the country. She then moved to Fort Sherman as Director of Ocean Breeze Recreation Center, where she worked until Murry retired in 1991.

After 27 years, Murry retired from the Atlantic side locks as Operation Supervisor (Lock Operations General Foreman). Many promotions were in line during this time for him to have achieved this position.

L-R: Murry, Larry, Lesa Walker Hirsch, Clayton, and Lessie. Christmas, 1997, Collierville, Tennessee.

The Joseph Anderson Wertz Family

WRITTEN BY THEIR BELOVED DAUGHTER, GINNY LEE ZORNES

From Memory and a Few Cherished Letters

I am forever grateful and honored for being chosen into a family that deeply loved me, my husband, my children and my grandchildren, a family where we recognized exceptional professional achievements, enjoyed many friendships, and were fortunate to work and live in the beautiful pristine Canal Zone and its picturesque host country of Panama. It is with pleasure that I share our story of how the Wertz family came to the Panama Canal Zone.

Joseph (Joe) Anderson Wertz was born in 1913, and raised in Milton, Pennsylvania. Joe was the older of two sons born to George and Vesta Wertz. After graduating from Milton High School, he studied drafting and architectural design at the University of New Mexico. In 1935, after receiving a two-year diploma, Joe returned to Milton at the request of his parents because his brother entered military service. After his return, Joe accepted a position as a draftsman for the Broscious Lumber Company in Sunbury, Pennsylvania.

As the years passed, Joe heard about more lucrative opportunities working for the U.S. Government in the Panama Canal Zone. He successfully passed the civil service exam, accepted an offer, and went there to work for the Engineering Division in the Panama Canal Zone in October 1940.

Joe, at that time, was engaged to his beautiful hometown sweetheart, Emma Margaret (Peggy) Chalfant. Peggy, born in 1919, was the third of six children born to George and Clara Chalfant, and her maternal heritage could be traced back to the 1700's to the Jerome Eckert family. She, too, graduated from Milton High School and, afterwards, Williamsport Business College.

Once established in the Canal Zone, Joe began sending his savings back to Milton in preparation to having Peggy join him. With his goal accomplished, Joe sent for Peggy in March 1941. Peggy, working then as a Secretary-Accountant for Smalls' Greenhouses, bravely said her goodbyes to friends and family and traveled to New York City, where she boarded a United Fruit Company ship and set sail for Panama via Havana, Cuba. Joe said "Bless her heart, she left a good home, her friends, and her family . . . to gamble on me!" People who knew and loved them remarked it was "a most fortunate gamble."

Peggy and Joe on their wedding day.

Peggy arrived in Cristobal, March 21, 1941, around 11 a.m. and found Joe anxiously waiting for her. Together they boarded the Panama Railroad for an exciting train trip across the isthmus to Balboa. That same afternoon, around 5 p.m. they were married in the Balboa Union Church. Joe quipped and said, "She had no chance at all; it [the marriage] had to be done and had to be done that day, the day that she got there; that was the law of the Canal Zone!" He said, "If I had not done that, I would not know what she would have done." They honeymooned in the beautiful historic Tivoli Guest

House, which was frequented by many prominent dignitaries, to include President Teddy Roosevelt, who in 1906 was a guest there while inspecting the progress on the construction of the Canal.

Joe and Peggy both worked in the Canal Zone until 1942, when a good friend of Joe's persuaded him to take a job in Puerto Rico. There he worked on the Roosevelt Roads project, at the request of the assistant to the resident officer in charge, and worked under the direction of the Navy design engineer from the Bureau of Yards and Docks. Peggy found living there tough, having to ride a ten-ton truck or hop on a labor truck to get back and forth for grocery shopping in town.

Peggy posing in pollera.

After a year in Puerto Rico, and fulfilling his contractual obligations, Joe and Peggy returned to the States to pursue Joe's intention of enlisting into the Armed Services. However, due to a punctured eardrum, he was turned down and unable to serve in World War II. They remained in Milton, Pennsylvania, through the winters of 1943 and 1944, after which Joe declared, "Too many winters!" They decided it was time to move on to a warmer climate. Thus began their return to the Canal Zone in 1945. Joe went ahead while Peggy followed once authorization for her travel was granted.

In planning a family, Joe and Peggy considered adopting in Costa Rica. In October of 1950, they received a wire from a doctor friend stating, "have cutest baby girl one month old come tomorrow if possible or maybe someone else will grab." Without hesitation, Joe and Peggy flew up to Costa Rica and within days Virginia "Ginny" Lee Wertz became theirs. She arrived in her new home in Curundu Heights within days. Ginny was their life. In addition to ballet and tap lessons, dance recitals, Brownie and Girl Scout troop meetings, Joe and Peggy took great notice of Ginny's fondness for horses. Because of her affection, Ginny's formable years focused on the care and riding of her beloved mares.

Joe and Peggy were very proud of their little girl, and her happiness was foremost in their lives. On June 6, 1970, they were elated when Ginny married her high school sweetheart Richard Sherman Zornes. Later, they were so proud and overjoyed upon the birth of their two grandsons, Jeffrey in 1972 and Joseph in 1976.

Joe cleaning lobster.

During their life in the Canal Zone, Joe and Peggy maintained households in Curundu Heights and on Albrook Air Force Base As in the typical Canal Zone home, many Panamanians, some from Barbadian descent, were entrusted with the normal household chores and the care and welfare of the children. For Joe and Peggy, Lica de Gabay, or Eunie as she preferred, became their loyal and faithful housekeeper and remained in their household until they departed the Zone.

Forever active and giving, Joe and Peggy participated in many local community activities and clubs. A favorite at Christmas was to join in with their friends and neighbors, contributing presents for sharing among Panamanian orphans who were bused into Curundu Heights at this special time of the year. A close friend and neighbor, Mr. Bill Bright, would play "Santa" and distribute the gifts to the awe and excitement of the children.

Throughout the years Joe was active in the Balboa Lions Club and the Balboa Elks Lodge. As president of the Lions, he held meetings at home. He was involved with and supported the local Boy Scouts. Peggy was a Past President of the Beta Sigma Phi Sorority, served as a Brownie and Girl Scout leader, taught Sunday school at the Curundu Protestant Church, and was the President of the Balboa Teen Club, which she and Joe managed every Friday night for five years. They are both remembered fondly by many of the Canal Zone teens who gathered there to dance to the tunes of the local live bands comprised of teens from both the military and Panama Canal communities.

For recreation, Joe and Peggy enjoyed traveling to and relaxing on the beaches of Santa Clara. Joe liked panning for gold in the streams of Panama's interior where he managed to pan enough to make two small pendants for Peggy and Ginny. Joe also enjoyed fishing and to the surprise of Ginny managed to take up riding one of her mares.

Career wise, Joe worked in the engineering offices of the Panama Canal Government and U.S. Corps of Engineers and for the U.S. Air Forces Southern Command. During his stay in Panama, his work and personal interests took him to all parts, from the jungles of the Darien to the high mountains of the Chiriqui Province and many South American countries. Joe retired as a general engineer with the U.S. Air Force. He was particularly proud of the fact that he was a key architect in the design of the Albrook Air Force Base swimming pool and its adjacent base exchange cafeteria/shopping center.

In 1941, Peggy had the unique experience of crossing the Canal on a ferry boat and hailing down a chiva bus every day in order to get to work for the U.S. Army engineers at Howard Air Force Base. In 1945, she began her long-term career working for the Signal Corps as a clerk stenographer, fiscal accountant, budget technician, administrative officer, and budget officer for 21 years and then transferring to the U.S. Army Strategic Communications Command-South, where she worked as a budget officer until her retirement.

Upon their joint retirement in 1971, Joe and Peggy returned to the United States to reside in Florida. Ginny and Richard remained in the Canal Zone to raise their sons with the help of Eunie, who became a lovable extension of their family.

Throughout their journey in the Canal Zone, Joe and Peggy built long time friendships with their neighbors and co-workers, many of whom they kept ties with well after retirement. Peggy and Joe cherished their life in the Canal Zone where friendships were endearing and everlasting.

PHOTO COURTESY OF WWW.CZIMAGES.COM

Balboa Teen Club, where weekend dances were held with local bands and managed and chaperoned by the Wertzes.

The John Emanuel Westberg Family

AS TOLD BY BETTY BARR-AUSNEHMER

John Emanuel Westberg was born July 5, 1870, in Stockholm, Sweden. He immigrated to New York on or about May 28, 1887, and became a naturalized citizen July 7, 1892, in Chicago, Illinois. John married Mary Barbara Bretz July 12, 1898, in Joliet, Illinois. They had four daughters: Dorothea Emma, Mary Frances, Dolores Esther, and Gurli Anna, who died in infancy. John traveled from New York to Panama in June 1906 arriving on the Isthmus aboard the *SS Advance*.

The men on the gold roll were encouraged to relocate their families to the Canal Zone as soon as living accommodations became available. The men were allotted living quarters based on salary, and the space allotment increased with marriage. The typical houses were two stories, and comprised of two to four apartments, with windows on all sides, high ceilings, and screened in verandas. (McCullough, David. *The Path Between the Seas: The Creation of the Panama Canal*. Simon and Schuster. New York, New York. 1977. (555-588).

John was joined by Mary and their three daughters on April 3, 1908. Mary was an organizer and president of Gorgas Auxiliary No. 2 of Chagres Camp No. 1, and John served as a police officer under the Isthmian Canal Commission from 1906 to 1914, for which he received the Roosevelt Medal No. 1508 with three bars. John died July 27, 1942, at Gorgas Hospital, Ancon, and is buried at Corozal Cemetery beneath a Spanish American War gravestone. Mary died November 14, 1946, and is also buried at Corozal Cemetery.

John's family remained connected to the Canal Zone for four generations. His daughter, Mary Frances, married Robert Warren Barr, who served as a fireman in Pe-

John E. Westburg and family.

dro Miguel and Balboa. Their children — John Robert, Barbara Frances, Peter James, Francis Richard, and Helen Elizabeth — worked or lived in the Canal Zone. Thomas Joseph, who was born and died in 1933, is also buried at Corozal Cemetery. John Robert served as Executive Director of the Boy Scouts of America, for which he received the V.F.W. Citizenship Medal. Barbara Frances entered the Foreign Mission Sisters of St. Dominic, commonly known as the Maryknoll Sisters. She celebrated her Golden Jubilee on May 4, 1997, for fifty years of service. Peter

James served with the Canal Zone fire department from October 13, 1952, to January 1, 1980, retiring as Assistant Fire Commander, Northern Branch, for which he received numerous medals and recognitions from the U.S. and Panamanian governments, including the Medal of Distinction. Francis Richard graduated from Balboa High School, class of 1952, and later joined the U.S. Air Force. Helen Elizabeth was employed from 1956 to 1984 by the Panama Canal Company as a supervisor in the Gorgas Hospital Admitting Office.

Peter James served in the U.S. Navy, during the Korean War, and met Betty Ruth Adams while stationed in Charleston, South Carolina. They were married in Saluda, South Carolina, in June 1951 and returned to the Canal Zone after Peter completed his tour of duty. He was employed as a security guard for the Gatun Locks in 1952, later transferring to the Balboa fire department. He served in all branches of the fire department uprooting his family with each new assignment. In his off duty time he claimed many awards and trophies for skeet and trap shooting; he was an avid outdoorsman. He often would share stories with his grandchildren of the experiences he had in the Canal Zone while hunting and fishing with his four sons: Robert Earl (Reb); James Anthony (Tony); Peter James, Jr. (Jimmy); and Sean Andrew. Betty Ruth taught English at the Cristobal Y.M.C.A., later working as sales supervisor of the Diablo Clubhouse from 1961 to 1963. She was president of the Gatun Civic Council and remained fervently involved with the Christian Women's Club for almost a decade, all while raising four wild boys in a jungle setting.

Robert Earl (Reb) graduated from Cristobal High School in 1971 and worked as a water safety instructor at Gatun pool. He was also involved with many of the sports leagues. James Anthony (Tony) graduated from Cristobal High School in 1973 and was also a water safety instructor. He was later stationed at Fort Davis in the Canal Zone as a water craft operator in the U.S. Army. Peter James, Jr., graduated from Cristobal High School in 1975, also worked as a water safety instructor, and went on to join the U.S. Navy. Sean Andrew attended Cristobal High School, playing multiple sports, but did not graduate from there due to relocation to the United States during the midterm of his junior year. All of the boys shared Peter, Sr.'s love for the outdoors. They also reminisce often of the experiences had while hunting pigeon and other exotic game and have many a tale, some taller than others, of fishing for peacock bass, snook, and tarpon in the lakes, rivers, and both oceans of the Canal Zone.

Peter J. Barr, Sr.

John Emanuel Westberg
Roosevelt Medal No. 1580 with Three Bars

The Joseph J. Wood Family

AS TOLD BY JOSEPH J. WOOD, JR.

A Story of Love at First Sight

In the early 1870's, James Moore, a young Englishman from Devon, England, was an engineer aboard a Pacific Steam Navigation Co. steamship that traveled between California and Panama. The company's ships docked at their Taboga Island coaling station, and on one of his trips Moore met Genarina Fernandez, a young lady from Panama City whose parents were vacationing in Taboga. Despite the language barrier, they soon married, had three children — Isabel, Anna and James — and settled in Taboga, where Moore built a house from redwood he brought from California. He continued sailing but several years later was lost at sea.

Moore's oldest daughter Isabel was born in 1877 and grew up during the unsuccessful French effort to build a canal in Panama. In 1903, after Panama gained its independence from Colombia, she met Carlos Fajardo, a young Colombian soldier from Cali, who had been stationed in Panama, but left the Colombian Army after the revolution. Isabel and Carlos were married in 1904, when the United States began construction of the Panama Canal.

After their marriage, Isabel, having been raised in a bilingual household, became a Spanish teacher and was giving lessons to Americans at the YMCA in Balboa, Canal Zone, in 1923, when Joseph "Joe" Wood, a young soldier stationed at Corozal Army Base, became one of her students. Joe left home in Kansas to "see the world"

Isabel F. Wood

U.S. Army Corporal Joseph J. Wood, Ft. Clayton, CZ, 1923.

and at age 22 enlisted in the U.S. Army on Governor's Island, New York. He was sent to the Canal Zone in February 1923 and immediately fell in love with Panama.

Joe wanted to learn Spanish more quickly than the once-a-week lessons offered at the YMCA, so he asked Isabel (Mrs. Fajardo) if she could give him private lessons. She agreed and invited him to her home in Panama City.

When Joe arrived for the lesson, Mrs. Fajardo was not quite ready, but commanded her 14 year-old daughter, also named Isabel, to sit in the "sala" to keep the gentleman company until she was ready. No conversation took place, and young Isabel, sitting with her arms crossed and scowling at this "Gringo," couldn't wait to get away to be with her friends. Joe's reaction apparently was quite different — for him, it was love at first sight — and he announced to his buddies when he returned to the barracks, "I just met the girl I'm going to marry!"

Years went by without Joe ever seeing Isabel. In the meantime, he was honorably discharged from the Army at which time his commanding officer wrote:

Joseph Wood has been a non-commissioned officer under my direct supervision for the past year. I have found him to be reliable, intelligent, a willing worker, possessed of an inordinate amount of good judgment and common sense. He has displayed tact, a remarkable ability to control the men under him, and a true sense of loyalty to his superiors.

He has been studious and sought to improve himself in various ways during this time, thus keeping himself occupied during off-duty hours, and free from any evidence of the many more or less petty vices that a man may become subject to in this country and climate.

In 1924, he joined the Canal's Supply Department and, while there, took accounting courses from the International Correspondence School (ICS). Upon getting a diploma in accounting, he transferred to the Accounting Department. His first assignment there was as "Inspector," which required him to stake out bars in Panama City to check up on unsuspecting Canal employees who were playing hooky from work!

In 1928, using the Spanish skills he learned from Mrs. Fajardo, Joe took a six-month sabbatical from his job with the Canal to serve with the American Electoral Mission in Nicaragua to monitor elections in that country. In a letter of commendation, Mission Chairman Brigadier General F. R. McCoy wrote:

I desire to express my personal appreciation of the splendid service rendered by you since accepting employment with the Electoral Mission on 3 July, 1928. Your coolheadedness in emergencies was especially commented upon, while your practical knowledge of the Spanish language was at all times of great help.

In 1929, Joe's language skills again were instrumental in his taking another leave of absence to serve with the American Mission in Peru that mediated the long-standing Tacna-Arica border dispute between Peru and Chile.

Upon returning to the Canal from Peru, Joe took up residence in a rooming house owned by none other than Mrs. Fajardo. Several other Americans were also living there and all had eyes for Mrs. Fajardo's then 21-year-old daughter, Isabel. She, wanting no part of them, was happily enjoying life and the active social scene in Panama, mainly with a group of friends who called themselves las treinta y dos, who got together often and had their own "comparsa" and float in the Carnival parades.

In the meantime, Joe was also enjoying life as a bachelor with his many friends as they frequented Kelly's Ritz, Happyland, The Stranger's Club, and other famous nightspots in Panama. On one occasion, after much partying, he and his buddies, Howard Sprague and Forrest Dunsmoor, commandeered a jitney convertible and drove it on the sidewalk from Kelly's Ritz to the Riviera bar on Central Avenue to continue their merrymaking.

Joe's persistent pursuit of Isabel paid off, and she finally agreed to go out with him. In 1933, after a courtship of several years — and ten years after he first declared his intention to marry her — they tied the knot. He was 32, and she was 24.

Pete Brennan, a well-known columnist for the *Star & Herald* newspaper in Panama at the time, wrote in his column about Joe and Isabel's upcoming marriage:

> *Joe's persistent pursuit of Isabel paid off, and she finally agreed to go out with him. In 1933, after a courtship of several years — and ten years after he first declared his intention to marry her — they tied the knot.*

Joe and I have been friends for many years, bachelor friends, and even at this writing, a bare fourteen hours before the wedding, I find it somewhat difficult to believe. I find myself in as acute a stage of the jitters as if I were being married myself. By tomorrow, I anticipate being as complete a wreck as Mr. Wood. They should make a genial couple, Miss Isabel Fajardo and Joe, for she is a beautiful and happy-spirited girl and he is—well, he's a good fellow and boon companion. I give them my benediction and best wishes for a long, happy and courageous life together.

Joe was an avid tennis player and participated in many tournaments in Panama, winning several. In 1938, he was selected to be the coach of Panama's men's and women's olympic tennis teams that participated in the IV Central American and Caribbean Games in Panama. The men's team consisted of prominent young Panamanians George Motta, Felipe Motta, Jack Pereira, George Westerman and Carlos Eleta.

Joe loved Panama and made many friends both in the Canal Zone and in the Republic. He returned to the United States only once after his first arrival in Panama when he was sent on a three-month assignment to New York in 1953 to audit the books of the Panama Line before its offices were transferred to New Orleans. Upon his return, he commented that "you can give Manhattan back to the Indians!" In 1957, while working as a Supervisory Auditor with the Panama Canal, Joe suffered a stroke and passed away at the age of 55. The love of his life, Isabel, later moved to New York to be with her daughters and then moved to Virginia, where she died at age 89 in 1997.

Joe and Isabel had three children, Isabel (1935); Joe, Jr. (1937); and Ann (1945). They were brought into the world at Panama Hospital by Dr. Frank Raymond, a well-known American doctor. As youngsters, they lived on 4th of July Avenue in Ancon across the street from "J" Street and the Cool Spot ice-cream parlor where they would go for nickel coke floats. All three graduated from Balboa High School and college in the United States. Isabel and Ann settled in New York, enjoying long careers in social work, and currently reside in Virginia. Isabel married Howard Egan, who became ill with multiple sclerosis and later passed away. Ann married Raymond Sanchez, and they have four children — Melissa, Raymond, Christina and Daniel, and four grandchildren — Dannah, Carol, Jesse and Jeremy.

After college and service in the U.S. Army, Joe, Jr., returned to the Canal Zone and, in 1971, married Beverly Bowman, whose parents L.D. and Audrey Bowman and grandparents Bert and Edna Benoit worked for the Panama Canal. Joe and Beverly also enjoyed long careers with the Canal and have three children — Craig, Brian and Scott, all graduates of Balboa High School — and three grandchildren — Cameron, Jordan and Adrian.

After retirement, Joe became one of the founders of the Panama Canal Museum in 1998, helping to preserve his family's heritage in Panama and the important role of the United States in Panama's history.

See also the Audrey Benoit Bowman family history.

PHOTO BY DON GOODE

Administration Building, Townsite of Balboa, Balboa Elementary School, and Balboa High School — places significant to the Wood family.

Robert C. Worsley

AS TOLD BY DAUGHTERS ROBERTA RICHARDS AND JACQUE WILLIAMS

Robert C. Worsley was born August 6, 1890, near Sharpsburg, North Carolina, the eldest of 13 children. When Bob was born, his father was a teacher who later moved the family to Rocky Mount, North Carolina, to become a partner in a small store.

Always entrepreneurial, Bob helped in the store at age 8. At age 10, he operated a scrap business from the store's loft. Because there was no local school until he was 11, Bob entered as a third grader, soon moved to fourth grade, and was quickly promoted to fifth grade, completing sixth grade the following year. The family moved to Wilson, North Carolina, where Bob opened the R.C. Worsley Company. He entered seventh grade at 14, and an older gentleman helped in the store while he attended school. Bob completed seventh grade with a 99+ average.

At 16, Bob moved to Battleboro, North Carolina, to work in a large farm department store as stenographer and general clerk. At 19, the Atlantic Coast Line Railroad Company hired and later promoted him to the main office in Wilmington, North Carolina. Three years later, he became secretary to the Chief Engineer of that railroad company and held that position until June 11, 1911.

In June 1911, the construction of the Panama Canal was about one half completed and in full swing. When adventurous Bob applied and was appointed by the U.S. Civil Service as stenographer for the Isthmian Canal Commission (ICC), Bob sailed for Panama on the *Allianca*. He walked the deck every morning with Colonel Goethals, Chief Engineer of the ICC, without knowing how important the Colonel was. When Bob arrived in the Canal Zone, he went to work in Culebra for Colonel Devol at the headquarters of the top ICC officials, doing occasional Sunday morning assignments for Colonel Goethals.

Robert C. Worsley

Canal employees worked eight hours a day, six days a week. It was typical for them to walk over eight miles to eat supper and return to quarters. They often walked from Culebra to Chorrera by trail on moonlit Saturday nights, the next day enjoying the Chorrera Falls and walking back to Culebra – about 22 miles each way. Bob Worsley and Glen Rose walked across the Isthmus from Colon to Panama in 14 hours, following the tracks of the Panama Railroad to Gamboa, down into the "cut" (then dry) to Las Cascadas, following the railroad through Empire, Culebra, Pedro Miguel, and Corozal to Panama.

John Warner inspired Bob's dedication to youth and community. John helped plan the Balboa Union Church; Bob and his wife, Anna, were among the first 100 charter members. Bob served six years as Superintendent of Sunday School and Financial Secretary and held almost every position on the church board.

On September 26, 1913, the tug boat *Gatun* was the first boat to pass through the Panama Canal locks and Bob Worsley was aboard!

In 1912, Bob was promoted to stenographer to Mr. Faure, Chief of Cost Accounting. In July 1914, Bob accepted the position of accountant for Union Oil Company of California and also served as stenographer and cashier. When Bob was appointed acting Special Agent, he undertook the challenge of building a commercial business in the Republic of Panama. Bob learned every detail of the business, from operating the pumping plant to maintenance and repair. That year, Union Oil started supplying fuel oil to Panamanian corporations. He also developed bulk oil terminals at Aguadulce, Chitre and Pedregal.

Bob took his first vacation to North Carolina, returning with his sister, Annie. He found her a job and introduced her to his friend, Bill De La Mater, Sr. Bill and Annie were married and had three children. Lois was secretary to the Engineer of Maintenance of the Canal and married Paul Bates, a World War II Army officer. Bill De La Mater, Jr., served as Protocol Officer to the Governor of the Canal Zone. Joanne moved to the U.S. for a career in music.

Bob's next vacation was in 1916. Bill De La Mater, Sr., made arrangements for his sister, Anna, to work for the Telephone Exchange in Panama. Bill asked Bob to meet her in New Orleans and escort her to Panama aboard ship, and Bob agreed. They fell in love and were married within the year. Anna did work at the exchange for a year. She was a very supportive and encouraging wife and a good nurse for tropical ailments and injuries.

Bob and Anna had four daughters. Marie, a nurse in the WAC in World War II, married a World War II veteran. Helen graduated from a business college in Tennessee, worked 21 years for a commercial real estate company, and married a brick mason. Roberta married a World War II Navy veteran who dedicated 25 years to the Boy Scouts, and she was an active community volunteer. Jacque performed in over 3,000 USO shows during World War II aboard aircraft carriers and at military outposts (including a War Bond show attended by General Eisenhower). She met and married her World War II veteran and actor husband (54 movies, 100 TV shows) at the Pasadena Playhouse.

Bob's daughters have fond memories of their home on Sosa Hill with a good view of the entrance to the Canal. Sosa Hill was often aflame during the dry season, so Bob bought a goat to eat the tall grass. On weekends, Bob would load up the car with family and friends to enjoy swimming at Amador Beach, later returning home to make homemade ice cream for them.

> *Bob's daughters have fond memories of their home on Sosa Hill with a good view of the entrance to the Canal. Sosa Hill was often aflame during the dry season, so Bob bought a goat to eat the tall grass.*

Two other sisters of Bob moved to the Canal Zone. Emma Leigh lived with Bob's family to finish high school and later worked for the Canal Zone Government. She married Jack Burke White. Their daughter, Arlene, was a champion swimmer, graduating from Balboa High School and Duke University. Mattie Ruth married Jack Clark and was known for her peanut business.

Bob joined the Boy Scouts of America, serving as a Scoutmaster for many years and continuing on the Executive Board for his lifetime. He and Colonel S. Wilson, successor to Lord Baden Powell, traveled to an Inter-American Scout Convention in Bogota, Colombia, as representatives of the Canal Zone and the U.S. Bob was recognized for contributing significantly to the Panama Scout Council and was awarded the Silver Beaver in 1963.

Bob's chosen sports for exercise and enjoyment in the earlier years were swimming, softball, handball, volleyball and marathon runs. He organized the inter-collegiate volleyball league in Panama City and served as Vice President on the Panama Council. He was awarded the Merit Medal from the Panama Department of Physical Education. Bob renewed competitions between Panamanian and Canal Zone teams.

Bob retired from Union Oil in 1950 but very quickly took over a small steamship agency which he named Agencia R.C. Worsley. He built up the business to 90 tuna clippers, and it continued successfully until Bob was 90.

Bob enjoyed dancing and swimming in his later years. He helped supervise dances and shows at the YMCA and Jewish Welfare Center. He joined the Pan-Amigos International Dance School, learning folk dances and the tango. He performed the tango several times on TV.

Bob celebrated his 86th birthday at a happy gathering given by his friends and various civic, community and religious organizations who honored him for his devotion to the spiritual and moral growth of the isthmian communities and his sterling qualities as a family man, church worker, civic and community leader and scout leader. He was given the title of "Grand Old Man" (saying Bob was "86 years young"). Bob's good friend, Governor Harold Parfitt, presented him with a Norman Rockwell plate depicting a Boy Scout.

Bob completed his honest and humble life with joy and respect, working for and with people to encourage spiritual and moral growth on the isthmus. Bob Worsley is a shining example of how to live a clean, healthful, active life, serving God first, mankind and doing what is right.

Bob's positions in Panama included: YMCA Secretary to Board of Directors (1924 – 1980); President, Panama Rotary Club (1936); member of the American Society (1925 – 1980); member of Masons (1924 – 1980) and Senior Past Master of Canal Zone Masons (1938 – 1980), holding position longer than any other; member of both the Executive Board of the Canal Zone (1975) and the Executive Board of Boy Scouts of America – Canal Zone Council (1980)

Bob's awards were many: Vasco Núñez de Balboa Silver Beaver Award (1963); Asociación de Scodets de Panama Award (1968); Norman Rockwell Boy Scouts "Gorham Plate" awarded by Governor of the Canal Zone (1976); Medal of Merit from Department of Education, Panama City (1977); Lions Club Certificate of Appreciation for Invaluable Services (1977); Panama Canal Public Service Award for Outstanding Service to the Community (1977); Balboa YMCA "Man of the Year" (1978); American Hemisphere International Boy Scout "Youth of Americas" medal (1980); El Club Rotario de Panama Merit Award (1980); and "Man of the Year Award" from Department of Latin American Reserve Officers Association (1980).

Robert C. Worsley
Roosevelt Medal No. 6712

William W. De La Mater
Roosevelt Medal No. 7028

PHOTO COURTESY OF PANAMA CANAL MUSEUM

Tugboat Gatun, first through the locks, with Bob Worsley on board, September 26, 1913.

The William Carl and Christine Peterson Zeeck Family

AS TOLD BY CHARLES E. ZEECK

For the Children and Those Yet to Be Born

The Great Depression of the 1930's was certainly a pivotal time in many people's lives. Such was the case for a young married couple, William Carl and Christine Peterson Zeeck.

Carl Zeeck was born October 14, 1908, in Lingleville, Erath County, Texas, to William Harry and Rhoda Hall Zeeck. His family migrated to farm in Dawson County, east of Lamesa, Texas, sometime around 1916. After graduation from Dawson County School in 1926, Carl worked at various menial jobs until attending Coyne Electrical School in Chicago, Illinois, graduating in 1929. Prior to attending school in Chicago, Carl married Christine Peterson at her parents' home in the McCarty Community, Lamesa, Texas, on June 24, 1928.

After Carl returned from Chicago, the young couple moved to Dallas, Texas, where he worked for Texas Construction Company and then Texas Power and Light, building power plants and installing electrical lines across East Texas. Carl's job came to an abrupt end in 1931 when all workers were laid off due to the depressed economy.

In the meantime, Evelyn Joyce Zeeck, had been born September 13, 1931, at the home of her Peterson grandparents in Dawson County. Several months passed while Carl tried to find work to support his small family before a close friend, Ted Foster, said he was going to Panama to work. He asked Carl if he would come to Panama if a job became available. "Send me a boat ticket, and I'll be there," Carl said. A ticket arrived in the mail shortly thereafter, and the family annals in Panama began.

Standing, L-R: Christine and Joyce. Seated, L-R: Richard, Carl and Charles.

William Carl Zeeck arrived in the Canal Zone in June 1932 aboard the United Fruit ship *Suriname*. His first day on the job at Madden Dam on the Chagres River was June 19, 1932, working for the Callahan Construction Company through the general contractors of Peterson, Shirley and Gunther. The beginning salary was 87 ½ cents per hour, a 12 cent increase over the Texas job. Family housing was provided, so Christine and Joyce arrived on November 11, 1932, aboard the *SS Matapan*.

At the Madden Dam project, Carl performed maintenance on equipment that had been purchased secondhand in the U.S. and had been out of service for many years. When the dam project was nearing completion, Carl ran the entire electrical conduit for the power house and main portion of the dam. His last day at Madden Dam was October 28, 1934. On October 30, he started work with the Locks Division and was transferred to the locks at Pedro Miguel where the family settled into four-family housing.

While still at Madden Dam, son Richard Carl was born at Gorgas Hospital, Ancon, Canal Zone, on August 30, 1933. A second son, Charles Ernest, was also born at Gorgas Hospital, Ancon, Canal Zone, on September 15,

1938.

Life was good, there was a steady income with benefits, and Carl was able to send money to family in Texas who were still in the throes of the depression and severe drought. But World War II would soon change life for the Zeeck family.

After Pearl Harbor was attacked, the Canal Zone was on high alert. Carl quickly secured passage on a ship bound for New Orleans for his family. They would live in Lamesa, Texas, until the summer of 1943 while Carl remained on the job in the Zone. When not performing his regular duties, Carl worked on defense jobs that were vital to the war effort including installation of power equipment for the Farfan Naval Radio Station and frequency changing stations for the U.S. Army.

During this period, Carl was accepted to the salvage diving school operated by the Mechanical Division. He then worked with the U.S. Navy to clear ships that had been sunk by German U-boats trying to block the entrance to the Canal, logging over 160 underwater hours. Since this type of diving required wearing a full diving suit with helmet, the shiny helmets were great targets for barracudas whose teeth marks often left dents in Carl's helmet.

When the threat of enemy attacks lessened, Christine and the children returned to wartime restrictions and regulations at Pedro Miguel. Families covered windows with black cloth at night, and barrage balloons were hoisted over the locks to prevent low-flying, enemy, aircraft bombing of the Canal. Smudge pots were also installed which would allow a complete blackout of each set of locks.

In 1944, Carl was promoted to electrical foreman on some wartime special item projects which included the SIP-3 project, a locks fire control system, as well as the highly-confidential SIP-7 project. From 1946 until retirement, he worked as a control house operator as well as at various overhauls at Pedro Miguel and Gatun Locks. Carl resigned from the Panama Canal Company in April 1949, at an annual salary of $5,075.20. His total time in Panama was seventeen years and ten months. The family returned to Lamesa, Texas, where Carl was a very successful farmer and electrician. Christine died in 1983, and Carl died in 1991. They are buried at Lamesa, Texas.

Life in the Canal Zone was not all work. The Zeeck family enjoyed times spent at Santa Clara beach where they owned two lots and planted mango trees. Carl and Christine were members of the Pedro Miguel Masonic Lodge and Eastern Star. He was a charter member of the Pedro Miguel Boat Club where his hand-hewn cayuco was a fixture and was great for navigating the jungle rivers. Carl, a skillful hunter and sportsman, also organized the Pedro Miguel Gun Club. In 1947, he qualified for his pilot's license and would later accompany his brother and wife, Jarvis and Beatrice Zeeck, when they flew Tony Arias's airplane to safety in Texas during a military coup. Christine worked at the movie theater at the Pedro Miguel clubhouse, played on the "Peppy Mommas" softball team, won medals for shooting archery, and enjoyed regular forays with friends into the jungle in search of orchids. The Zeecks were very active members of the Pedro Miguel Union Church.

> *At Christmas, everyone eagerly awaited the arrival of the Christmas tree ship. After the holidays, the children competed by going door-to-door collecting the trees for the annual Christmas tree bonfire held at the baseball field near the fire station.*

Pedro Miguel was a wonderful place for growing children. They had the clubhouse with a movie theater, bowling alley, swimming pool and a food grill that served great milkshakes. At Christmas, everyone eagerly awaited the arrival of the Christmas tree ship. After the holidays, the children competed by going door-to-door collecting the trees for the annual Christmas tree bonfire held at the baseball field near the fire station.

Joyce Zeeck was a champion swimmer at Balboa High School and competed in the 1947 nationals in Florida. She was popular and often took her brother to school events instead of a date. Joyce graduated from Balboa High School in 1949. Richard played flag football and ran track at BHS for two years. He also played baseball and was a member of the Pedro Miguel boxing team where he won recognition as the "Outstanding Boxing Student." Richard attended Canal Zone schools from first through tenth grade. Charles played on the "D" league baseball team coached by Henry Leisy and challenged brother, Richard, for recognition on the Pedro Miguel boxing team. During World War II, Charles and friends often climbed Pedro Miguel Hill where they exchanged homemade sandwiches for K rations with soldiers on duty at the Army gun emplacements. He attended Pedro Miguel Elementary School from first through fifth grade.

Other family members who worked in Panama and for the Canal were: Jarvis R. Zeeck and Harry L. Zeeck (brothers of Carl), Beatrice Thrapp Zeeck (wife of Jarvis), Lunelle Zeeck (sister of Carl, Jarvis and Harry), Opal Peterson Durham (sister of Christine), and Eugene Slover (cousin of Christine).

Life often travels full circle. Carl and Christine Zeeck left Texas due to a severe drought and economic depression only to return to several years of similar conditions. But the years spent in the Panama Canal Zone are forever ripe with memories of living quarters, clubs and organizations, church activities, career positions, the lush jungle terrain, travels and always . . . "the lasting friendships."

PHOTO COURTESY OF DENNIS WHITE

This big old Cuipo tree was a landmark at the Silver town of Red Tank for many years. Red Tank had its largest population of over 2,200 in the decade between 1931 and 1941 when it was finally deserted and torn down. This photo was taken at the end of a causeway over an arm of Miraflores Lake heading toward Pedro Miguel.

The Llewellyn "Lew" Zent Family

AS TOLD BY MARGE ZENT GARNER

Years 1943–1963 in the Canal Zone

The family lived in Santa Fe, New Mexico, before arriving in the Canal Zone. Llewellyn "Lew" Zent's job took him there in 1942, while his family remained in Santa Fe. In 1943, Lorraine and daughters Lou, Liz, Martha and Marge traveled to Panama, with stops in Mexico City, Guatemala City, and Managua. They arrived in Balboa on November 14, 1943.

Their first home was at Fort Kobbe/Howard Field. It was a nice place to start in a new location and to meet new friends. While living there, there was an addition to the family. Llewellyn "Butch" Zent II was born in December 1944. A military ambulance took Lorraine to the hospital, which was a good thing as the Miraflores Locks bridge was open to allow a ship to enter the locks, causing a traffic delay. They arrived at Gorgas Hospital in time. The family was happy to have a son and a brother, who was well cared for by his four sisters.

The family moved to Pedro Miguel where their new home was a four-family dwelling, and their new neighbors were the Rosans, Mitchells, and Dawsons. Home was on the main highway across from the railroad tracks. On the other side of the tracks was Pedro Miguel Locks. They had no problem adjusting to the sounds of the trains and sounds of the locks as ships were passing through. Since it was World War II, security around all the locks was very tight. At times, as troop ships were passing through, the main entrance gate was opened, allowing people to enter. There was a lot of waving and shouting back and forth with the troops. Some sailors tossed their caps, and the marines tossed their brass insignias to the crowd. It was fun for everyone.

L-R: Lew holding Butch, Lorraine, Margaret standing in front of her, Elizabeth, Lou Ellen, and Martha, at Summit Gardens.

Lew had worked for the U.S. Employees Compensation Commission, but in 1948 went to work for the U.S. Army as housing manager. This meant a move to Curundu, where his office was located and civilians who worked for the military lived. Their new home was located on 2nd Street, and the office was on 4th Street at one end of the post office. In 1956, he went to work for the Inter-American Geodetic Survey at Fort Clayton. He was a member of the Elks, Scottish Rite, and Masons, and Lew and Lorraine were Eastern Stars. Lorraine was a member of the women's clubs in Pedro Miguel and Curundu. In 1949 the family moved to Curundu Heights to a duplex on the second level. Among the many friends and neighbors were the Frangionis, Hoopes, Lavallees, Russells, Halls and Levys. In 1963, after 20 years in the Canal Zone, Lew retired. He and Lorraine returned to Santa Fe, New Mexico, where he remained active in many organizations.

He died in 1978, and Lorraine remained there until her death in 1983.

Lou Ellen graduated from Balboa High School in 1945 and recalls her teachers and many good friends. She keeps in touch with some, such as Jim and Dulie Coffey. Lou recalls having fun in Mr. Turbyfill's drama class and plays that they performed. Her classmate, Betty Gaines, introduced her brother Bill to Lou; he had returned to the Canal Zone serving with the Navy Air Corp. Bill and Betty's parents were Bill and Della Gaines. Lou and Bill were married in the Pedro Miguel Church in 1947; daughter Sharon was born in Gorgas Hospital. They left the Canal Zone in 1948 and settled in Peoria, Illinois, where Bill finished his last two years of college at Bradley University. There were other Canal Zone friends there, such as Charlie and Charlotte Norris. In 1950, son Bill was born. After completing ROTC, Bill joined the Air Force and the family travels began. When Bill was stationed in Greenland, Lou and the children returned to the Canal Zone. They stayed with her parents, and Lou went to work for the Army. When Bill retired from the Air Force, they settled in Zephyrhills, Florida. Bill died in 1990, and Lou still resides there.

Liz recalls her teachers and many friends at Balboa High School to include Jim O'Donnell and Jackie West. After moving to Pedro Miguel, her friends, Pat and Vee Hatchett, made Liz a part of their family. Following graduation in 1946, she went to work at the Administration Building while attending Balboa Junior College. She met Pat Beall, whose family lived in Panama. They were married in Ancon in 1949 and their three children — Carol, Richard and Robert — were born at Gorgas Hospital. They lived on the Pacific side and then moved to the Atlantic side. The three children graduated from Cristobal High School. In 1976, following a divorce, Liz moved to Washington, D.C., where she worked for the Veterans Administration and attended Northern Virginia Junior College. While there, a friend suggested Liz work for her husband. Liz applied and got a job with the U.S. Senate Budget Committee, working for Senator Dominici from New Mexico. She retired and moved to Largo, Florida. She has many memories and lasting friendships from the Canal Zone.

Martha remembers and describes the school bus from Fort Kobbe/Howard Field to Cocoli as an Army truck with canvas over the back and with wooden benches. A regular school bus then took the junior high and high school students on to Balboa. The Standifers, Wrights, Joneses, and Zents made up most of the school bus riders. In Pedro Miguel, more friendships were formed, such as the Zeecks, Dedeauxs, Powells, Hatchetts, Coffeys, Dawsons and Underwoods. After graduating from Balboa High School in 1949, Martha went to Bradley University, Peoria, Illinois, along with Dorothy Dedeaux and George Bull. Mr. Zip Zeirten, the senior class sponsor, encouraged and helped many Canal Zone students get accepted at Bradley University. In 1950, Martha and George were married. George graduated and entered the U.S. Air Force pilot training program in 1953. In his 27-year Air Force career, Martha and George welcomed three children — Susan, Linda and Michael. They also moved 14 times in those 27 years. When he retired from military service, George and Martha moved to Yalaha, Florida, near his parents, Lelia and Earl Freund. George passed away in 1999; Martha lives in Leesburg, Florida.

Marge found the Canal Zone a great place to grow up. She attended Cocoli, Pedro Miguel and Diablo Heights elementary schools and Balboa Junior/Senior High School. She graduated in 1955 and went to work at the Administration Building in the wage and classification section. During this time, she met SP-3 Howard Garner, who was with the 534th Military Police Company at Fort Clayton. They were married in the Curundu Protestant Church in 1956 and lived in Los Rios, Gamboa and Panama City. They left the Canal Zone in 1957 and following How-

> *Martha remembers and describes the school bus from Fort Kobbe/Howard Field to Cocoli as an Army truck with canvas over the back and with wooden benches. A regular school bus then took the junior high and high school students on to Balboa. The Standifers, Wrights, Jones, and Zents made up most of the school bus riders.*

ard's discharge settled in California where they raised four children — Keith, Greg, Leslie and Donna. Howard retired in 1998 after 32 years with the Los Angeles City Fire Department. They now reside in Prescott Valley, Arizona. Marge formed long lasting friendships and keeps in touch with many, such as Kay Frangioni Pierce, Judy Hoopes, Glenda Lee Migliozzi, Sadie Williams Turlington and Rose Marie Wildsmith. Other friendships were renewed at the Balboa High School 55-year class reunion and at various Panama Canal reunions.

Butch attended Diablo, Balboa and Kobbe elementary schools and Balboa Junior/Senior High School. While in high school he was active in the math club and was class president in his freshman, sophomore and junior years. He attended the Curundu Protestant Church and was active in community baseball and scouting. Following graduation in 1962 he attended the U.S. Air Force Academy in Colorado Springs, Colorado. In 1966, his parents, sisters and other family members were there to celebrate his graduation. Butch had a 23-year career with the Air Force as a pilot, flying F-4 fighter aircraft during the Vietnam War. During his second tour there he met his wife, Raklay, in Thailand. They were married in 1973 and have two sons — Eddie and David. He retired as a colonel in 1989. He went back to school and received his teaching degree; in 1990 he became a math teacher in Tampa, Florida, where he continues to this day.

There were so many things to see and do in the Canal Zone and in Panama. The Zent family enjoyed their time there from 1943 to 1963 and will always think of it as a second home. Thanks to all their Canal Zone friends, there are many wonderful memories and friendships.

PHOTO COURTESY OF MARGE ZENT GARNER

Curundu, Canal Zone, circa 1950. The "X" designates the movie theater (front left), and the "✓" (front right) designates the former commissary which became a community center (taken from hill behind Curundu Clubhouse).

The American Era Revisited — A Pictorial Journey Through Time

We Built It...

President Theodore Roosevelt sits at the controls of a 95-ton Bucyrus steam shovel. William F. Ashton, Roosevelt Medal holder, shows President Roosevelt "the ropes." November 1906.

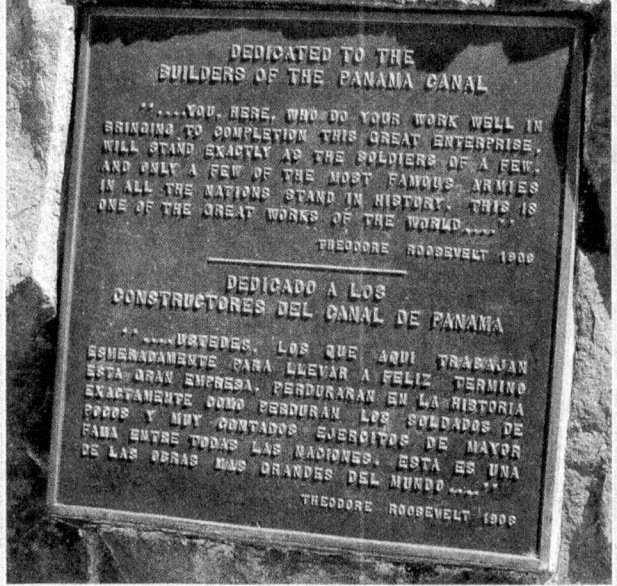

Theodore Roosevelt's 1906 dedication to the builders of the Panama Canal.

Col. George W. Goethals, appointed Chief Engineer of the Panama Canal by President Theodore Roosevelt in 1907.

We Opened It...

PHOTO COURTESY OF PANAMA CANAL MUSEUM
Opening of the Panama Canal. SS Ancon approaching Cucaracha slide looking north, August 15, 1914.

PHOTO BY DON GOODE
Miraflores Locks and Pedro Miguel Locks.

PHOTO BY DON GOODE
Gatun Locks.

We Guided It...

George W. Goethals, first Governor of the Panama Canal.
PHOTO COURTESY OF PANAMA CANAL MUSEUM

First Panamanian Administrator Gilberto Guardia (left) and first U.S. Deputy Administrator Raymond P. Laverty, of the Panama Canal Commission.
PHOTO COURTESY OF PANAMA CANAL AUTHORITY

Alberto Aleman (left), last Administrator of the Panama Canal Commission and current Administrator of the Panama Canal Authority, at the ceremony turning over stewardship of the Canal to Panama, December 1999.
PHOTO COURTESY OF DEPT. OF THE ARMY

First and only U.S. Administrator and Panamanian Deputy Administrator of the Panama Canal Commission, D.P. McAuliffe (right) and Fernando Manfredo (left), with former U.S. President Gerald Ford.
PHOTO COURTESY OF PANAMA CANAL AUTHORITY

Maj. Gen. Harold R. Parfitt, last Governor of the Canal Zone, in his second-floor office overlooking the town of Balboa. George Goethals was the first governor to use this office.
PHOTO COURTESY OF PANAMA CANAL REVIEW

We Defended It...

Panama Canal Police Capt. Wall talks with a Panamanian student delegation about raising the Panamanian flag at Balboa High School, January 1964 riots.

Life Magazine featuring coverage of the 1964 riots.

Operation Just Cause (left and above), December 1989.

We Made It Our Home...

Ft. Clayton, Miraflores Locks in background, Pacific side.

Gatun townsite and locks.

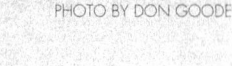

Cocoli, Pacific side.

Ft. Sherman, Atlantic side.

Amador, La Boca and Gavilan, with causeway, Pacific side.

Cristobal, Atlantic side.

We Worked...

Canal pilot on bridge of transiting ship.

Locks control house.

Locks maintenance.

Canal Zone police officers.

We Grew Up...

PHOTO COURTESY OF PANAMA CANAL MUSEUM
Flag ceremony with Scouts at Balboa Train Station.

PHOTO COURTESY OF PANAMA CANAL MUSEUM
Crossing guard at Balboa Elementary School.

PHOTO BY SUE ROBBINS
No weekend at the beach was complete without a trip to the market in El Valle.

PHOTO COURTESY OF JEAN MEDINGER MCGUIRE
Fun at the Pacific Saddle Club.

We Worshipped...

St. Mary's Catholic Church, Balboa, built in 1916.

Gatun Union Church.

First Baptist Church, Balboa Heights.

PHOTO COURTESY OF PANAMA CANAL AUTHORITY
Redeemer Lutheran Church, Balboa.

PHOTO COURTESY OF PANAMA CANAL AUTHORITY
Sacred Heart Chapel, Ancon.

PHOTO BY CAPT. WILBUR VANTINE
Holy Family Church, Margarita.

We Learned...

Balboa Elementary School.

Cristobal High School, 1965.

Curundu Junior High School with geodesic dome at left.

Margarita Elementary School looking from Brazos Boulevard.

We Traveled...

Tocumen Airport, Panama City.

Home leave charter flight, Tocumen Airport.

Engine 901 of the Panama Railroad, operated by the Panama Canal Company, rounds the bend near Pedro Miguel Locks.

SS Cristobal, 1939-1981.

We Played...

Farewell Golf Tournament for Panama Canal Administrator D.P. McAuliffe, Amador, 1989. Kneeling L-R: Bob Hauser, Bob Rupp, Burt Mead, Gus De La Guardia, Fred Sapp, Bill Kerivan, Fernando Martinez, Hubert Jordan, Rolo Winberg. Standing L-R: Joe Wood, Gary Anderson, Ted Lucas, Dick Burgoon, Bill Van Hoorde, Bill De La Mater, Ken Underwood, Bob Best, D.P. McAuliffe, Jack Hern, Lou Vogel, Bob Barnes, Ken Morris, Bill Coffey, Sandy Hinkle, Bill Cofer, Ray Laverty, Tim Corrigan, Jay Seilman, Bill Joyce, Jim Farnsworth, Ed Donahue.

Pacific Little League baseball game.

Ft. Amador Base and Golf Course.

Paddlers take a break during Ocean-to-Ocean Cayuco Race.

Camping, Chagres River.

Two of the famous fishing Schmidt Brothers, John (left) and Louis, Jr. (right), are shown with record 1,006 lb. black marlin; June 1949.

Fishing from the bridge at the Gamboa Boat Club.

We Gathered...

Gatun Yacht Club and vicinity.

Pedro Miguel Yacht Club.

The Pacific Sailfish Club was earlier known as the Naval Officers' Club at Ft. Amador. In its last iteration, it was the Ft. Amador American Legion Club. The popular Balboa Yacht Club was located below. The building was destroyed by fire in February 1999.

Hotel Contadora, Contadora Island, Pacific side.

Tivoli Hotel.

We Were Cared For...

Colon Hospital.

Gorgas Hospital.

Coco Solo Hospital.

We Shopped...

PHOTO COURTESY OF PANAMA CANAL MUSEUM
Balboa Commissary.

PHOTO BY CAPT. WILBUR VANTINE
Margarita Clubhouse.

PHOTO COURTESY OF PANAMA CANAL MUSEUM
Coco Solo Commissary.

PHOTOS COURTESY OF PANAMA CANAL AUTHORITY
Commissary shopping in the Canal Zone.

We Embraced the Local Culture...

PHOTO COURTESY OF CAPT. WILBUR VANTINE
Chinese Garden near Margarita.

PHOTO COURTESY OF DEPT. OF THE ARMY
Street vendor selling Panamanian watercolors at Plaza de Francia in Panama City.

PHOTO COURTESY OF DENNIS WHITE
Lottery vendor.

PHOTOS COURTESY OF PANAMA CANAL MUSEUM
Kuna Indian and indigenous mola artwork.

We Celebrated...

Goethals Memorial dedication ceremony, Balboa, March 31, 1954.

Fireworks over Sosa Hill.

Fourth of July Parade, Balboa.

Independence Day Celebration, Balboa, July 4, 1942.

We Remember...

PHOTO BY DON GOODE
Canal Zone Police fold U.S. flag, Last Day of the Canal Zone ceremony, Sept. 30, 1979.

PHOTO BY DON GOODE
Governor Harold R. Parfitt presiding at the Panama Canal Company flag lowering ceremony, Sept. 30, 1979.

Lowering and folding of the U.S. flag (left and above), December 1999.

Flag transfer ceremony, December 1999.

Canal Zone

by Celina Barkema Vargas

The memories that we have of you will always be with us.
You were our home, but now you're gone, remember you we must.
We won't forget the lush green grass we ran through every day,
The mango trees, the majestic palms, and places we would play.
We won't forget the doodle bugs, the giant beetles too,
The skating rink, the clubhouses, and most of all the schools.
We won't forget the fishing trips, and going to the beach,
And picking mangos off the trees, as high as we could reach.
We won't forget the cardboard sleds, and playing on the street.
We won't forget the theaters, and bowling alleys too.
The Post Office and Commissary, or where we bought our shoes.
You see, there is so much of you, that we will always cherish.
You were our home, our home sweet home, to us you'll never perish.
You now belong to someone else, but the memories of you we own.
I'll say goodbye but I'll never forget, the good old Canal Zone.

The Panama Canal Museum Celebrates Roosevelt Medal Holders and Their Descendants

The Panama Canal Museum

Write of Passage

PHOTO COURTESY OF PANAMA CANAL MUSEUM

A February 2008 Panama Canal Museum luncheon commemorated the 150th birthday of President Theodore Roosevelt. Pictured at the event are Teddy Roosevelt (actor Michael O. Smith) standing next to Joe Wood, PCM President, and flanked by Tampa Bay Rough Riders and PCM members Dan Ulrich (left) and Rick Gayer (right). The First U.S. Volunteer Cavalry, known as "Rough Riders," was organized by President Roosevelt to fight during the Spanish-American War of 1898.

PHOTO COURTESY OF PANAMA CANAL MUSEUM

Roosevelt Medal descendants attending the luncheon included: Back row, L-R: Carol (Peterson) Krueger, Diane (Peterson) Pearson, Jack Hern, Dick Cunningham, Jean (Medinger) McGuire, Janet Sutherland, Kathy Egolf, and Jason Ohman. Middle row, L-R: Christine Heintz, Shirley (Million) Muse, Joan (McCullough) Ohman, Peggy (Hale) Huff, Cheryl (Peterson) Russell, Judi McCullough, and Lois (Hollowell) Jones. Front row, L-R: Gerry DeTore, Michelle Pearson, Kristle Pearson, Michael O. Smith (Teddy Roosevelt), Blanche (Adler) Browne, and Chuck Hummer.

At the museum following the luncheon, PCM President Joe Wood (right) and actor Michael O. Smith blow out the candles on a cake celebrating the museum's 10th anniversary and Teddy Roosevelt's 150th birthday.

Peggy (Hale) Huff, the museum's Musings and Review editor, appearing in the guise of her great-grandfather, William F. Ashton (a Canal construction worker pictured in the top right corner of the wall hanging showing TR on the Bucyrus steam shovel), surprises Teddy Roosevelt (Michael O. Smith).

Panama Canal Society President Bob Russell (left) and museum trustee Bob Karrer view a TR exhibit at the museum with Michael O. Smith.

NASA Astronaut and Gaillard descendant Scott Parazynski, M.D. (front, center), Panama Canal Society President Bob Russell (left) and Panama Canal Museum President Joe Wood (right) with Roosevelt Medal holder descendants at Panama Canal Society Commemorative Ceremony held in their honor, July 5, 2008.

Parazynski honored his ancestors by taking a Roosevelt Medal into space aboard the shuttle Discovery on October 23, 2007.

PHOTO BY DAVID WRIGHT PHOTOGRAPHY

PHOTO COURTESY OF PANAMA CANAL MUSEUM

Roosevelt Medal back on earth after historic journey into space. Circumnavigating the earth in 238 orbits, the medal traveled 6.2 million miles in 15 days, 2 hours, and 23 minutes. The medal was issued to Peter J. Sundberg (No. 5000 for 1909-11 with one bar No. 3103 for 1911-13) and donated by Evelyn Brunswick to the Panama Canal Museum.

PHOTO COURTESY OF PANAMA CANAL MUSEUM

NASA Certificate of Authenticity attesting that Roosevelt Medal was aboard the shuttle Discovery during NASA's STS-120 mission into space.

Panama Canal Society President Bob Russell (right) and Panama Canal Museum President Joe Wood (left) present a special commemorative plaque to NASA Astronaut Scott Parazynski.

NASA montage presented to the Panama Canal Museum depicting the Discovery space shuttle (STS-120) that took the Roosevelt Medal into space on October 23, 2007, with a photograph of the Discovery crew, personally autographed by crew members Pam Melroy, Doug Wheelock, Stephanie Wilson, George Zamka, Dan Tani, Scott Parazynski, and Paolo Nespoli (Italian Astronaut).

NASA Astronaut Scott Parazynski, descendant of Roosevelt Medal holders Col. David DuBose Gaillard and David St. P. Gaillard. As lead spacewalker, Parazynski accumulated more than 27 hours in 4 spacewalks, including an unplanned and dangerous spacewalk to successfully repair a damaged solar panel.

NASA Astronaut Scott Parazynski presents STS-120 mission montage to Panama Canal Museum President Joe Wood.

> "Far better it is to dare mighty things, to win glorious triumphs, even though checkered by failure, than to take rank with those poor spirits who neither enjoy much nor suffer much, because they live in the gray twilight that knows not victory or defeat."
>
> — Theodore Roosevelt

... running across stickers in your bare feet to see who wins ... waving at the guys on ships a Golden Altar ... dancing at the Panama Hilton ... chasing the rain ... eating michas ... ¡Carn at Summit Gardens ... Saturday afternoon at Amador Beach ... riding the Panama Railro ... submarine races out on the Causeway ... a "real" tropical rainstorm ... dances at the Ti French fries and gravy at the Gatun Clubhouse ... ring-a-levio at dusk ... heading home wh Margarita hill on the sides of the tennis courts ... sunset over the Caribbean ... the Gatu Conjunto ... sleeping to the sound of rain on a tin roof ... riding the train to football game races ... Christmas tree burns ... battleball in the Margarita Gym ... jeeps full of G.I.'s whis Bravo's band floating across the bay from Colon ... onion rings at the American Legion ... t ... banyan trees on Roosevelt Avenue ... steak sandwiches at Gamboa Golf Club ... swimmin ... Boy Scout Camp Chagres ... meat on a stick ... the lottery ... Balboa Police Station (end ditches after a big rain ... Chorrera Falls ... BHS letters on Sosa Hill ... cases of soda deli Canal Zone at the Civil Affairs Building ... Draft Board Local #1 ... collecting snakes ... Frida in the big sleigh at Mr. Townsend's garage on Santa Claus Lane ... the American Legion floor Lake ... swimming races for movie tickets during summer vacation at Gatun Pool ... walki the pond ... fishing on the Chagres River ... skydiving over CocoSolo/France Field ... ceviche docked at Pier 1 ... visiting the Leper Colony ... spaghetti night at Amador Officers' Club .. carnivalito ... midnight mass at St. Mary's ... drag races at France Field ... meatball sandv ...taking your life into your hands on the Transisthmian Highway ... Jamboree parties at th trails of Mindi Acres and Ft. Davis ... fried corvina — anywhere ... swimming parties at the through Balboa whenever Sosa Hill burned ... the old guy with the swagger stick who walked a Panama at the beach ... hot feet on the black sand at San Carlos beach ... 5-cent ice crea low-flying planes over BHS ... plantain chips ... blocks of ice delivered to your home ... sea the Seventh Fleet at the Atlantic breakwater ... Chinese plums ... the morning flag-raising — anywhere ... driving up the Admin. Building hill during your driver's license test ... drinkin dinners at Balboa Yacht Club ... fishing trips to Isla Perlas ... gathering around the plaque mud ... beer and pizza at Ft. Davis snack bar after Friday night football games ... putting wall parties, Coco Solo ... hunting lobster at night on the reefs outside the breakwater wal Whiskey-a-Go-Go in Panama City ... the natural slide at Goofy Falls ... weekends at Santa Gatun Lake ... watching the flying fish from the Taboga launch ... tree frogs in El Valle ... top of the hill in Curundu ... watching the U2s take off from Albrook Field ... picking berries Curundu ...the manual pinsetters at Balboa Bowling Alley ... eating raspadilla snow cones the best pizza in the world from the Napoli restaurant ... snorkeling around Taboga Island . the flats at Albrook AFB ... sound of the bullfrogs ... cruising the Causeway at night ... the

... riding the chivas with your friends ... a kiss under the palm trees at Rio Del Mar ... picnics
... wading in the Goethals Memorial ... sunrise from Gold Hill the morning after graduation
... skinny-dipping at Cocoli Lake ... hiking the Las Cruces Trail ... weekends at Santa Clara ...
the streetlights turned on ... sliding down the Admin. Building hill on palm fronds—and the
spillway with all gates open ... ginnup season ... eating fresh limes with sugar ... Lucho y Su
"the other side" ("the other side" being the opposite side from where you lived) ... cayuco
at the girls ... warm rain ... mango season, from green to ripe ... the sound of Colegio Able
Auto Cine ... panning for gold in Rio Pecorar ... Fourth of July celebrations at Balboa Stadium
K-9 ... the DDT truck ... leaf-cutter ants ... weekends at Taboga Island ... 25-cycle electricity
said) ... green police cars ... Nancy trees ... Farfan Beach ... teen clubs ... swimming in the
to your home for a dollar ... Chinese Gardens ... Cocoli Gun Club ... the rubber map of the
at the skating rink — then walking fast to make the Owl Show at Diablo Theatre ... Santa
ping with the music ... Saturday speedboat races at the Cristobal Yacht Club and Miraflores
the railroad tracks from the Aids Building to the Gatun Yacht Club ... catching guppies at
the Police Lodge ... 25-cent rum and cokes at the drive-in theater ... sneaking aboard ships
day nights at El Rancho ... Johnny Mazetti from the clubhouse ... skipping school to go to
es from the Napoli ... toad hunting at night ... Margarita's "Snob Hill" ... France Field crabs
Docks ... chop suey and wontons from the Balboa YMCA ... riding horses through the back
tel Washington saltwater pool ... catching the ferry to Ft. Sherman ... wild animals running
nd wearing a WWII Nazi uniform ... the Blue Angels flying over Panama Bay ... drinking Cerveza
nes at the YMCA ... veterans of the Spanish American War at Memorial Day ceremonies ...
es taking off and landing behind Cristobal High School ... watching the sun come up behind
Guardia Nacional stations ... gooey cinnamon buns at Margarita Clubhouse ... empanadas
rple Passion at the drive-in ... throwing up Purple Passion on the way home ... $1.50 lobster
HS, trying not to step on it ... Little Theatre plays ... football games in torrential rains and
ies on the train tracks ... hidden beaches past Piña Beach ... Mine Dock keg parties ... sea
seeing people off on the S.S. Cristobal ... The Snake Pit in Curundu ... the Sombrero and the
ra ... the spider monkey at Rio Mar ... camping at the Taboga cross ... fishing for bluegill in
horses at Santa Clara ... camping with the scouts at Rio Hato ... the rope swing at the
rro Punta ... watching Carnival parades on Via España in Panama City ... chasing iguanas in
the Canal Zone bus terminal ... the sound of an old palm branch falling from a palm tree ...
ing monkey meat on a stick ... the sound of the approaching rain coming from the jungle ...
wooden Balboa Clubhouse ... the GAP (later the BRIDGE) Clubhouse for Balboa teenagers ...

www.ingramcontent.com/pod-product-compliance
Lightning Source LLC
Chambersburg PA
CBHW060328240426
43665CB00048B/2851